Protein Secretion Pathways in Bacteria

Protein Secretion Pathways in Bacteria

Edited by

Bauke Oudega

Vrije Universiteit Amsterdam,
Department of Molecular Microbiology,
The Netherlands

KLUWER ACADEMIC PUBLISHERS
DORDRECHT / BOSTON / LONDON

A C.I.P. Catalogue record for this book is available from the Library of Congress.

ISBN 1-4020-1255-1

Published by Kluwer Academic Publishers,
P.O. Box 17, 3300 AA Dordrecht, The Netherlands.

Sold and distributed in North, Central and South America
by Kluwer Academic Publishers,
101 Philip Drive, Norwell, MA 02061, U.S.A.

In all other countries, sold and distributed
by Kluwer Academic Publishers,
P.O. Box 322, 3300 AH Dordrecht, The Netherlands.

Printed on acid-free paper

TABLE OF CONTENTS

PREFACE

The passed decades have seen an explosive growth in our understanding of the properties and functions of living cells. Especially our knowledge of the various components and complex structures of bacterial cells increased enormously. The genomes of over 50 different bacterial species have been sequenced. We have not yet identified all the different gene products of, for instance, *Escherichia coli* or *Bacillus subtilis*, but it will not be to long before we know most of the gene products, their cellular location, structure and function, and their specific role in one or more of the cellular processes. Some people even begin to dream about a complete picture of the functioning of a simple, living, unicellular organism, to understand life itself! This book plays a part in that dream, that beautiful future perspective of Biology. The book provides an broad overview of how bacterial cells are organized, how proteins are targeted out of the bacterial cytosol and end up in the inner membrane, the periplasm or even in the outer membrane. In addition, several chapters deal with mechanisms by which proteins are released into the extracellular environment.

Many inspiring scientists have contributed to the book. Each of them is well known is his or her field of research. I like to thank them all for their invaluable contribution and their continuing research efforts in my area of research, the bacterial cell surface.

In my work as a teacher and researcher so far, I have been looking for a book like this one, a book dealing with all different aspects of protein targeting, membrane insertion, folding and secretion. I hope that this first edition can inspire numerous students and young scientist to use the knowledge presented, to continue investigating all the marvelous aspects of the bacterial cell and the intriguing mechanisms and molecular machines that are used by bacterial cells. I also hope that teachers can use the book in their courses, especially in the area of molecular microbiology, medical microbiology and biotechnology.

Bauke Oudega

Chapter 1

PROTEIN TARGETING TO THE INNER MEMBRANE

Joen Luirink and Bauke Oudega

Department of Molecular Microbiology, IMBW/BioCentrum Amsterdam
Faculty of Biology, Vrije Universiteit Amsterdam
De Boelelaan 1087
1081 HV Amsterdam, NL

1. INTRODUCTION

Bacteria appear to be simply organized organisms, but bacterial cells do have different subcellular compartments: the cytoplasm, the cytoplasmic or inner membrane (IM), the cell wall and in the case of gram-negative bacteria, the periplasm and the outer membrane (OM). Proteins are synthesized in the cytosol, and a large number of these proteins fold in the cytoplasm and play their role in this compartment. However, another large number of proteins (about 40 % of the total amount of proteins) is destined for functioning in one of the extra-cytoplasmic environments. These proteins have to be targeted to the IM, during or after completion of translation, and they have to be inserted into or translocated across this membrane. The mechanisms by which extra-cytoplasmic proteins are inserted into or translocated across the IM, transported to their final destination, fold and function, are described in more detail in one of the following chapters of this book. This chapter will deal with the first steps in these translocation pathways: the early recognition of membrane proteins and of secreted proteins by various targeting factors, and their targeting to generic or specific protein translocation sites (translocons) in the IM of *Escherichia coli*.

1

B. Oudega (ed.), Protein Secretion Pathways in Bacteria, 1–21.
© 2003 *Kluwer Academic Publishers. Printed in the Netherlands.*

2.   TRANSLOCONS AS RECEPTORS FOR TARGETING COMPLEXES

The *E. coli* IM contains different types of translocons (Fig. 1). The best studied and most widely used is the so-called Sec-translocon which appears to function both in secretion and in membrane integration of proteins. It consists of several integral membrane proteins that are thought to form a proteinaceous channel, and a peripheral bound ATPase to drive the translocation reaction (reviewed in Fekkes and Driessen, 1999). The core of this multisubunit complex is formed by SecY, SecE and SecG together with the peripherally bound SecA. This core complex is also referred to as translocase to underscore its catalytic activity. In addition, under certain experimental conditions three accessory proteins can be co-purified with the core translocon. These proteins are SecD, SecF and YajC. They apparently increase the efficiency of the translocation reaction although their exact role is not clear (Fekkes and Driessen, 1999).

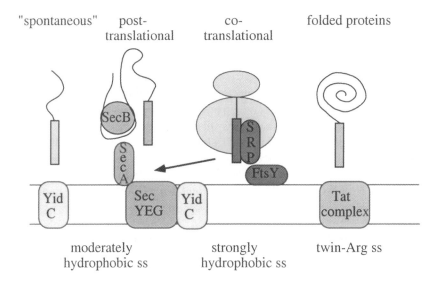

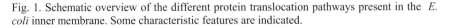

Fig. 1. Schematic overview of the different protein translocation pathways present in the *E. coli* inner membrane. Some characteristic features are indicated.

SecA is an important component of the Sec translocase. It binds to almost all components that are involved in the translocation process, exhibits ATPase activity, and regulates its expression by binding to its own mRNA (reviewed in Driessen *et al.*, 1995). Most importantly for this chapter, it also functions in the reception of secretory proteins that are being targeted to the translocase (see below as well as chapter 2). When bound to the membrane at the SecYEG complex, SecA is "activated" for the high-affinity recognition of the SecB export chaperone (Hartl *et al.*, 1990), and of targeting signals in secreted proteins.

The Sec-translocon is the common target of two distinct targeting pathways that pilot IM and secreted proteins to the membrane. IM proteins are in general targeted co-translationally to the Sec-translocon by the so-called SRP (Signal Recognition Particle) targeting route (detailed below). The signal that is recognized in this route is a hydrophobic targeting peptide that is most often found at the N-terminus of the protein. In contrast, most periplasmic and OM proteins as well as some proteins that are secreted into the extracellular environment are targeted post-translationally (or late co-translationally) to the Sec-translocon via the so-called SecB targeting route (detailed below). These latter proteins are synthesized with an N-terminal cleavable signal peptide or signal sequence.

Recently, a new accessory translocon component has been identified, YidC, that appears to operate exclusively in the integration of membrane proteins perhaps as an intermediate beween the aqueous Sec-translocon and the hydrophobic lipid bilayer (Scotti, 2000; see also chapter 4). Interestingly, YidC also acts independent of the Sec translocon either as a separate entity or in a different, yet unknown translocon. This "independent" YidC translocon is thought to facilitate membrane insertion of a subset of IM proteins (Samuelson *et al.*, 2000; Fröderberg *et al.*, 2001).

In addition to the Sec-machinery, another protein translocation pathway has been discovered. This pathway is used by proteins that are characterized by a so-called twin-arginine motif in their N-terminal targeting signal (TAT pathway, reviewed in Berks *et al.*, 2000). The TAT pathway appears to function specifically in the translocation of folded proteins like enzymes that contain a cofactor. So far, specific cytoplasmic targeting factors that play a targeting role in this route have not been identified. Details of this pathway will be described in chapter 3.

Besides the Sec and TAT translocons, other specific targeting and secretion routes exist that are used by proteins that have to be translocated across the entire cell envelope. Examples are the so-called type I, type III and IV routes, the flagellum biosynthesis route as well as the bacteriocin release route. These routes will be described in more detail in one of the following chapters of this book.

Finally, it can not be excluded that some IM proteins integrate into the membrane spontaneously, i.e. without the aid of a protein machinery.

3.   TARGETING SIGNALS

The targeting signals present in IM proteins as well as in secreted proteins, that mediate targeting by the main SRP and SecB targeting pathways to the Sec-translocon will be described in this section. Signals for targeting of most secreted proteins are located both in a cleavable, hydrophobic N-terminal extension of the pre-secretory protein (preprotein) as well as in the mature region. The N-terminal extention is called signal sequence, signal peptide or leader peptide. The signal sequence is cleaved (processed) during passage of the IM by a signal peptidase (see chapter 2). IM proteins are sometimes targeted by a cleavable signal sequence just like secreted proteins. More often, however, targeting is achieved by a signal sequence that is not cleaved and also serves to anchor the protein in the IM. This type of targeting signal is called "signal anchor sequence" (SA) to emphasize its dual function.

3.1   Signal sequence

The N-terminal signal sequence of preproteins ranges in length from about 20 to 30 amino acid residues. It is characterized by a positively charged N-terminal region (about 6 residues), a hydrophobic central part or core region (10-15 residues), and a polar C-terminal region (about 6 residues) that precedes the cleavage site of the signal peptidase (von Heijne, 1985) (Fig. 2). Signal sequences show no primary sequence similarity but have similar physical characteristics and are often functionally interchangeable between different organisms (von Heijne, 1990). These shared features are reflected by common mechanisms for recognition and functioning of signal sequences both in targeting and membrane insertion.

Fig. 2. Schematic presentation of a signal sequence of a preprotein. N, amino-terminal, positively charged region; H, hydrophobic, central region; C, carboxyl-terminal region containing rather well conserved amino acids with short neutral side chains at position –1 and –3. The arrow indicates the cleavage site.

A net positive charge in the N region (arginines and/or lysines) enhances the processing and translocation rates of a preprotein but is not essential. Preproteins with signal sequences that carry a neutral or even negatively charged N region can be targeted and secreted, albeit at reduced rates (Gennity *et al.*, 1990). With increasing positive charge at the N region of the signal sequence, the SecA requirement for translocation is reduced while the interaction of the preprotein with SecA is enhanced (Akita *et al.*, 1990). Moreover, a secretion deficiency caused by a reduction in the number of positive charges can be restored by mutations in *secA* (Puziss *et al.*, 1989 and 1992). These data suggest that the N region plays a role in the targeting of the preprotein to the translocase. The N region has also been suggested to bind the negatively charged surface of the lipid bilayer of the membrane (de Vrije *et al.*, 1990). In concept, this property might contribute both to targeting and to fix the N-terminus of the preprotein at the cytoplasmic side of the membrane. This would help to orient the C region towards the periplasmic side of the membrane where the cleavage by the signal peptidase takes place. A reduction in the number of positive charges in the N region can be compensated for by an increased hydrophobicity of the H region suggestive of distinct but cooperative functions of these subdomains (Phoenix *et al.*, 1993).

The H domain is the core of a signal sequence and varies in length from 10 to 15 amino acid residues. The hydrophobicity of this region is the most important and distinguished feature of the signal sequence both in targeting and translocation (Hikita and Mizushima, 1992; MacFarlane and Müller, 1995; Phoenix *et al.* 1993). In particular, the hydrophobicity plays a crucial role in the choice between a co-translational and post-translational targeting mechanism (see below). Mutations in the H region that disrupt the hydrophobic character usually result in defective signal sequences. On the other hand, increasing the total hydrophobicity of the H region increases the translocation efficiency of a preprotein to a certain maximum (Chou and Kendall, 1990). The length of the H region has to fall between certain limits to allow proper positioning of the C region relative to the catalytic site of the signal peptidase at the periplasmic side of the membrane (Dalbey *et al.*, 1997). Structurally, the H region has a tendency to adopt an α–helical conformation in a non-polar environment. A "helix breaking" residue, i.e., a glycine or proline residue, is often found in the center of the H region. This may allow the signal sequence to form a hairpin-like structure or at least a kink in the helix which appears important during the targeting and insertion process (for reviews, see Pugsley, 1993; Izard and Kendall, 1994). A proline or glycine is also predominant at the border between the H- and C regions.

The C-terminal region of the signal sequence contains the signal peptidase cleavage site, but is not important for targeting and translocation.

It is the only part of the signal sequence that demands some primary sequence specificity, since this part interacts with the signal peptidase which cleaves the signal sequence at the periplasmic side of the IM (for a review, see Dalbey and von Heijne, 1992). Amino acid residues in this region have small neutral side chains, such as alanine, glycine, serine, and threonine, with a preference for alanine (von Heijne, 1984). The cleavage site of a preprotein can be omitted or mutated, which often results in a protein that remains anchored to the membrane by the uncleaved signal sequence.

3.2 Signals in the mature protein

Secretory proteins that lack a signal sequence can be translocated under certain conditions, which indicates that these proteins have additional targeting information (Prinz *et al.*, 1996). On the other hand, not all proteins that are artificially equipped with a signal sequence are translocated (Lee *et al.*, 1989). This also suggests that specific information in the mature structure is required. Secretory proteins often have binding sites for the cytosolic targeting factor SecB at several positions in the mature region. The exact nature of the SecB binding site is not clear: no consensus sequence has been identified that is responsible for conferring SecB dependence although short clusters of both basic and hydrophobic amino acids appear important (see below). It should be noted that the signal sequence may affect the folding of the mature region of the preprotein and thus the accessibility of targeting information in this region (Collier *et al.*, 1988).

Targeting information is also located within the mature part of IM proteins and secreted proteins. IM proteins possess one or more hydrophobic transmembrane segments that function in targeting and anchoring the protein in the lipid bilayer (see also chapter 4).

4. THE SRP TARGETING PATHWAY

Having some idea on the destination of targeted proteins and the signals in secreted and membrane proteins that are essential for targeting and membrane insertion, we can now discuss the two major targeting pathways in *E. coli* that involve the targeting factors SRP and SecB. SRP and SecB recognize targeting information in the signal or signal anchor sequence and in the mature domain, respectively, and guide the bound proteins to the Sec-translocon. The SecB pathway appears specific for gram-negative bacteria whereas the SRP pathway is ubiquitous. In fact, the realization that *E. coli* has a, be it simple, SRP is of relatively recent date. Therefore, most of the relevant evidence on the functioning of SRPs comes from the literature on eukaryotic SRPs, that will be briefly discussed below.

4.1   SRP pathway in eukaryotes

The mammalian SRP-targeting pathway targets both secretory and membrane proteins to the membrane of the endoplasmic reticulum (ER). The mammalian SRP is a large ribonucleoprotein complex that consists of six polypeptides of 9, 14, 19, 54, 68 and 72 kDa, respectively, arranged on a 7S RNA moiety (Lütcke, 1995) (Fig. 3). The core of the eukaryotic SRP consists of a GTP-binding protein, the 54 kDa subunit (SRP54), and the 7S RNA. SRP54 binds to the hydrophobic core of the signal sequence or anchor sequence of secretory and membrane proteins, when the nascent polypeptide emerges from the ribosome. Further translation of the nascent chain is then inhibited. SRP9 and SRP14 form a heterodimer that is involved in this translation arrest. Together with the 5' and 3' ends of the 7S RNA it forms the so-called Alu-domain of theis SRP (see Fig. 3.). Translation arrest is relieved when the SRP contacts its receptor (SR) at the ER membrane. The SR consists of two subunits, SRα which is the actual receptor, and SRβ which functions to tether SRα to the ER membrane. Both SRα and SRβ are GTP-binding proteins. Upon targeting of the ribosome-nascent chain-SRP complex to the SR, the nascent chain is released from the SRP in a GTP dependent reaction and is transferred to the translocon (the Sec61 complex) that is homologous to the *E. coli* SecYEG translocon (Matlack *et al.*, 1998). As a result, translation of the protein is continued and the protein is inserted co-translationally (driven by translation) into the translocon. The ribosome contributes to the targeting reaction by having affinity for both SRβ and the translocon.

The targeting reaction is regulated and proceeds in one direction by conformational changes in the key components, SRP54, SRα and SRβ that take place upon binding and hydrolysis of GTP.

4.2   The *E. coli* SRP

Genetic screens employed to identify components involved in secretion of proteins have never pointed to the existence SRP route components in *E. coli*. Cloning and sequencing of mammalian SRP components actually gave the first hints for a similar, yet simpler targeting

8

system in bacterial cells. The *E. coli* SRP consists of only a 4.5S RNA and a 48 kDa GTPase designated Ffh (for Fifty-four homologue), that show homology to the eukaryotic 7S RNA and SRP54, respectively (reviewed in Herskovits *et al.*, 2000) (Fig. 3).

Mammalian SRP

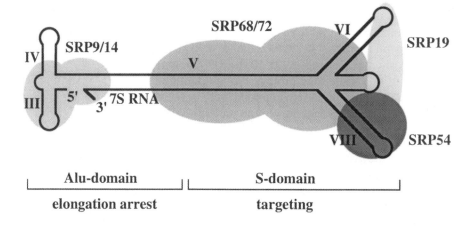

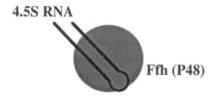

E. coli SRP

Fig. 3. Schematic presentation of the mammalian and *E. coli* SRPs. The RNA moieties are indicated by the black lines, whereas the proteins are indicated in grey shades with the molecular mass in kDa in the labels. The function of the Alu-domain and the S-domain of mammalian SRP is indicated below the top part of the figure.

Ffh and 4.5S RNA are essential for cell viability indicative of an important function. The conservation of targeting functions is underscored by the fact that human 7S RNA can functionally replace the 4.5S RNA of *E. coli* SRP (Ribes *et al.*, 1990). In turn, *E. coli* Ffh can replace SRP54 in a reconstituted mammalian SRP without affecting signal sequence recognition (Bernstein *et al.*, 1993). It should be noted that depletion of 4.5S RNA, but not of Ffh, affects protein translation (reviewed in Brown, 1991). This suggests that 4.5S RNA which is present in excess over Ffh, may have an additional cellular activity in protein synthesis. Conditional strains in which the cellular content of Ffh and 4.5S RNA can be regulated revealed a relatively mild but significant effect of SRP depletion on secretion of a subset of proteins (Poritz *et al.*, 1990; Ribes *et al.*, 1990; Phillips and Silhavy, 1992). Apparently, the SRP is involved in the targeting of some secretory proteins perhaps depending on growth conditions and the availability of other targeting factors. Consistently, the *E. coli* SRP has been shown to interact with the signal sequence of nascent secretory proteins in an *in vitro* cross-linking approach (Valent et al., 1995 and 1997).

Despite the mild effects of depletion of SRP on secretory proteins, the preferred substrates for targeting by the *E. coli* SRP appear to be IM proteins (reviewed in de Gier and Luirink, 2001). Biogenesis of most IM proteins is strongly hampered upon the depletion of Ffh, 4.5S RNA and FtsY. Furthermore, a recent genetic screen designed to identify components involved in the biogenesis of IM proteins in *E. coli,* yielded several mutations in the ffs gene that encodes 4.5S RNA (Tian *et al.*, 2000). Finally, cross-linking studies have demonstrated that the *E. coli* SRP binds with higher affinity to particularly hydrophobic targeting signals present in the N-terminus of nascent IM proteins (Valent *et al.*, 1995; Valent *et al.*, 1997). Intuitively, a co-translational mode of insertion of IM proteins would seem beneficial and prevent the exposure of hydrophobic domains that could result in undesired intra- and inter-molecular interactions (Newitt *et al.*, 1999). Interestingly, depletion of the SRP does not affect the targeting of all IM proteins. Some IM proteins are completely unaffected, whereas others show a variable degree of SRP dependence suggesting they can use alternative mechanisms for targeting and maintaining insertion competence. In conclusion, the exact substrate specificity of the SRP, i.e. the subset of IM proteins and perhaps secretory proteins that preferentially use the SRP pathway, is still a not completely resolved issue.

It has been suggested that the essential nature of the SRP and its receptor is related to their participation in the targeting of essential IM proteins. However, it was recently shown that the toxic accumulation of misfolded IM proteins is the primary cause of cell death under these conditions. Cells depleted of SRP try to counteract this effect by increasing the expression of heat shock regulated proteases (Bernstein and Hyndmann, 2001).

4.3 The SRP receptor FtsY

Initially, FtsY has been identified as a homologue of SRα based on sequence similarity (Bernstein *et al.*, 1989; Römisch *et al.*, 1989). FtsY is essential for cell viability (Gill and Salmond, 1990; Luirink *et al.*, 1994) and depletion of FtsY results in a mild accumulation of preproteins that also accumulate upon over-expression of Ffh. Based on these observations, FtsY has been proposed to be the functional *E. coli* homologue of the mammalian SRP receptor (Luirink *et al.*, 1994). *In vitro* studies have shown that FtsY interacts with reconstituted SRP (Miller *et al.*, 1994; Kusters *et al.*, 1995) as well as with purified ribosome-nascent-chain-SRP complexes (RNC-SRP complex) (Valent *et al.*, 1998). In addition, FtsY is required for the release of SRP from a nascent substrate in an *E. coli* cell-free homologous protein targeting assay (Valent *et al.*, 1998), consistent with its proposed function as receptor for the SRP. Using *in vitro* synthesized translocation intermediates, FtsY was shown to be one of the components of a soluble targeting complex that also contained SRP (Valent *et al.*, 1998). Based on these observations, it has been suggested that FtsY recognises and binds RNC-SRP in the cytosol. Hence, FtsY might play an important role in the targeting of RNC-SRP complexes, rather than acting as a fixed membrane bound receptor for the SRP.

FtsY was found to be evenly distributed between the cytosol and the inner membrane. The mechanism of association of FtsY with the cytoplasmic membrane appears to be unusual and complex perhaps reflecting the necessity of a dynamic, flexible and coordinated targeting mechanism (reviewed in Bibi *et al.*, 2001). FtsY does not contain hydrophobic α–helices that could form permanent transmembrane anchor sequences (Gill *et al.*, 1986). Besides, SRβ homologues, that might tether FtsY to the inner membrane, have not been identified in the *E. coli* genome sequence.

FtsY contains two distinct domains: a highly acidic amino-terminal domain (called the A-domain) and a carboxy-terminal GTP-binding domain (called the NG-domain). Both the A domain and the NG domain contribute to the targeting process (Zelazny *et al.*, 1997; de Leeuw, 2000). Two distinct mechanisms have been proposed for the membrane association of FtsY. In the first mechanism, FtsY associates with IM via a direct protein-lipid interaction, involving primarily the NG domain. In the second mechanism, FtsY associates with IM via a protein-protein interaction, involving primarily the A domain.

The interaction of FtsY with phospholipids has been examined in more detail (de Leeuw *et al.*, 2000). FtsY has a general affinity for phospholipids. In addition to the association with lipids, FtsY can insert into a phospholipid monolayer. Upon interaction with membrane lipids, the GTPase activity of FtsY (and probably also of the bound SRP) is stimulated, which might result in the dissociation of the targeting complex. This indicates that lipid binding of FtsY is important for the regulation of SRP-mediated protein targeting. So far, a protein that serves as a membrane-anchor for FtsY has not be identified.

Interestingly, a role for FtsY in the general targeting of ribosomes to the *E. coli* inner membrane has also been postulated, but the fate and function of these targeted ribosomes awaits further analysis (Herskovits *et al.*, 2000; Herskovits and Bibi, 2000; Bibi *et al.*, 2001).

4.4   Structural aspects

Ffh and FtsY, as well as their eukaryotic homologues SRP54 and SRα, share a conserved GTP binding domain (G domain) and a domain, that precedes the G domain, the so-called N domain (Herskovits *et al.*, 2000). In FtsY and SRα, a charged amino-terminal domain, the A domain, precedes the NG part of the protein. In SRα, this A domain is involved in co-translational membrane targeting to SRβ (Young *et al.*, 1995; Young and Andrews, 1996). Ffh and SR54 contain in addition to the N and G domains, a methionine-rich domain (the M domain). The M domain is located C-terminally from the NG domain.

The crystal structure of the NG domains of *T. aquaticus* Ffh (Freymann *et al.*, 1997) and *E. coli* FtsY (Montoya *et al.*, 1997; Fig. 4) have been resolved without bound nucleotide. The structures are almost identical and reveal the existence of three subdomains: the α-helical N domain, the G domain that appears related to the Ras-like GTPases and a surface exposed insertion box (I-box). This I-box is an α–β–α insertion in the G domain, that is unique to SRP-type GTPases. It is postulated to play a role in the interaction with regulatory proteins (Montoya *et al.*, 1997). Alternatively, it has been proposed that the I-box acts as a built-in nucleotide-exchange factor (Moser *et al.*, 1997). The crystal structure of the NG domain of *T. aquaticus* Ffh in complex with GDP has also been resolved (Freymann, 1999). This structure shows minimal rearrangement in the I-box upon the binding of GDP, indicating that this structural subdomain is not primarily involved in the nucleotide occupancy of the GTP binding pocket. The overall structure of the NG domain of Ffh and of FtsY reveals a wide open GTP binding site which explains the low affinity of both proteins for GTP (Freymann *et al.*, 1997, Freymann, 1999; Montoya *et al.*, 1997).

As described above, so far the crystal structure of only a number of isolated NG domains have been determined (Jovine *et al.*, 2000).

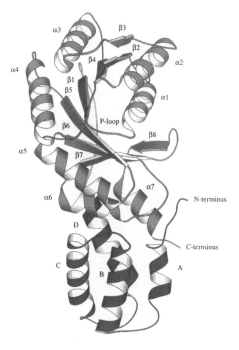

Fig. 4. Cα ribbon representation of the NG domain of FtsY. The three segments are indicated at the right. In the N domain, helices are labeled alphabetically in capital letters (A-D). In the G domain, helices and strands are labeled with numbers according to their appearance in the sequence.The P loop, in volved in GTP binding (Montoya *et al.*, 1997), and the N and C termini (close in space) are indicated.

Therefore, the effect of the M domain of Ffh on the structure of the NG domain and the effect of the FtsY A domain on the NG domain is not known. The first 10 amino acid residues in the structure of FtsY NG domain are not well defined (Montoya *et al.*, 1997). This may be due to the absence of stabilising interactions with the acidic A domain.

The M domain interacts with SRP RNA and with the signal or signal anchor sequence of nascent secretory and membrane proteins (Römisch *et al.*, 1990; Lütcke *et al.*, 1992). The crystal structure of the M domain from bacterial and human origin has also been resolved (Keenan *et al.*, 1998; Clemons *et al.*, 1999). The conserved structure shows a helix-turn-helix motif containing an arginine-rich helix that is required for RNA binding. Furthermore, the structure reveals a surface exposed, deep hydrophobic groove that is formed by four amphipatic α-helices. The dimensions of the groove are sufficient to accommodate the core of a signal sequence (approximately 17 amino acid residues in α–helical conformation). The groove surface is rich in methionine residues that have flexible side chains. This could provide some structural flexibility needed for the interaction with different signal sequences. Next to the groove lies a conserved hydrophobic loop that was postulated to close the groove in the absence of a functional signal sequence.

Recently, the M domain has been co-crystallized with a conserved part of the 4.5S RNA (Batey *et al.*, 2000) (Fig. 5).

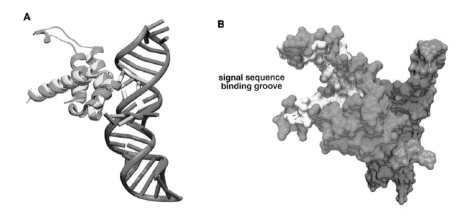

Fig. 5. Structure of the *E. coli* 4.5S RNA-M domain. **A**, ribbon representation of the complex with the RNA at the right and the protein part at the left.; **B**, molecular surface representation of the complex oriented to show the signal peptide binding groove.

The RNA was shown to change its structure drastically upon association with the M domain resulting in a very intimate contact. Surprisingly, a looped structure of the RNA lies adjacent to the groove in the M-domain forming a combined interaction platform that is two-third protein and one-third RNA. It has been suggested that the RNA participates directly in signal sequence binding, perhaps by electrostatic contact with the positively charged N region of the signal sequence. The combined hydrophobic and hydrophilic interactions might explain why mutations in the N region of the signal sequence can be compensated for by in creased hydrophobicity of the

H region (see above). Likewise, the signal anchor sequence of a membrane protein is often flanked by a positively charged region which might interact with the RNA in the groove. Direct proof for these assumptions has to await the co-crystallization of the M-domain, SRP RNA and a signal sequence.

How do the M and NG domains of Ffh communicate with each other? The interaction between *E. coli* SRP and FtsY at the membrane stimulates GTP binding to both the NG domain of Ffh and FtsY. The conformational change in the Ffh NG domain is probably sensed by the M domain which alters its conformation to trigger the release of the signal sequence. As a result, the signal sequence is free to enter the translocon. The conformational change in the M domain upon release of the signal sequence is in turn sensed by the NG domain. This initiates the mutual stimulation of GTPase activity by the SRP and FtsY which results in dissociation of the SRP-FtsY complex and recycling of the individual components for a new round of targeting. Likewise, the A and NG domain of FtsY must talk to each other. For instance, interaction of the A domain with membrane lipids stimulates the GTPase activity of the NG domain. More insight into this intra- and inter-molecular cross-talk requires structures of the full length Ffh (complexed with RNA, signal sequence and different nucleotides), the full length FtsY and of SRP-FtsY complexes. Even then, it should be realized that interactions with upstream (ribosome) and downstream (translocon) elements of the targeting pathway are not yet considered.

5.   SECB: FUNCTION AND STRUCTURE

SecB is a molecular chaperone which is required for the efficient transport of a subset of secretory proteins (reviewed in Pugsley, 1993; Fekkes and Driessen, 1999). It functions by binding to nascent and full length secretory proteins as they emerge from the ribosome in the cytosol (Kumamoto and Francetic, 1993). SecB interacts with the mature region of the preprotein (Randall *et al.*, 1990; Topping and Randall, 1994; Khisty *et al.*, 1995) to prevent its premature folding and to target the preprotein to the membrane in a transport competent state (Collier *et al.*, 1988; Hartl *et al.*, 1990). The SecB-preprotein complex binds SecA at the translocon and this interaction is thought to facilitate the transfer of the preprotein to SecA with the subsequent release of SecB (Hartl *et al.*, 1990; Fekkes *et al.*, 1997). The rules which govern a requirement for SecB assistance in protein transport are unknown. A variety of outer membrane and periplasmic proteins require SecB, while others do not. Moreover, SecB exhibits very broad substrate specificity involving either positively charged or hydrophobic regions of preproteins (Randall, 1992; Knoblauch *et al.*, 1999). This suggests that SecB utilization is not dictated by a requirement for specific high affinity binding sites. Nor does the mere presence of SecB-binding sites determine whether a preprotein is inherently SecB-dependent. Modifications in SecB-independent proteins can lead to SecB utilization, indicating that at least some proteins are endowed with the ability to employ SecB whether they need to or not (Kim *et al.*, 1992; Kim and Kendall, 1998). Furthermore, no consensus sequence or region consistent among proteins has been found responsible for conferring SecB dependence (Gannon *et al.*, 1989; Altman *et al.*, 1990). One model suggests that the requirement for SecB depends on the partitioning of the preprotein between transport productive and aggregation-prone, nonproductive pathways (Hardy and Randall, 1991; Diamond and Randall, 1997). A functional signal sequence does not directly bind SecB (Randall *et al.*, 1990), but it functions to impede the folding of the preprotein so that SecB can bind (Randall and Hardy, 1986; Diamond and Randall, 1997), whereas a defective signal sequence does not. This model has been called into question, because SecB can also bind fully folded proteins (Stenberg and Fersht, 1997) and because the rate of SecB-ligand association can be much faster than the rate of folding of SecB-independent proteins (Fekkes *et al.*, 1995).

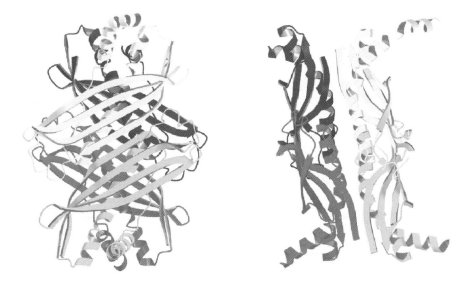

Fig. 6. Crystal structure of the SecB tetramer at the left and of two SecB dimers at the right, as determined by Xu *et al.*, 2000.

SecB functions as a tetramer of identical 16 kDa subunits, which appears to be a dimer of dimers that are in a dynamic equilibrium (Topping *et al.*, 2001). Recently, the crystal structure of SecB has been determined without bound ligand (Fig. 6) (Xu *et al.*, 2000). As in Ffh, there is a large surface exposed groove that presumably constitutes the ligand binding region. The tetramer contains four grooves, two on each side of the tetramer. Two grooves on the same side of the tetramer form one long continuous channel. The mid section of the channel is deep and rich in conserved aromatic amino acid residues with flexible side chains that might provide the necessary plasticity for promiscuous hydrophobic ligand binding. The end of the channel is more open and shallow and is proposed to interact with more polar regions in the substrate molecule. To fill the channels at both sides of the tetramer, a long stretch of the substrate molecule has to wrap around the tetramer. This might explain the complexity of SecB binding regions in preproteins as determined in biochemical assays. It remains to be determined how ligand binding influences the structure of different regions of the ligand binding site and other regions of the tetramer. SecB targets its substrate secretory protein to SecA via a direct interaction with the basic extreme C-terminal region of SecA (see Fekkes and Driessen, 1999). Consistently, two acidic regions in SecB that were previously proposed to be involved in an electrostatic interaction with SecA are highly solvent exposed in the crystal structure.

6. PATHWAY SPECIFICITY

Which features of the *E. coli* secretory and membrane proteins determine the preferred mode of targeting via the distinct SRP/FtsY or via the SecB/SecA pathway? The hydrophobicity of the targeting signal appears to be the major factor in the choice for a specific targeting route. The *in vitro* affinity of the targeting signal for the SRP increases with its hydrophobicity to a certain maximum (Valent *et al.*, 1997). In addition, secretory proteins that normally follow the SecB pathway or membrane proteins that are targeted spontaneously can be shunted into the SRP route *in vivo* by simply increasing the hydrophobicity of the targeting signal (Lee and Bernstein, 2001). Other factors that might influence SRP binding like translational pausing to increase the time window for co-translational SRP binding, have been proposed, but have not been addressed experimentally. Finally, SecB binding regions in the early mature region may play an important role in pathway specificity (Kim *et al.*, 2000).

In vitro cross-linking studies point to a role of trigger factor as an important decision maker (Beck *et al.*, 2000). Trigger factor is a cytosolic, partly ribosome-bound chaperone that interacts with nascent chains of various cytosolic and secreted proteins (Valent *et al.*, 1995; Hesterkamp *et al.*, 1996). Interaction of trigger factor with the early mature region of nascent pro-OmpA (an outer membrane protein) appears to prevent SRP binding to the signal sequence and opens the way to SecB binding upon prolongation of the nascent chain or upon release from the ribosome after completion of translation (Beck *et al.*, 2000). This proposed pivotal role of trigger factor is somewhat difficult to reconcile with the *in vivo* data described above and with the fact that trigger factor is not essential *in vivo* (Guthrie and Wickner, 1990).

SecA appears not to be essential per se for targeting via the SRP/FtsY pathway to the SecYEG translocon, but it may act at later stages in the insertion of IM proteins (see chapter 4).

Recent evidence identifies YidC as a possible novel component of the IM translocon (see above, and see chapter 4). Targeting to YidC may occur spontaneously, without the involvement of cytosolic targeting factors (Samuelson *et al.*, 2000). Alternatively, some artificial hybrid proteins depend both on SRP and on YidC, but not on the Sec-machinery, for efficient assembly into the IM, suggesting the possibility of a direct SRP/YidC targeting pathway (Fröderberg *et al.*, unpublished results).

It should be noted that the pathway specificity is not absolute for most substrate proteins. Both secretory and membrane proteins may use alternative targeting mechanisms when their preferred targeting route is compromised. In addition, the affinity of ribosomes for the Sec-translocon appears to be a conserved feature that also applies to *E. coli* (Prinz *et al.*, 2000). This affinity together with an exposed targeting signal might be sufficient to target a subpopulation of IM proteins to the translocon in the absence of SRP.

7. MODEL FOR SRP- AND SECB-MEDIATED TARGETING AND FUTURE PERSPECTIVES

A tentative model for targeting via the main SRP- and SecB targeting pathways is given in Fig. 7. SRP binds to a particularly hydrophobic targeting sequence that emerges from the ribosome thus forming a RNC-SRP complex. The interaction is dependent on the context of the ribosome (i.e. it only occurs co-translationally) and is not dependent on any nucleotide. The targeting complex may contact the SRP receptor FtsY in the cytosol, again in a nucleotide-independent process. The complex diffuses to the membrane and docks by a direct FtsY-lipid interaction. This docking is possibly assisted by an interaction between FtsY and a receptor protein yet to be identified. GTP is then bound to probably both FtsY and Ffh, which results in the release of the nascent chain form the SRP.

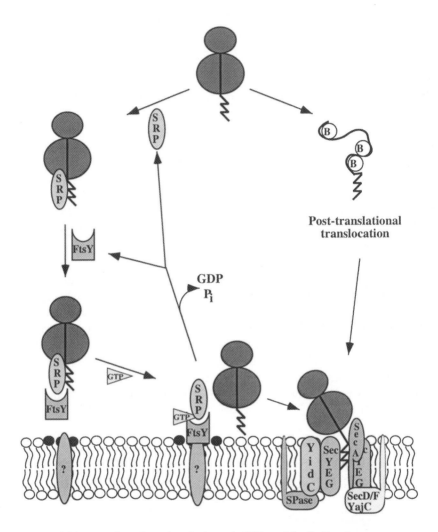

Fig. 7. Model for targeting of proteins via the main SRP- and the SecB pathway.

The nascent chain is transferred into the Sec-translocon in a poorly defined mechanism. Possibly, the affinity of the nascent chain and ribosome for the Sec-translocon contributes to this transfer step although the involvement of dedicated, yet unidentified factors can not be excluded at present. Upon hydrolysis of GTP, the SRP-FtsY complex dissociates and the individual components are recycled for another round of targeting.

Although not explicitly depicted in the model, targeting via SRP/FtsY to YidC seems also possible. Secretory proteins that have less hydrophobic targeting signals and which possess binding motifs for SecB in their mature region, bind SecB in a late co-translational or post-translational fashion. These proteins are received at the Sec-translocon via a direct SecB-SecA interaction.

This is the general picture that has emerged over the last years. Several fundamental questions need more work. What is the role of trigger factor in targeting? Does the SRP or components of the SRP play any role in translation? How is the prokaryotic SRP cycle regulated by GTP binding and hydrolysis? What is the mechanism of interaction between FtsY and the IM and how is the nascent chain transferred from the SRP to the translocon? What is the molecular basis for substrate selection by SecB? And finally, does YidC function in the reception of targeting complexes?

The study on these issues will benefit from the recently elucidated structures of targeting components, the genetic screening methods that have been developed, and the partial *in vitro* reconstitution of the targeting reactions.

8. REFERENCES

Akita, M., Sasaki, S., Matsuyama, S. and Mizushima, S. (1990) J Biol Chem 265: 8164-8169.
Altman, E., Bankaitis, V.A. and Emr, S.D. (1990) J Biol Chem 265: 18148-18153.
Batey, R.T., Rambo, R.P., Lucast, L., Rha, B. and Doudna, J.A. (2000) Science 287: 1232-1239.
Beck, K., Wu, L.-F., Brunner, J. and Müller, M. (2000) EMBO J 19: 134-143.
Berks, B.C., Sargent, F. and Palmer, T. (2000) Mol Microbiol 35: 260-274.
Bernstein, H.D., Poritz, M.A., Strub, K., Hoben, P.J., Brenner, S. and Walter, P. (1989) Nature 340: 482-486.
Bernstein, H.D., Zopf, D., Freymann, D.M. and Walter, P. (1993) Proc Natl Acad Sci USA 90: 5229-5233.
Bernstein, H.D. and Hyndman, J.B. (2001) J Bacteriol 183: 2187-2197.
Bibi, E., Herskovits, A.A., Bochkareva, E.S., Zelazny, A. (2001) Trends Biochem Sci. 26: 15-16.
Brown, S. (1991) The New Biologist 3: 430-438.
Clemons, W.M., Gowda, K., Black, S.D., Zwieb, C and Ramakrishnan, V. (1999) J Mol Biol 292: 697-705.
Collier, D.N., Bankaitis, V.A., Weiss, J.B. and Bassford, P.J., Jr. (1988) Cell 53: 273-283.
Cou, M.M. and Kendall, D.A. (1990) J Biol Chem 265: 2873-2880.
Dalbey, R.E. and von Heijne, G. (1992) Trends Biochem Sci 17: 474-478.
Dalbey, R.E., Lively, M.O., Bron, S and van Dijl, J.M. (1997) Protein Sci 6: 1129-1138.
De Gier, J.W. and Luirink, J. (2001) Mol Microbiol 40: 314-322.
De Leeuw, E.P.H., te Kaat, K., Moser, C., Menestrina, G., Demel, R., de Kruyff, B., Oudega, B., Luirink, J. and Sinning, I. (2000) EMBO J 19: 531-541.
De Leeuw, E.P.H. (2000). Thesis, Vrije Universiteit Amsterdam

20

De Vrije, T., Batenburg, A.M., Killian, A. and de Kruijff, B. (1990) Mol Microbiol 4: 143-150.

Diamond, D.L. and Randall, L.L. (1997) J Biol Chem 272: 28994-28998.

Driessen, A.J.M., de Wit, J.G., Kuiper, W., van der Wolk, J.P.W., Fekkes, P, van der Does, C., van Wely, K., Manting, E.H. and den Blaauwen, T. (1995) Biochem Soc Trans 23; 981-985.

Fekkes, P., den Blaauwen, T and Driessen, A.J.M. (1995) Biochemistry 34: 10078-10085.

Fekkes, P., van der Does, C. and Driessen, A.J.M. (1997) EMBO J 16: 6105-6113.

Fekkes, P. and Driessen, A.J.M. (1999) Microbiol Mol Biol Rev 63: 161-173.

Freymann, D.M., Keenan, R.J., Stroud, R.M. and Walter, P (1997) Nature 385: 361-364.

Freymann, D.M., Keenan, R.J., Stroud, R.M. and Walter, P (1999) Nature Strc Biol 6: 793-800.

Fröderberg, L., Samuelson, J.C., Chen, M., Luirink, J., Dalbey, R., von Heyne, G. and de Gier, J.-W. (2001).

Gannon, P.M., Li, P. and Kumamoto, C.A. (1989) J bacteriol 171: 813-818.

Gennity, J., Goldstein, J. and Inouye, M. (1990) Biomemr. 22: 233-269.

Gill, D.R., Hatfull, G.F. and Salmond, G.P.C. (1986) Mol Gen Genet 205: 134-145.

Gill, D.R. and Salmond, G.P.C (1990) Mol Microb 4:575-583.

Guthrie, B. and Wickner, W. (1990) J Bacteriol 172: 5555-5562.

Hardy, S.J.S. and Randall, L.L. (1991) Science 251: 439-443.

Hartl, F.-U., Lecker, S., Schiebel, E., Hendrick, J.P. and Wickner, W. (1990) Cell 63: 269-279.

Hesterkamp, T., Hauser, S., Lütcke, H. and Bukau, B. (1996) proc Natl acad sci USA 93: 4437-4441.

Herskovits, A.A. and Bibi, E. (2000) Proc Natl Acad Sci USA 97: 4621-4626.

Herskovits, A.A., Bochkareva, E.S. and Bibi, E. (2000) Mol Microbiol. 38: 927-39.

Hikita, C. and Mizushima, S. (1992) J Biol Chem 267: 12375-12379.

Izard, J.W. and Kendall, D.A. (1994) Mol Microbiol 13: 765-773.

Jovine, L., Hainzl, T., Oubridge, C., Scott, W.G., Li, J., Sixma, T.K., Wonacott, A., Skarzynski, T. and Nagai, K. (2000) Structure Fold Des 8: 527-540.

Keenan, R.J., Freymann, D.M., Walter, P. and Stroud, R.M. (1998) Cell 94: 181-191.

Khisty, V.J., Munske, G. and Randall, L.L. (1995) J Biol Chem 270: 25920-25927.

Kim, J., Lee, Y., Kim, C. and Park, C. (1992) J Bacteriol 174: 5219-5227.

Kim, J. and Kendall, D.A. (1998) J Bacteriol 180: 1396-1401.

Kim, J., Luirink, J. and Kendall, D.A. (2000) J Bacteriol 182: 4108-4112.

Knoblauch, N.T., Rudiger, S., Schonfeld, H.J., Driessen, A.J.M., Schneider-Mergener, J. and Bukau, B. (1999) J Biol Chem 274: 34219-34225.

Kumamoto, C.A. and Francetic, O. (1993) J. Bacteriol 176: 2184-2188.

Kusters, R., Lentzen, G., Eppens, E, van Geel, A., van der Weijden, C.C., Wintermeyer, W. and Luirink, J. (1995) FEBS Lett 372: 253-258.

Lee, C., Li, P., Inouye, H., Brickman, E.R. and Shanmugan, K.T. (1989) J Bacteriol 171: 4609-4616.

Lee, H.C. and Bernstein, H.D. (2001) Proc Natl Acad Sci USA 98: 3471-3476.

Luirink, J., ten Hagen-Jongman, C.M., van der Weijden, C.C., Oudega, B., High, S., Dobberstein, B. and Kusters, R. (1994) EMBO J 13: 2289-2296.

Lütcke, H.,High, S., Römisch, K., Ashford, A.J. and Dobberstein (1992) EMBO J 11: 1543-1551.

Lütcke, H. (1995) Eur J Biochem 228: 531-550.

MacFarlane, J. and Müller, M. (1995) Eur J Biochem 133: 766-771.

Matlack, K.E., Mothes, W. and Rapoport, T. (1998) Cell 92: 381-390.

Miller, J.D., Bernstein, H.D. and Walter, P. (1994) Nature 367: 657-659.

Montoya, G., Svensonn, C., Luirink, J. and Sinning, I. (1997) Nature 385: 365-368.

Moser, C., Mol, O., Goody, R.S. and Sinning, I (1997) Proc. Natl. Acad. Sci. USA 94: 11339-11344.

Newitt, J.A., Ulbrandt, N.D. and Bernstein, H.D. (1999) J Bacteriol 181: 4561-4567.

Phillips, G.J. and Silhavy, T.J. (1992) Nature 359: 744-746.

Phoenix, D.A., Kusters, R., Hikita, C., Mizushima, S. and de Kruijff, B. (1993) J Biol Chem 268: 17069-17073.

Poritz, M.A., Bernstein, H.D., Strub, K., Zopf, D., Wilhelm, H. and Walter, P. (1990) Science 250: 1111-1117.

Prinz, W.A., Spiess, C., Ehrmann, M., Schierle, C. and Beckwith, J (1996) EMBO J 15: 5209-5217.

Prinz, A., Behrens, C., Rapoport, T.A., Hartmann, E. and Kalies, K.U. (2000) EMBO J. 19: 1900-1906.

Pugsley, A.P. (1993) Microbiol Rev 57: 50-108.

Puzzis, J.W., Fikes, J.D. and Bassford, P.J., Jr. (1989) J Bacteriol 171: 2303-2311.

Puzzis, J.W., Strobel, S.M. and Bassford, P.J., Jr. (1992) J Bacteriol 174: 92-101.

Randall, L.L. and Hardy, S.J.S. (1986) Cell 46: 921-928.

Randall, L.L., Topping, T.B. and Hardy, S.J.S. (1990) Science 248: 860-863.

Randall, L.L. (1992) Science 257: 341-345.

Ribes, V., Römisch, K., Giner, B., Dobberstein, B. and Tollervey, D. (1990) Cell 63: 591-600.

Römisch, K., Webb, J., Herz, J., Prehn, S., Vingron, M. and Dobberstein, B. (1989) Nature 340: 478-482.

Römisch, K., Webb, J., Lingelbach, K., Gausepohl, H. and Dobberstein, B. (1990) J Cell Biol 111: 1793-1802.

Samuelson, J.C., Chen, M., Jiang, F., Moller, I., Wiedmann, M., Kuhn, A., Phillips, G.J. and Dalbey, R.E. (2000) Nature 406: 637-641.

Scotti, P.A., Urbanus, M.L., Brunner, J., de Gier, J.W., von Heijne, G., van der Does, C., Driessen, A.J., Oudega, B. and Luirink, J. (2000) EMBO J 19: 542-549.

Stenberg, G. and Fersht, A.R. (1997) J Mol Biol 274: 268-275.

Tian, H., Boyd, D. and Beckwith, J. (2000) Proc Natl Acad Sci USA 97: 4730-4735.

Topping, T.B. and Randall, L.L (1994) Protein Sci 3: 730-736.

Topping, T.B., Woodbury, R.L., Diamond, D.L., Hardy, S.J.S. and Randall, L.L. (2001) J Biol Chem 276: 7437-7441.

Valent, Q.A., Kendall, D.A., High, S., Kusters, R., Oudega, B. and Luirink, J. (1995) EMBO J 14: 5494-5505.

Valent, Q.A., de Gier, J.W.L., von Heijne, G., Kendall, D.A., ten Hagen-Jongman, C.M., Oudega, B. and Luirink, J. (1997) Mol Microbiol 25: 53-64.

Valent, Q.A., Scotti, P.A., High, S., de Gier, J.W., von Heijne, G., Lentzen, G., Wintermeyer, W., Oudega, B. and Luirink, J. (1998) EMBO J 17: 2504-2512.

Von Heijne, G. (1984) J Mol Biol 173:243-251.

Von Heyne, G. (1985) J Mol Biol 184: 99-105.

Von Heyne, G. (1990) J Membr Biol 115:195-201.

Xu, Z., Knafels, J.D. and Yoshino, K. (2000) Nature Struct Biol 7: 1172-1177.

Young, J.C., Ursini, J., Legate, K.R., Miller, J.D., Walter, P. and Andrews, D.W. (1995) J Biol Chem 270: 15650-15657.

Young, J.C. and Andrews, D.W. (1996) EMBO J 15: 172-181.

Zelazny, A., Seluanov, A., Cooper, A. and Bibi, E. (1997) Proc Natl Acad Sci USA 94: 6025-6029.

Chapter 2

THE SEC TRANSLOCASE

Chris van der Does, Nico Nouwen and Arnold J.M. Driessen

Department of Microbiology, Groningen Biomolecular Sciences and Biotechnology Institute,
University of Groningen
Kerklaan 30
9751 NN Haren, NL

1.   INTRODUCTION

In bacteria, the major route of protein translocation across the cytoplasmic membrane is the so-called "Sec-pathway". Secretory proteins that use this route are synthesized as precursors (preproteins) with an amino-terminal extension, the signal peptide. The signal peptide consists of three domains: a positively charged N-terminus (N-domain), a hydrophobic central region (H-domain) and a C-terminal domain containing the signal peptidase cleavage site. The signal peptide directs the preprotein to the translocation machinery and is needed for initiation of the translocation process. It also slows down the folding of the mature domain of a secretory preprotein, and thereby increases the time window for molecular chaperones like SecB to interact. SecB is a molecular chaperone that stabilizes preproteins in a translocation-competent (non-aggregated, loosely folded) state and that targets them to the membrane (Fig. 1, route b) (reviewed by Fekkes and Driessen, 1999). Preproteins with a very hydrophobic signal peptide and most membrane proteins are targeted to the cytoplasmic (inner) membrane via the signal recognition particle (SRP)-pathway (Fig. 1, route c). SRP binds to the H-domain of the signal peptide when it emerges from the ribosome, and subsequently targets the ribosome-bound nascent chain complex via the SRP receptor FtsY at the membrane to the translocation machinery. Preprotein translocation is mediated by a multicomponent membrane complex that consists of the membrane-associated ATPase SecA,

B. Oudega (ed.), Protein Secretion Pathways in Bacteria, 23–49.
© 2003 *Kluwer Academic Publishers. Printed in the Netherlands.*

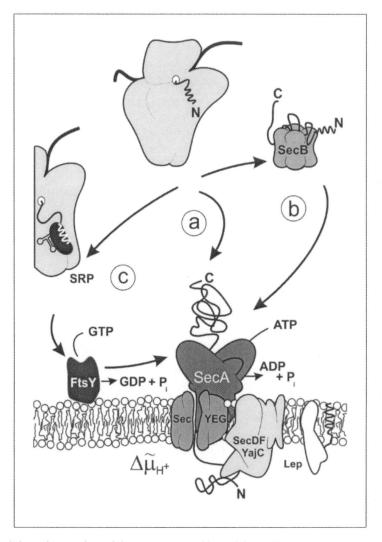

Fig. 1. Schematic overview of the components of bacterial translocase. A preprotein with a signal sequence can be targeted to the translocase by three different routes. (**a**) The preprotein post-translationally associates with the translocase, or (**b**) utilizes the molecular chaperone SecB as a targeting factor, or (**c**) the signal peptide of the nascent preprotein associates with the signal recognition particle (SRP), and via FtsY is targeted to the translocase. In the latter process, the transfer of the preprotein from SRP and FtsY to the translocase requires the hydrolysis of GTP. After targeting, the preprotein is translocated across the membrane via an integral membrane complex that consists of SecY, SecE and SecG. Energy in the form of ATP hydrolysis by SecA drives the translocation across the membrane, a process that is stimulated by the pmf. During or shortly after translocation the signal sequence is removed by the enzyme signal peptidase (Lep) and the mature protein is released into the periplasm. SecD, SecF and YajC are other membrane proteins involved in protein translocation, but their exact function is not known.

and a large integral membrane domain consisting of SecY, SecE, SecG, SecD, SecF and YajC as separate subunits. The essential (also termed '*core*') subunits of this complex are the SecY, SecE and SecA proteins. Energy in the form of ATP and the proton motive force (pmf) is used to drive the translocation of preproteins across the cytoplasmic membrane. During or shortly after the translocation process the signal peptide of the preprotein is removed by signal peptidase, and the mature domain is released into the periplasm (Gram-negative bacteria) or medium (Gram-positive bacteria). In the case of Gram-negative bacteria some proteins are subsequently targeted to the outer membrane and integrated into or translocated across this membrane (see chapter 5 and the following chapters). In this chapter, the different subunits of the bacterial preprotein translocase will be described and their function will be discussed.

2. SUBUNITS OF THE TRANSLOCASE

2.1 SecA

SecA is a motor protein that couples the hydrolysis of ATP to the stepwise translocation of preproteins across the membrane. The *secA* gene was discovered in a screen for proteins that cause a conditionally pleotropic defect in preprotein secretion (Oliver and Beckwith, 1981). The *secA* gene encodes a 102 kDa protein of 901 amino acid residues (Oliver and Beckwith, 1982). In the cell, SecA distributes in the cytosol and at the membrane. SecA binds with low affinity to phospholipids, but recognizes the SecYEG complex with high affinity. SecA binds the binary SecB-preprotein complex and interacts directly with the signal sequence of the preprotein. Therefore, it is an essential component in the targeting of preproteins to the SecYEG-complex. SecA is the only ATPase involved in preprotein translocation. It has a low endogenous ATPase activity that is increased by its interaction with the SecYEG complex and preproteins. This ATPase activity is termed "SecA Translocation ATPase" as it is coupled to preprotein translocation (Lill *et al.*, 1989).

SecA functions as a homodimer (Akita *et al.*, 1991; Driessen, 1993) in which each monomer contains two thermodynamically independently folding domains (N-and C-domain) (den Blaauwen *et al.*, 1996). In addition, the monomer contains two essential nucleotide-binding sites (NBSs) (Mitchell and Oliver, 1993). NBS-1 is located in the N-domain and is responsible for high-affinity ATP binding ($K_d = 0.13$ μM). It contains the typical Walker A and Walker B motifs. The second NBS (NBS-2) is proposed to function as a low-affinity ATP binding (K_d = 340 µM) and is located in the C-domain. This region, however, does not seem to function as an independent nucleotide binding site, but instead acts as a regulatory domain that controls the hydrolysis of ATP at NBS-1 (Nakatogawa et al., 2000a; Sianidis et al. 2001). Both regions are essential for the preprotein-stimulated SecA ATPase activity and preprotein translocation (Klose et al., 1993; van der Wolk et al., 1993; Mitchell and Oliver, 1993; Sato et al., 1996). ADP-binding to SecA promotes the interaction between the N-and C-domains of SecA (den Blaauwen et al., 1996) (Sianidis et al., 2001) yielding a compact SecA conformation (den Blaauwen et al., 1996; den Blaauwen et al., 1999). Various biochemical methods indicate that the soluble SecA undergoes conformational changes upon nucleotide binding (den Blaauwen et al., 1996; Shinkai et al., 1991; Sianidis et al., 2001), but these could not be detected by small angle X-ray scattering (Shilton et al., 1998). Unfortunately, the latter study did not access the SecA activity after the exposure to the high-energy irradiation.

The signal sequence of the preprotein has been crosslinked to a region of SecA just adjacent to NBS-1, i.e., amino acid residues 267 to 340 (Akita et al., 1990). Mutagenesis of Tyr-326 in this region results in strong protein translocation defect and a loss of SecA translocation ATPase activity, and this has been attributed to a defect in preprotein release (Kourtz and Oliver, 2000). Mutations in SecA that suppress the export defect caused by signal sequence mutations, i.e. the so-called prlD suppressors (protein localization) are located throughout the entire primary sequence of SecA (for overview of SecA mutations see den Blaauwen and Driessen, 1996). Remarkably, many prlD and azi mutations coincide (Huie and Silhavy, 1995). Azi mutations render SecA resistant to azide, an inhibitor of the translocation ATPase (Oliver et al., 1990). Azi- and PrlD-SecA proteins have a reduced affinity for ADP, an elevated membrane ATPase activity, and an altered conformation (Schmidt et al., 2000). These findings suggest that alterations in the turnover of SecA may lead to signal sequence suppression.

The extreme C-terminus of SecA has been implicated in various catalytic activities, such as SecB binding (Breukink et al., 1995; Fekkes et al., 1997; Fekkes et al., 1999), and lipid interaction (Breukink et al., 1995). In conjunction with N-terminal and central regions of SecA, the C-terminus of the SecYEG-bound SecA protein is accessible to membrane-impermeable reagents and proteases added from the periplasmic face of the membrane (van der Does et al., 1996; Ramamurthy and Oliver, 1997; Eichler and Wickner, 1998). This indicates a complex membrane-topology of the SecYEG-bound SecA and suggests that domains of SecA penetrate the membrane. Alternatively, the proteases/membrane impermeable reagents may gain access to regions of SecA via the protein conducting pore formed by SecYEG complex.

2.2   SecY

The localization of the gene encoding SecY was discovered in a screen for mutations, which could suppress signal sequence mutations. This genetic locus was termed *prlA* (Emr *et al.*, 1981), and shown to be the first of two open reading frames downstream of the *spc* ribosomal operon followed by the gene encoding the adenylate kinase protein (Cerretti *et al.*, 1983). The genetic organization of this locus is extremely well conserved in prokaryotes. Hydropathy analysis predicts that SecY harbors 10 hydrophobic transmembrane segments (TMS) (Akiyama and Ito, 1987) (Fig. 2). SecY is present in all eubacteria, and homologues of SecY are even present in archaea, in the endoplasmic reticulum (Sec61α) of eukaryotes and in plant thylakoids. When the sequences of different SecY proteins are aligned, only the N- and C-terminal cytoplasmic domains, TMS 3, 4 and 6, and the periplasmic loops except P1 (periplasmic loop 1) are found not to be conserved. Dominant loss of function mutations have been identified in cytosolic loop 5, and this region, together with C6 has been proposed to be important for a functional SecY-SecA interaction (Nakatogawa *et al.*, 2000b). The conditionally lethal mutations are all located in the conserved domains. In addition, SecY contains a large numbers of strong *prl* mutations (Flower *et al.*, 1995; and references therein), and these are mainly found in the conserved regions. Related to *prl* mutations are a number of mutations in TMS 7 (*prlA4-1*, F286Y, *prlA401*, S282R, *secY121*, I290T, and *secY161*, P287L) that specifically block staphylokinase translocation (Sako and Iino, 1988; Iino and Sako, 1988).

The protein translocation activity associated with SecY was first purified and reconstituted from octylglucoside-solubilized *E. coli* inner membranes (Driessen and Wickner, 1990). Two other proteins, SecE and SecG, were found to co-purify with SecY (Brundage *et al.*, 1990; Akimaru *et al.*, 1991; Driessen *et al.*, 1991). When purified separately, the translocation activity of reconstituted SecY increases proportional with the amount of co-reconstituted SecE (Tokuda *et al.*, 1991). SecY and SecE are interacting proteins. In the absence of SecE, SecY is unstable and becomes degraded by FtsH (Kihara *et al.*, 1995). Overproduction of SecY in FtsH mutant cells has a deleterious effect on cell growth and protein export suggesting that elimination of uncomplexed SecY is important for cell viability (Kihara *et al.*, 1995). Pulse chase experiments demonstrated that there is no exchange of SecE molecules between SecYE complexes (Joly *et al.*, 1994; Taura *et al.*, 1993) indicating that SecY and SecE are tightly interacting proteins.

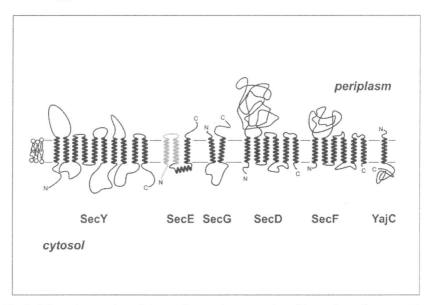

Fig. 2. Schematic overview of the topology and conservation of integral membrane proteins involved in protein translocation.

2.3 SecE

SecE has first been identified in a screen for mutations that increase the expression of the *secA* gene. The basis of this screening method is that mutations that cause protein secretion defects in general lead to elevated levels of SecA in the cell. Further screens for extragenic suppressors of a LamB signal sequence mutation also showed that dominant suppressor mutations (called *prlG*) were linked to the *secE* gene (Stader *et al.*, 1989). The gene is located at approximately 90 min on the *E. coli* chromosome and forms an operon with the *nusG* gene (Riggs *et al.*, 1988). The *secE-nusG* genes are co-transcribed (Downing *et al.*, 1990), and the genetic region surrounding *secE* seems well conserved in prokaryotes. Since *secE* encodes a rather small integral inner membrane protein, it is often overlooked in the annotation of bacterial genomes. In contrast to what has been suggested in the literature (Yang *et al.*, 1997), *secE* is also present in the smallest bacterial genome of *Mycoplasma genitalium* where it is localized downstream from *nusG*.

The *E. coli* SecE contains three membrane-spanning domains (Schatz *et al.*, 1989) (Fig. 2). Most bacterial SecE proteins, however, are much smaller and contain the conserved C2 region and the third TMS of *E. coli* SecE. Also in *E. coli*, this region suffices to support growth (Schatz *et al.*, 1991) and protein translocation (Nishiyama *et al.*, 1992). SecE proteins of various bacteria with only one TMS domain can complement the lethal *secE* deletion in *E. coli*. Replacement of the third TMS domain of the *E. coli* SecE for an unrelated TMS of the MalF protein yields a SecE protein that can complement the Δ*secE* strain at 37° C but not at 30 or 42° C (Murphy and Beckwith, 1994). This suggests that this TMS mainly fulfills a passive function in maintaining the proper topological arrangement of the conserved cytoplasmic domain (C2) that precedes TMS3. However, two mutations in TMS3 (L111R and D112P) have been found that completely abolish the SecE function (Kontinen *et al.*, 1996a; Pohlschroder *et al.*, 1996). In addition, TMS3 like the P2 loop, contains strong suppressors of defective signal sequences (*prlG1*, L108R *prlG2*, S105P; *prlG3*, S120F; *prlG8*, ΔV116-R117 (Flower *et al.*, 1994)), suggesting that TMS3 is involved in a catalytic function of SecE.

The eukaryotic homologue of SecE is Sec61γ which interacts with Sec61α (Bairoch *et al.*, 1997). The C2 domain is the most conserved sequence between SecE and Sec61γ. Most single amino acid substitutions in the C2 domain cause mild secretion defects and have little effect on growth at 37° C (Murphy and Beckwith, 1994). However, double substitutions or the deletion of the conserved proline residues cause severe growth and translocation defects (Pohlschroder *et al.*, 1996; Kontinen *et al.*, 1996a). Several regions contribute to the formation of a stable SecE-SecY complex (Fig. 3). The C-terminal part of SecE is essential for the interaction with SecY (Nishiyama *et al.*, 1992), and mutants in this region destabilize the SecE-SecY complex (Pohlschroder *et al.*, 1996). Sites of direct contact between SecY and SecE have also been suggested by the synthetic lethality of certain combinations of *prlA* and *prlG* mutations. Combination of *prl* mutations in the P1 loop of SecY (F67C, S68P) and the P2 loop in SecE (S120F), or in TMS10 (L407R, I408N) or TMS7 (I278N) in SecY with mutations in TMS3 of SecE (L108R) are lethal, suggesting that these regions interact (Flower *et al.*, 1995) (Fig. 3). By means of cysteine-scanning mutagenesis, indeed, interactions between the P1 loop of SecY and P2 loop of SecE (Harris and Silhavy, 1999), between TMS2 of SecY and TMS3 of SecE (Kaufmann *et al.*, 1999) and between TMS7 of SecY and TMS3 of SecE have been demonstrated. Cytoplasmic loop C4 of SecY also plays an important role in the SecY-E interaction. This was discovered using a dominant negative *secY* allele, *secY*^{d}, which contains a major deletion (R372-T374) at the interface of C5 and TMS9 of SecY (Shimoike *et al.*,

1992). The deletion inactivates SecY but preserves its ability to interact with SecE and thereby sequesters the SecE in an inactive complex. This lethal effect can be overcome by over-expressing SecE (Baba *et al.*, 1994) or by second site mutations that disrupt the SecE binding. All found mutations clustered in the C4 loop of SecY. Cysteine scanning mutagenesis of TMS3 of SecE revealed that SecE of one SecYEG subunit is in close proximity to a SecE molecule in another SecYEG complex (Kaufmann *et al.*, 1999). These contacts are found all along the SecE TMS3 suggesting that SecYEG complexes assemble into a higher oligomeric state.

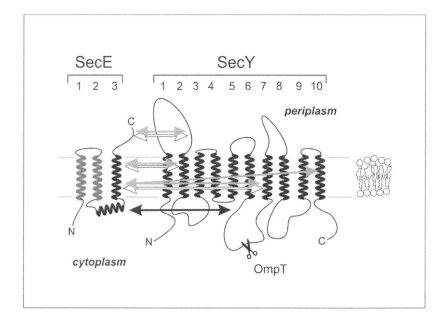

Fig. 3. Sites of interactions between the SecY and SecE proteins. Synthetically lethal combinations of *prlA* and *prlG* mutants suggest interactions between the P1 loop of SecY with the P2 loop of SecE, and interactions between transmembrane segment number 7(TMS7)and TMS10 of SecY with TMS3 of SecE (Grey arrows). An interaction between the P1 loop of SecY and the P2 loop of SecE and between TMS2 and TMS7 of SecY and TMS3 of SecE was confirmed with cysteine directed crosslinking (White arrows). Mutations in the C4 loop of SecY diminish the interaction between SecY and SecE, probably by lessening the interaction with the C2 loop op SecE (Black arrow).

2.4   SecG

SecG was not identified through a genetic screen, but found via co-purification with the SecYE complex (Brundage *et al.*, 1990; Nishiyama *et al.*, 1993). SecG is not essential for protein translocation, as authentic translocation could be reconstituted with purified SecA, SecY, SecE and acidic phospholipids (Akimaru *et al.*, 1991). However, when SecG is co-reconstituted with SecYE, it strongly enhances the protein translocation activity (Nishiyama *et al.*, 1993). The *secG* gene is located at 69 min on the *E. coli* chromosome, and encodes a hydrophobic protein of 11.4 kDa. Deletion of *secG* renders cells cold sensitive (Cs) for growth and protein export but this phenomena is depend on the genetic background of the *E. coli* strains (Bost and Belin, 1995) and was found to relate to a defect in phospholipid biosynthesis (Flower, 2001). SecG therefore seems to be important for the efficiency of protein translocation, particularly at lower temperature (Nishiyama *et al.*, 1994). *In vitro* studies also indicate that SecG is important in the absence of the pmf (Hanada *et al.*, 1996). SecG does not affect the binding of SecA to SecYE (Hanada *et al.*, 1994).

SecG contains two TMSs (Nishiyama *et al.*, 1996). Proteolysis studies suggest that the topological of these two TMSs completely reverses upon the membrane insertion of SecA (Nishiyama *et al.*, 1996). It was speculated that this remarkable event is needed to facilitate membrane insertion of SecA, in particular at lower temperature when the microviscosity is high. The latter suggestion is based on a genetic screen for suppressors of the Cs growth defect of a ΔsecG strain that yielded genes involved in phospholipid biosynthesis (Kontinen and Tokuda, 1995) (Shimizu *et al.*, 1997) (Kontinen *et al.*, 1996b). By increasing the total amount of anionic phospholipids, protein translocation becomes elevated and this would render cells less dependent on SecG (Suzuki *et al.*, 1999). Mutations in the translocase membrane subunits often result in a Cs growth phenotype. It has been suggested that this reflects the cold sensitivity of protein export itself (Pogliano and Beckwith, 1993) which would become exacerbated by any mutations that interfere with the activity of the translocase. The mechanism underlying thermal sensitivity of this process is not known, but potential cold-sensitive steps could be the insertion of the signal sequence and/or SecA into the membrane, the topology inversion of SecG and/or the oligomerization of the SecYEG complex into a protein conducting channel (see section on translocase structure).

Bacterial SecG homologues have been identified by sequence analysis, but functional evidence has been obtained only in *B. subtilis* (van Wely *et al.*, 1999). The *B. subtilis* ΔsecG strain shows a mild Cs phenotype for growth, which is exacerbated by high-level production of secretory proteins. Sequence comparison reveals that conserved amino acid residues are present throughout the entire SecG sequence, while the C-terminal periplasmic domain of Gram-negative SecG proteins is lacking in Gram-positive bacteria. The genetic localization of SecG is not conserved and so far no homologues have been found in eukaryotes and archaea. Although no suppressors of signal sequence mutations were mapped to *secG* by the original screens, a direct selection of SecG mutants yielded *prl* alleles of *secG* (*prlH*). These mutations are distributed along the SecG sequence but do not map to the three residues that yield a defective SecG protein. Some of the *prlH* alleles are as strong as the strongest known *prlG* alleles of *secE*, and it has been suggested that SecG also contributes to signal sequence recognition (Bost and Belin, 1997).

Little is known about the interaction between SecG and the SecYE complex. SecG increases the stability of SecY (Nishiyama *et al.*, 1995), the SecYE complex (Homma *et al.*, 1997) and of an unstable truncate of SecE (Nishiyama *et al.*, 1995). Other studies suggest that SecG interacts mainly with SecY (Homma *et al.*, 1997; Bost and Belin, 1995) and SecA (Bost and Belin, 1995), and not with SecE. Although binding of SecA to the SecYE complex is not affected by SecG (Hanada *et al.*, 1994), the topology inversion of SecG is hampered when a partially functional SecA truncate is used that lacks eight N-terminal amino acid residues (Mori *et al.*, 1998). SecA is thought to interact with both SecY and SecG since a SecA mutant (SecA36) suppressed both the Cs sensitive phenotype of the SecY205 and Δ*secG* mutant (Matsumoto *et al.*, 1998). Various lines of evidence suggest that the SecG and SecA conformational changes are coupled (Matsumoto *et al.*, 1997), but the interaction has not been demonstrated biochemically.

2.5   SecD, SecF and YajC

SecD was identified in a screen for mutants altered in export of PhoA-LacZ and LamB-LacZ fusion proteins (Gardel *et al.*, 1987). These mutants have a Cs phenotype and accumulate precursors in the cell. Analysis of the genomic region around the *secD* gene showed that the presence of another gene, *secF*, that when mutated exhibits the same phenotype as *secD* mutants. The first gene in the *secDF* operon, *yajC* or *orf12*, encodes for a small membrane protein with one TMS and a large cytosolic domain. YajC is not essential for growth nor does it affect secretion (Gardel *et al.*, 1990a). However, co-immunoprecipitation studies showed that YajC forms a complex with both SecYEG and SecDF (Duong and Wickner, 1997a). SecD, SecF and double depletion strains exhibit a severe export defects at 37°C and are barely viable. SecD and SecF are integral membrane proteins with 6 TMSs and a large periplasmic domain (Gardel *et al.*, 1990b; Pogliano and Beckwith, 1994) (Fig. 2). In *B. subtilis* and some other bacteria, SecD and SecF form a single polypeptide with 12 putative TMS domains (Bolhuis *et al.*, 1998). YajC is also conserved among bacteria, and its gene is often located in an operon together with *secD* and *secF*.

The exact role of SecD, SecF, and YajC in protein translocation is not clear. In *E. coli*, SecD has been suggested to be involved in the release of proteins from the periplasmic side of the cytoplasmic membrane (Matsuyama *et al.*, 1993). This hypothesis is based on *in vivo* studies that showed that an antibody against SecD resulted in preprotein accumulation and prevented the release of mature MBP and OmpA from spheroplasts into the medium. The large periplasmic loops of SecD and SecF have also been suggested to act as valve that prevents passage of protons/ions through the protein-conducting channel that is formed by the SecYEG complex. Indeed, inner membrane vesicles from a strain in which SecD and SecF are depleted are unable to maintain a pmf (Arkowitz and Wickner, 1994). Since protein translocation is strongly dependent on the pmf, it has been argued that the secretion defect in SecDF depletion strains indirectly resulted from the lack of a pmf. SecDFYajC also has been suggested to control protein translocation by regulating the ATP-driven cycle of SecA membrane insertion and de-insertion (Duong and Wickner, 1997a). Overproduction of SecD and SecF stabilizes SecA in the membrane-inserted state (Kim *et al.*, 1994; Economou *et al.*, 1995). By stabilizing the SecA membrane inserted form, the SecDFYajC complex slows the movement of preprotein in transit against both reverse and forward translocation (Duong and Wickner, 1997b). However, it is questionable that SecD and SecF directly affect the membrane cycling of SecA as the latter is absent in archaea whereas these organisms contain SecD and SecF proteins (Tseng *et al.*, 1999). Finally, SecD and SecF show some structural similarity to certain multidrug resistance RND (resistance, nodulation and division) type pumps with 12 TMSs (i.e. AcrF and ActII-3) (Saier *et al.*, 1998; Tseng *et al.*, 1999). Based on this similarity, it is possible that the SecDF complex fulfills a transport function, such as the removal of the signal peptide or phospholipids from the aqueous protein-conducting pore formed by the translocase.

3.   REGULATION OF PROTEIN TRANSLOCATION

In *E. coli*, none of the integral membrane subunits of the translocase seems to be regulated in relation to the secretion demand of the cell (Rensing and Maier, 1994;   Schatz *et al.*, 1991; Bost and Belin, 1995;

Pogliano and Beckwith, 1994), but the relative amounts seem well-balanced. For instance, SecY that is not associated with SecE is readily removed by proteolysis by FtsH, while over-expression of one of the membrane components results in increased amounts of the other subunits (Taura *et al.*, 1993; Matsuyama *et al.*, 1990a; Sagara *et al.*, 1994). Estimates for the number of translocase subunits in the cell vary tremendously and the following numbers have been reported: SecY, 100 to 400 (Seoh and Tai, 1997; Matsuyama *et al.*, 1992; Matsuyama *et al.*, 1990a); SecE, 100-600 (Seoh and Tai, 1997; Schatz *et al.*, 1991; Matsuyama *et al.*, 1992; Matsuyama *et al.*, 1990a); SecG, 100-1000 (Nishiyama *et al.*, 1995; Matsuyama *et al.*, 1990a); SecD, less than 30 (Pogliano and Beckwith, 1994) up to 450-900 (Matsuyama *et al.*, 1992), and SecF, 30-60 (Pogliano and Beckwith, 1994; Matsuyama *et al.*, 1992). SecD and SecF thus appear to be present in sub-stoichiometric amounts compared to SecY, SecE and SecG.

SecA expression is tightly regulated in relation to the protein secretion demand. Normally, SecA is present at about 500 to 5000 copies per cell (Oliver and Beckwith, 1982; Matsuyama *et al.*, 1992; Seoh and Tai, 1997), but mutations that interfere with the activity of the translocase, the high level expression of secretory proteins or expression of poorly translocated preproteins, may cause a 10-20 fold increase in the cellular SecA (Riggs *et al.*, 1988; Rollo and Oliver, 1988). SecA regulates its own translation (Dolan and Oliver, 1991; Rollo and Oliver, 1988) by binding to a large secretion-responsive element that spans a 96 nucleotide long region (Salavati and Oliver, 1997) located near the 5' end of the *secM-secA* mRNA (Schmidt *et al.*, 1991; Schmidt and Oliver, 1989). The ribosome and SecA binding site on the *secM-secA* mRNA are overlapping, indicating that a simple competition mechanism regulates the translation initiation step (Salavati and Oliver, 1995). *SecM* (before termed *geneX*) encodes a non-essential secretory protein (Rajapandi *et al.*, 1991; Oliver *et al.*, 1998) that is rapidly degraded by the periplasmic tail-specific protease (Nakatogawa and Ito, 2001). The nascent SecM undergoes self-translation arrest at a position close to its C-terminus. It has been proposed that this event results in an altered secondary structure of the *secM-secA* messenger RNA which in turn enhances the translation of *secA*. In this manner, SecM thus functions as a secretion sensor. The Walker B motif of NBS-1 of SecA shows some similarity to the DEAD-box found in RNA binding proteins such as helicases (Koonin and Gorbalenya, 1992). SecA indeed possess RNA helicase activity (Park *et al.*, 1997), which may function in the autogenous regulation of SecA translation (Salavati and Oliver, 1997). It has been suggested that this activity serves to hairpin unwind the *secM-secA* mRNA thereby leading to autogenous regulation of its translation (Koonin and Gorbalenya, 1992).

4.   ROLE OF LIPIDS IN TRANSLOCATION

Phospholipids play an important role in protein translocation. The *E. coli* inner membrane contains about 70% phosphatidylethanolamine (PE) and 30% anionic phospholipids, mainly phosphatidylglycerol (PG) and cardiolipin (CL). With the aid of an *E. coli* mutant in which the synthesis of the major anionic membrane phospholipids could be controlled, it has been demonstrated that the rate of preprotein translocation is directly proportional to the PG content (Kusters *et al.*, 1991). Only the negative charge is essential, as any anionic phospholipids can restore protein translocation in PG-depleted inner membrane vesicles (Kusters *et al.*, 1991). Anionic phospholipids are also needed for functional reconstitution of the SecYE(G) complex into liposomes (Hendrick and Wickner, 1991; Akimaru *et al.*, 1991; Tokuda *et al.*, 1990; van der Does *et al.*, 2000). Elevated levels of SecA can counteract the reduced translocation activity caused low levels of anionic phospholipids (Kusters *et al.*, 1992). On the other hand, anionic phospholipids stimulate the binding and penetration of SecA into the membrane (Breukink *et al.*, 1992) (Ulbrandt *et al.*, 1992). Unlike the binding of SecA to the SecYEG complex, binding of SecA to the lipid surface occurs with low affinity. The catalytic function of the lipid-bound SecA in protein translocation remains to be elucidated since the SecYEG-bound SecA is shielded from lipids, as it cannot be labeled with photo-reactive lipid probes (van Voorst *et al.*, 1998; Eichler *et al.*, 1997).

Numerous experiments showed that anionic phospholipids are needed for a functional interaction of the signal sequence with the membrane (Phoenix *et al.*, 1993; Keller *et al.*, 1992; Keller *et al.*, 1996). Since signal sequences contain at least one or two positive charges at their N-terminus it has been postulated that negatively charged phospholipid head groups interact electrostatically with the positive charges in the signal sequence. Indeed, removal of the positive charges in the signal sequence reduces the translocation efficiency of preproteins and this correlates with the ability of synthetic signal peptides to bind to membranes composed of anionic phospholipids. The loss of activity with the reduction of positive charges in the N-domain can be compensated by an increase in the hydrophobicity of the H-region (Hikita and Mizushima, 1992). Strikingly, this phenomenon is paralleled by a reduction in the requirement for anionic phospholipids, suggesting that the signal sequence-membrane interaction is governed both by electrostatic and hydrophobic interaction. Proteins without a cleavable signal sequence, like integral membrane proteins (van Klompenburg et al., 1997) or the M13 phage coat protein (Kusters et al., 1994) also require anionic phospholipids for insertion. This suggests that lipids may affect protein translocation and membrane insertion by similar mechanisms. In conclusion, the data gathered thus far points to an important role of anionic phospholipids in the initial stages of translocation/membrane insertion (van Klompenburg and de Kruijff, 1998).

Non-bilayer lipids, such as phosphatidylethanolamine, are needed for efficient protein transport (Rietveld et al., 1995). Depletion of PE in the E. coli cell is accompanied by an increase in the amount of PG and CL, but in the presence of high concentrations of Mg^{2+} or other divalent cations, this phenomenon is not lethal. Mg^{2+} forces the type II lipid CL into a non-bilayer conformation (Rand and Sengupta, 1972), thus compensating for the loss of PE. In vitro, preprotein translocation into PE-depleted inner membrane vesicles can be re-activated by addition of Mg^{2+} or by re-introduction of PE (Rietveld et al., 1995). In contrast, protein translocation can be reconstituted into liposomes that do not contain non-bilayer lipids (Hendrick and Wickner, 1991), but inclusion of non-bilayer lipids like PE or diacylglycerol strongly enhances the protein translocation efficiency (van der Does et al., 2000). Unlike the anionic phospholipid requirement, non-bilayer lipids only stimulate translocation. Although speculative, they may facilitate insertion of the signal sequence into the membrane or affect the conformation or oligomeric state of the SecYEG complex.

5.   ENERGETICS OF PROTEIN TRANSLOCATION

Preprotein translocation is driven by ATP hydrolysis (Chen and Tai, 1986) and the pmf (Daniels et al., 1981; Bayan et al., 1993; Enequist et al., 1981). In vivo, both the pmf and ATP hydrolysis are essential for protein translocation (Geller et al., 1986; Bakker and Randall, 1984). Both components of the pmf, the membrane potential ($\Delta\psi$) and the transmembrane gradient of protons (ΔpH) participate in the translocation process (Bakker and Randall, 1984; Yamane et al., 1987; Driessen and Wickner, 1991; Driessen, 1992). In vitro, however, ATP hydrolysis alone can drive protein translocation (Yamane et al., 1987), whereas the pmf is only stimulatory (Yamada et al., 1989; Geller and Green, 1989; Schiebel et al., 1991).

ATP and the pmf function at different stages of protein translocation (Tani et al., 1989; Geller and Green, 1989). ATP is essential for the initiation of preprotein translocation, while the pmf can drive translocation at the later stages and provides directionality to the process (Driessen and Wickner, 1991; Schiebel *et al.*, 1991; Tani *et al.*, 1990).

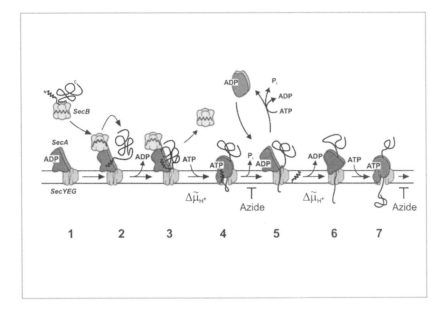

Fig. 4. Model for targeting and stepwise translocation of precursor proteins. (1) SecB targets the preprotein to the translocase where it binds the SecYEG bound SecA with high affinity. (2) Binding of the SecB/preprotein complex to SecA results in a conformational change in SecB, and in a transfer of the signal sequence of the bound preprotein to SecA. The latter elicits another conformational change in SecA that stimulates the binding of SecB to SecA and results in dissociation of preprotein from SecB. (3) Upon binding of the precursor, SecA is stimulated for the exchange of ADP for ATP. Binding of ATP to SecA results in a release of SecB into the cytosol, and (4) causes a conformational change of SecA that allows translocation of approximately 2.5 kDa of preprotein. (5) Upon the hydrolysis of ATP, SecA reverses its conformational change and de-inserts from the membrane-integrated state, a step that is inhibited by azide. Simultaneously, SecA releases the preprotein to the SecYEG complex. (6). Hydrolysis of another ATP is needed for the release of SecA from the membrane. De-inserted SecA can rebind to the preprotein, and this binding reaction allows the translocation of another 2.5 kDa of preprotein. Next, exchange of ADP for ATP takes place, leading to a new round of conformational changes in SecA and further translocation of the preprotein (7). Consecutive steps of ATP driven SecA insertion and de-insertion drive translocation of the preprotein in a stepwise fashion. After the initial insertion of the preprotein, the pmf can drive translocation provided that SecA is no longer associated with the translocating preprotein.

5.1   Mechanism of ATP-driven translocation

The last decade, the mechanism by which ATP hydrolysis drives the movement of a preprotein across the cytoplasmic membrane has been partially resolved. ATP-driven translocation is a stepwise process (Fig. 4). First, the preprotein is targeted to SecA bound with high affinity to SecYEG. Preprotein binding to SecA promotes the exchange of ADP for ATP (Lill *et al.*, 1989; Matsuyama *et al.*, 1990b). Upon ATP-binding the conformation of SecA changes from a compact ADP-bound state into a more extended ATP-bound state (den Blaauwen *et al.*, 1996; den Blaauwen *et al.*, 1999) and this may lead to insertion of SecA domains into the membrane (Economou and Wickner, 1994; Economou *et al.*, 1995). Concomitantly with ATP binding, the chaperone SecB is released from its association with SecA (Fekkes *et al.*, 1997) and a topology inversion of SecG takes place (Nishiyama *et al.*, 1996). This initial event results in translocation of a loop of the signal sequence and early mature domain to the extent that the signal sequence can be processed by signal peptidase (Schiebel *et al.*, 1991). Subsequent hydrolysis of ATP releases the bound preprotein (Schiebel *et al.*, 1991) and the SecA conformation reverses into its compact ADP-bound state leading to membrane de-insertion. Re-binding of SecA to the partially translocated preprotein results in the translocation of an another segment of about 2-2.5 kDa. Exchange of ADP for ATP leads to a new cycle of conformational changes allowing the translocation of another of 2-2.5 kDa of polypeptide mass (Schiebel *et al.*, 1991; van der Wolk *et al.*, 1997; Uchida *et al.*, 1995). According to this model, a complete catalytic cycle of the preprotein translocase permits the stepwise translocation of 5 kDa polypeptide segments by two consecutive events, i.e. ~2.5 kDa upon binding of the polypeptide by SecA, and another 2.5 kDa upon binding of ATP to SecA (van der Wolk *et al.*, 1997). The translocation poise azide blocks the de-insertion step (van der Wolk *et al.*, 1997). Although it is generally accepted that conformational changes in SecA drive protein translocation, the hypothesis that these conformational changes lead to a membrane insertion-deinsertion cycle of SecA is controversial (Economou and Wickner, 1994; Economou *et al.*, 1995). Insertion of SecA domain into the membrane has been concluded from results that showed the formation of membrane-protected 30 kDa and 65 kDa N- and C-terminal protease resistant fragments (Eichler and Wickner, 1997). These fragments cover almost the entire mass of SecA, and it has been argued that SecA penetrates the membrane to expose these domains to the periplasmic face of the membrane. However, provided that the SecA-SecY interaction is maintained, the 30 kDa fragment can also be found in detergent solution (van der Does *et al.*, 1998). Also a strict correlation between translocation and the formation of these fragments does not seem to exist. For instance, overproduction of SecDFYajC, stabilizes the 30 kDa fragment (Economou *et al.*, 1995; Kim *et al.*, 1994) and slows down release of SecA from the membrane (Eichler and Wickner, 1997; Duong and Wickner, 1997b) without affecting the catalytic activity of the translocase. The other way around, PrlA mutants of SecY bind SecA more tightly and enhance the translocation activity (van der Wolk *et al.*, 1998) but a reduced amount of 30 kDa fragment is formed (Nishiyama *et al.*, 1999). Alternatively, the conformational changes in the SecA protein may drive membrane insertion of the preprotein domain only.

5.2 Mechanism of proton motive force driven translocation

The mechanism by which the pmf drives or stimulates protein translocation remains obscure. Early models postulated that the pmf acts on the preprotein. In this model, the membrane potential, $\Delta\psi$, via an electrophoretic mechanism, drives the translocation of negatively charged residues in the preprotein across the membrane. Indeed, the charge distribution of the mature domain, especially around the signal sequence, strongly affects the pmf requirement for protein translocation (Cao *et al.*, 1995). However, other studies showed that translocation of a completely uncharged preprotein is still stimulated by the pmf (Yamada *et al.*, 1989; Kato *et al.*, 1992). Electrophoretic mechanisms have also been implicated in the membrane insertion of small phage coat proteins (Schuenemann *et al.*, 1999) and in the determination of membrane protein topology (Cao *et al.*, 1995; Andersson and von Heijne, 1994; von Heijne, 1989). The membrane potential or charge distribution across the membrane ($\Delta\psi$), inside negative, would prevent translocation of loops with excess positive charge according to the so-called 'positive-inside' rule. However, it is questionable if the $\Delta\psi$ is a major factor in this process, as its orientation does not seem to determine the topology of polytopic membrane proteins. For instance, in acidophilic bacteria and archaea in which the $\Delta\psi$ has the opposite orientation, i.e. outside negative, the 'positive-inside' rule still applies (van de Vossenberg *et al.*, 1998).

Other experimental data suggest that the pmf acts on the membrane components of the translocation machinery. First, the pmf stimulates SecA membrane deinsertion and release of SecA from the membrane (Nishiyama *et al.*, 1999). The latter corroborates with the observation that pmf-driven translocation occurs when SecA is no longer bound to the translocating preprotein (Schiebel *et al.*, 1991), and is consistent with the observation that the pmf reduces the amount of ATP needed to translocate preproteins (Driessen, 1992). The pmf lowers the apparent K_m value of the translocation reaction for ATP (Shiozuka *et al.*, 1990) The latter has been attributed to a stimulatory effect of the pmf on the release of ADP from SecA, a step that may be rate determining in the SecA catalytic cycle. Second, in PrlA strains translocation of preproteins becomes less dependent on the pmf (Nouwen *et al.*, 1996). From the latter, one has postulated that the pmf acts on SecYEG and modulates the opening or formation of the translocation channel. There is still some debate on this hypothesis since the *prlA4* mutation of SecY also causes an increased affinity for SecA (van der Wolk *et al.*, 1998). Tighter binding of SecA to the mutant SecYEG protein therefore will favor ATP-dependent translocation and in this way reduce the requirement for the pmf. However, whereas the pmf reduces the amount of membrane inserted SecA (as monitored by a protease protected fragment) in the wild-type protein, the amount of membrane inserted SecA is low under all conditions (Nishiyama *et al.*, 1999). This suggests that the mutant SecYEG complex has a conformation that is similar to the conformation of the wild-type complex in the presence of a pmf and that this causes the altered interaction with SecA.

6. DYNAMICS IN THE PREPROTEIN TRANSLOCASE

Both the idle surface-bound state and the active "membrane inserted" state of SecYEG-bound SecA are largely shielded from the phospholipid acyl chains (van Voorst *et al.*, 1998; Eichler *et al.*, 1997). Since dimeric SecA is a very large molecule that is 15 nm long and 8 nm wide (Shilton *et al.*, 1998), this remarkable finding can only be understood if the SecA is part of an oligomeric assembly of SecYEG heterotrimers. Indeed, high-resolution electron microscopy of *B. subtilis* SecYE (Meyer *et al.*, 1999) suggests an oligomeric assembly of translocase subunits. Using *E. coli* translocase, it has been shown that SecA recruits multiple SecYEG complexes upon binding of a nonhydrolyzable ATP analogue to form a large oligomeric structure (Manting *et al.*, 2000). This large oligomeric structure has a diameter of 10.5 nm and exhibits an approximately 5 nm central cavity. Mass measurements and quantitative immunoblotting indicates that oligomeric structure consists of a SecA dimer and a SecYEG tetramer, and that a similar structure is formed when the translocase is actively involved in preprotein translocation (Manting *et al.*, 2000). Such an oligomeric organization of the SecYEG complex explains why SecY and SecE do not dissociate *in vivo* (Joly *et al.*, 1994) whereas they appear to interact dynamically in a genetic screen based on suppressor directed inactivation (Bieker-Brady and Silhavy, 1992). Within a single SecYEG heterotrimer, SecY and SecE will remain physically associated, but within the oligomeric assembly, they may appear as dissociable when the oligomer disassembles. The conditions that elicit oligomer formation also induce a pair wise contact between SecE subunits of the translocase (Kaufmann *et al.*, 1999). The presence of a large pore in the translocase may be deleterious to the cells as for its size one would predict that it allows passage of small solutes and protons. Since the oligomerization of the SecYEG complex requires an active SecA, one may speculate the large channel structure accommodates part of the SecA protein. In addition, the large periplasmic domains that are present in SecD and SecF also might act as a valve that prevents leakage of small solutes and protons during translocation.

The translocase is a dynamic complex. When translocation is slow, i.e. for instance at low temperature or at low ATP concentration, distinct preprotein translocation intermediates accumulate (Schiebel *et al.*, 1991; van der Wolk *et al.*, 1997; Uchida *et al.*, 1995). However, deletion or relocation of hydrophobic segments in the mature domain of the preprotein can significantly alter the pattern of translocation intermediates (Sato *et al.*, 1997a). In this respect, intermediate stages of translocation are reversible (Schiebel *et al.*, 1991), and in the absence of an energy source reverse movement may take place (Driessen, 1992). This reverse translocation may also uncouple the SecA ATPase from translocation progress, which is particularly notable when a non-translocatable polypeptide moiety is bound to the preprotein. The presence of hydrophobic segments in preproteins may result in a translocation arrest (Saaf *et al.*, 1998; Duong and Wickner, 1998). Also the presence of positive charges around a hydrophobic domain may cause a translocation arrest (Yamane *et al.*, 1990). Such polypeptide regions may elicit the de-insertion of SecA from the membrane, and thereby allow the hydrophobic domain to partition into the lipid bilayer (Sato *et al.*, 1997b). Small and moderate hydrophobic regions may escape membrane insertion as a result of rapid translocation (Duong and Wickner, 1998), while more hydrophobic regions may laterally leave the translocation pore formed by the SecYEG complex (von Heijne, 1997). It is not clear if this is a passive or active process that requires additional proteins. The recently discovered integral membrane protein YidC (see chapter 4) that can be crosslinked to hydrophobic transmembrane segments (Scotti *et al.*, 2000; Samuelson *et al.*, 2000; Houben *et al.*, 2000) would be a good candidate for the removal of hydrophobic segments from the SecYEG channel. Alternatively, hydrophobic segments may be simply released into the lipid bilayer by dissociation of the oligomeric SecYEG pore complex once SecA has de-inserted.

7. CONSERVED PROTEIN TRANSLOCATION MECHANISM IN BACTERIA, EUKARYOTES AND ARCHEA; FUTURE PERSPECTIVES

In recent years, homologues of the bacterial SecY and SecE have been identified in eukaryotes and archaea (Hartmann *et al.*, 1994; Schatz and Dobberstein, 1996; Pohlschroder *et al.*, 1997; Jungnickel *et al.*, 1994)). In eukaryotes, these homologues form a trimeric complex, called Sec61p, which is involved in protein translocation into the endoplasmatic reticulum. The subunits of Sec61p are called Sec61α, Sec61β and Sec61γ. Sec61α and Sec61γ are homologous to the bacterial SecY and SecE, respectively. Homologues of Sec61β have only been identified in eukaryotes (Esnault *et al.*, 1993). However, although it shows no sequence similarity, Sec61β is probably functionally homologous to SecG. The endoplasmic reticulum of *S. cerevisiae* contains two trimeric complexes, Sec61p and Ssh1p, which are involved in preprotein translocation. In this case, Sec61p consist of Sec61α, Sec61β and the Sss1p protein. In the Ssh1p complex, the α- and β -subunits, Ssh1p and Sbh2p are close homologues to Sec61α and Sec61β, while the γ-subunit, Sss1p, is common for both trimeric complexes (Finke *et al.*, 1996). In archaea, only Sec61α and Sec61γ have been identified. These subunits are more closely related to the eukaryotic counterparts than to the bacterial SecY and SecE subunits (Arndt, 1992). SecE and SecY homologues have also been identified in chloroplasts of higher plants, cyanobacteria, plastids and algae (Vogel *et al.*, 1996; Muller *et al.*, 1994; Scaramuzzi *et al.*, 1992a). Higher plants, like *Arabidopsis thaliana* contain both a Sec61α and a SecY homologue. Sec61α is present in the endoplasmic reticulum whereas the SecY protein is present in the thylakoid membrane of the chloroplast (Laidler *et al.*, 1995).

As discussed in the previous sections, preprotein translocation in eubacteria is driven by the hydrolysis of ATP by SecA which '*pushes*' the protein in a stepwise manner across the membrane (Fig. 5, b). In mammals, protein translocation occurs mainly co-translationally and is driven by chain elongation in the ribosome (pushing force) (Fig. 5, e). During this process, the Sec61α subunit of the Sec61p complex associates directly with the large subunit of the ribosome. Such interaction has also been demonstrated for the eubacterial SecY protein (Fig. 5, a) (Prinz *et al.*, 2000), but it is not clear if chain elongation at the ribosome suffices to drive preproteins across the membrane (Neumann-Haefelin *et al.*, 2000). In contrast, protein translocation in *S. cerevisiae* can occur post-translational and is than driven

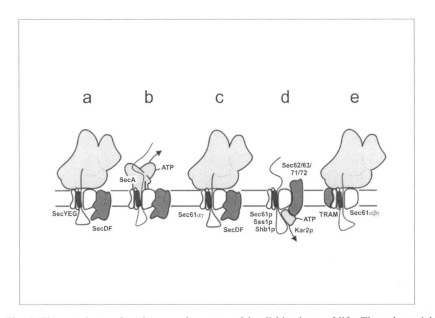

Fig. 5. The protein translocating pore is conserved in all kingdoms of life. The eubacterial SecY and SecE proteins (**a** and **b**) are homologous to the Sec61α and Sec61γ proteins of the protein-conducting channel of the endoplasmic reticulum of eukarya (**d** and **e**) and archaea (**c**). SecG and Sec61β are found in eubacteria and eukarya, respectively but bear no significant sequence similarity. Although the heterotrimeric organization of the translocation pore is conserved, post-translational translocation can be driven by different mechanisms. Hydrolysis of ATP by SecA (eubacteria, cyanelles, plastids and plant thylakoids) pushes the preprotein in a stepwise fashion through the pore (**b**), while hydrolysis of ATP by Kar2p in yeast (**d**) pulls the preprotein through the channel, a process that is assisted by Sec62, Sec63, Sec71 and Sec72. SecD and SecF are accessory proteins that are found only in eubacteria and archaea (**c**), while the TRAM protein is found only in mammals (**e**). During co-translational translocation of proteins into the endoplasmic reticulum, the ribosome directly associates with the Sec61p complex (**e**). Such interaction also exists in bacteria (**a**), but it is not clear if the chain elongation at the ribosome suffices to drive preprotein translocation. The model for the archaeal translocase is speculative and based on the identfication of homologous proteins in genetic database.

by a lumenal chaperone called Kar2p/Bip ('*pulling*') that associates with the Sec62/63p complex (Fig. 5, d). Translocation in archaea may be co-translational and driven by chain elongation since (Fig. 5, c). Although a large number of complete genome sequences of archaea are available, no homologues of SecA or BiP/Kar2p have been identified. In thylakoids, translocation can be both co-and post-translational. Both homologues of SecA (Nakai *et al.*, 1994; Nohara *et al.*, 1995; Scaramuzzi *et al.*, 1992b) and components of SRP (Franklin and Hoffman, 1993; Packer and Howe, 1996; Kogata *et al.*, 1999) have been identified in these organelles, which suggests that translocation occurs mainly according to the mechanisms defined in bacteria. It therefore appears that the basic structure of the translocation channel is remarkably conserved among all three kingdoms of life, whereas the energetic mechanism by which protein translocation takes place may differ. Further detailed analyses of the various translocases and of the structure and dynamics of the constituents will be a challenge for the future.

8.   REFERENCES

Akimaru, J., Matsuyama, S., Tokuda, H. and Mizushima, S. (1991) Proc Natl Acad Sci USA 88: 6545-6549.

Akita, M., Sasaki, S., Matsuyama, S. and Mizushima, S. (1990) J Biol Chem 265: 8164-8169.

Akita, M., Shinaki, A., Matsuyama, S. and Mizushima, S. (1991) Biochem Biophys Res Commun 174: 211-216.

Akiyama, Y. and Ito, K. (1987) EMBO J 6: 3465-3470.

Andersson, H. and von Heijne, G. (1994) EMBO J 13: 2267-2272.

Arkowitz, R.A. and Wickner, W. (1994) EMBO J 13: 954-963.

Arndt, E. (1992) Biochim Biophys Acta 1130: 113-116.

Baba, T., Taura, T., Shimoike, T., Akiyama, Y., Yoshihisa, T. and Ito, K. (1994) Proc Natl Acad Sci. USA 91: 4539-4543.

Bairoch, A., Bucher, P. and Hofmann, K. (1997) Nucleic Acids Res 25: 217-221.

Bakker, E.P. and Randall, L.L. (1984) EMBO J 3: 895-900.

Bayan, N., Schrempp, S., Joliff, G., Leblon, G. and Shechter, E. (1993) Biochim Biophys Acta 1146: 97-105.

Bieker-Brady, K. and Silhavy, T.J. (1992) EMBO J 11: 3165-3174.

Bolhuis, A., Broekhuizen, C.P., Sorokin, A., van Roosmalen, M.L., Venema, G., Bron, S., Quax, W.J. and van Dijl, J.M. (1998) J Biol Chem 273: 21217-21224.

Bost, S. and Belin, D. (1995) EMBO J 14: 4412-4421.

Bost, S. and Belin, D. (1997) J Biol Chem 272: 4087-4093.

Breukink, E., Demel, R.A., de Korte-Kool, G. and de Kruijff, B. (1992) Biochemistry 31: 1119-1124.

Breukink, E., Nouwen, N., van Raalte, A., Mizushima, S., Tommassen, J. and de Kruijff, B. (1995) J Biol Chem 270: 7902-7907.

Brundage, L., Hendrick, J.P., Schiebel, E., Driessen, A.J.M. and Wickner, W. (1990) Cell 62: 649-657.

Cao, G., Kuhn, A. and Dalbey, R.E. (1995) EMBO J 14: 866-875.

Cerretti, D.P., Dean, D., Davis, G.R., Bedwell, D.M. and Nomura, M. (1983) Nucleic Acids Res 11: 2599-2616.

Chen, L. and Tai, P.C. (1986) J Bacteriol 168: 828-832.

Daniels, C.J., Bole, D.G., Quay, S.C. and Oxender, D.L. (1981) Proc Natl Acad Sci USA 78; 5396-5400.

Den Blaauwen, T. and Driessen, A.J.M. (1996) Arch Microbiol 165: 1-8.

Den Blaauwen, T., Fekkes, P., de Wit, J.G., Kuiper, W. and Driessen, A.J.M. (1996) Biochemistry 35: 11994-12004.

Den Blaauwen, T., van der Wolk, J.P., van der Does, C., van Wely, K.H. and Driessen, A.J.M. (1999) FEBS Lett 458: 145-150.

Dolan, K.M. and Oliver, D.B. (1991) J Biol Chem 266: 23329-23333.

Downing, W.L., Sullivan, S.L., Gottesman, M.E. and Dennis, P.P. (1990) J Bacteriol 172: 1621-1627.

Driessen, A.J.M. (1992) EMBO J 11: 847-853.

Driessen, A.J.M. (1993) Biochemistry 32: 13190-13197.

Driessen, A.J.M., Brundage, L., Hendrick, J.P., Schiebel, E. and Wickner, W. (1991) Methods Cell Biol 34:147-165.

Driessen, A.J.M. and Wickner, W. (1990) Proc Natl Acad Sci USA 87: 3107-3111.

Driessen, A.J.M. and Wickner, W. (1991) Proc Natl Acad Sci USA 88: 2471-2475.

Duong, F. and Wickner, W. (1997a) EMBO J 16: 2756-2768.

Duong, F. and Wickner, W. (1997b) EMBO J 16: 4871-4879.

Duong, F. and Wickner, W. (1998) EMBO J 17: 696-705.

Economou, A., Pogliano, J.A., Beckwith, J., Oliver, D.B. and Wickner, W. (1995) Cell 83: 1171-1181.

Economou, A. and Wickner, W. (1994) Cell 78: 835-843.

Eichler, J., Brunner, J. and Wickner, W. (1997) EMBO J 16: 2188-2196.

Eichler, J. and Wickner, W. (1997) Proc Natl Acad Sci USA 94: 5574-5581.

Eichler, J. and Wickner, W. (1998) J Bacteriol 180: 5776-5779.

Emr, S.D., Hanley-Way, S. and Silhavy, T.J. (1981) Cell 23: 79-88.

Enequist, H.G., Hirst, T.R., Harayama, S., Hardy, S.J. and Randall, L.L. (1981) Eur J Biochem 116: 227-233.

Esnault, Y., Blondel, M.O., Deshaies, R.J., Scheckman, R. and Kepes, F. (1993) EMBO J 12: 4083-4093.

Fekkes, P., de Wit, J.G., Boorsma, A., Friesen, R.H. and Driessen, A.J.M. (1999) Biochemistry 38: 5111-5116.

Fekkes, P. and Driessen, A.J.M. (1999) Microbiol Mol Biol Rev 63: 161-173.

Fekkes, P., van der Does, C. and Driessen, A.J.M. (1997) EMBO J 16: 6105-6113.

Finke, K., Plath, K., Panzner, S., Prehn, S., Rapoport, T.A., Hartmann, E. and Sommer, T. (1996) EMBO J 15: 1482-1494.

Flower, A.M. (2001) J Bacteriol 183: 2006- 2012.

Flower, A.M., Doebele, R.C. and Silhavy, T.J. (1994) J Bacteriol 176: 5607-5614.

Flower, A.M., Osborne, R.S. and Silhavy, T. J. (1995) EMBO J 14: 884-893.

Franklin, A.E. and Hoffman, N.E. (1993) J Biol Chem 268: 22175-22180.

Gardel, C., Benson, S., Hunt, J., Michaelis, S., and Beckwith, J. (1987) J Bacteriol 169: 1286-1290.

Gardel, C., Johnson, K., Jacq, A. and Beckwith, J. (1990a) EMBO J 9: 4205-4206.

Gardel, C., Johnson, K., Jacq, A. and Beckwith, J. (1990b) EMBO J 9: 3209-3216.

Geller, B.L. and Green, H.M. (1989) J Biol Chem 264: 16465-16469.

Geller, B.L., Movva, N.R. and Wickner, W. (1986) Proc Natl Acad Sci USA 83: 4219-4222.

Hanada, M., Nishiyama, K. and Tokuda, H. (1996) FEBS Lett 381: 25-28.

Hanada, M., Nishiyama, K.I., Mizushima, S. and Tokuda, H. (1994) J Biol Chem 269: 23625-23631.

Harris, C.R. and Silhavy, T.J. (1999) J Bacteriol 181: 3438-3444.

Hartmann, E., Sommer, T., Prehn, S., Gorlich, D., Jentsch, S. and Rapoport, T.A. (1994) Nature 367: 654-657.

Hendrick, J.P. and Wickner, W. (1991) J Biol Chem 266: 24596-24600.

Hikita, C. and Mizushima, S. (1992) J Biol Chem 267: 12375-12379.

Homma, T., Yoshihisa, T. and Ito, K. (1997) FEBS Lett 408: 11-15.

Houben, E.N., Scotti, P.A., Valent, Q.A., Brunner, J., de Gier, J.L., Oudega, B. and Luirink, J. (2000) FEBS Lett 476: 229-233.

Huie, J.L. and Silhavy, T.J. (1995) J Bacteriol 177: 3518-3526.

Iino, T. and Sako, T. (1988) J Biol Chem 263: 19077-19082.

Joly, J.C., Leonard, M.R. and Wickner, W.T. (1994) Proc Natl Acad Sci USA 91: 4703-4707.

Jungnickel, B., Rapoport, T.A. and Hartmann, E. (1994) FEBS Lett 346: 73-77.

Kato, M., Tokuda, H. and Mizushima, S. (1992) J Biol Chem 267: 413-418.

Kaufmann, A., Manting, E.H., Veenendaal, A.K., Driessen, A.J.M. and van der Does, C. (1999) Biochemistry 38: 9115-9125.

Keller, R.C., Killian, J.A. and de Kruijff, B. (1992) Biochemistry 31: 1672-1677.

Keller, R.C., ten Berge, D., Nouwen, N., Snel, M.M., Tommassen, J., Marsh, D. and de Kruijff, B. (1996) Biochemistry 35: 3063-3071.

Kihara, A., Akiyama, Y. and Ito, K. (1995) Proc Natl Acad Sci USA 92: 4532-4536.

Kim, Y.J., Rajapandi, T. and Oliver, D. (1994) Cell 78: 845-853.

Klose, M., Schimz, K.L., van der Wolk, J., Driessen, A.J.M. and Freudl, R. (1993) J Biol Chem 268: 4504-4510.

Kogata, N., Nishio, K., Hirohashi, T., Kikuchi, S. and Nakai, M. (1999) FEBS Lett 447: 329-333.

Kontinen, V.P., Helander, I.M. and Tokuda, H. (1996b) FEBS Lett 389: 281-284.

Kontinen, V.P. and Tokuda, H. (1995) FEBS Lett 364: 157-160.

Kontinen, V.P., Yamanaka, M., Nishiyama, K. and Tokuda, H. (1996a) J Biochem (Tokyo) 119: 1124-1130.

Koonin, E.V. and Gorbalenya, A.E. (1992) FEBS Lett 298: 6-8.

Kourtz, L. and Oliver, D. (2000) Mol Microbiol 37: 1342-1356.

Kusters, R., Breukink, E., Gallusser, A., Kuhn, A. and de Kruijff, B. (1994) J Biol Chem 269: 1560-1563.

Kusters, R., Dowhan, W. and de Kruijff, B. (1991) J Biol Chem 266: 8659-8662.

Kusters, R., Huijbregts, R. and de Kruijff, B. (1992) FEBS Lett 308: 97-100.

Laidler, V., Chaddock, A.M., Knott, T.G., Walker, D. and Robinson, C. (1995) J Biol Chem 270: 17664-17667.

Lill, R., Cunningham, K., Brundage, L.A., Ito, K., Oliver, D. and Wickner, W. (1989) EMBO J 8: 961-966.

Manting, E.H., van der Does, C., Remigy, H., Engel, A. and Driessen, A.J.M. (2000) EMBO J 19: 852-861.

Matsumoto, G., Mori, H. and Ito, K. (1998) Proc Natl Acad Sci USA 95: 13567-13572.

Matsumoto, G., Yoshihisa, T. and Ito, K. (1997) EMBO J 16: 6384-6393.

Matsuyama, S., Akimaru, J. and Mizushima, S. (1990a) FEBS Lett 269: 96-100.

Matsuyama, S., Fujita, Y. and Mizushima, S. (1993) EMBO J 12: 265-270.

Matsuyama, S., Fujita, Y., Sagara, K. and Mizushima, S. (1992) Biochim Biophys Acta 1122: 77-84.

Matsuyama, S., Kimura, E. and Mizushima, S. (1990b) J Biol Chem 265: 8760-8765.

Meyer, T.H., Ménétret, J.F., Breitling, R., Miller, K.R., Akey, C.W. and Rapoport, T.A. (1999) J Mol Biol 285: 1789-1800.

Mitchell, C. and Oliver, D. (1993) Mol Microbiol 10: 483-497.

Mori, H., Sugiyama, H., Yamanaka, M., Sato, K., Tagaya, M. Mizushima, S. (1998) J Biochem (Tokyo) 124: 122-129.

Muller, S.B., Rensing, S.A., Martin, W.F. and Maier, U.G. (1994) Curr Genet 26: 410-414.

Murphy, C.K. and Beckwith, J. (1994) Proc Natl Acad Sci USA 91: 2557-2561.

Nakai, M., Goto, A., Nohara, T., Sugita, D. and Endo, T. (1994) J Biol Chem 269: 31338-31341.

Nakatogawa, H. and Ito, K. (2001) Mol Cell 7: 185-192.

Nakatogawa, H., Mori, H. and Ito, K. (2000a) J Biol Chem 275: 33209-33212.

Nakatogawa, H., Mori, H., Matsumoto, G. and Ito, K. (2000b) J Biochem (Tokyo) 127: 1071-1079.

Neumann-Haefelin, C., Schafer, U., Muller, M. and Koch, H.G. (2000) EMBO J 19: 6419-6426.

Nishiyama, K., Hanada, M. and Tokuda, H. (1994) EMBO J 13: 3272-3277.

Nishiyama, K., Mizushima, S. and Tokuda, H. (1992) J Biol Chem 267: 7170-7176.

Nishiyama, K., Mizushima, S. and Tokuda, H. (1993) EMBO J 12: 3409-3415.

Nishiyama, K., Mizushima, S. and Tokuda, H. (1995) Biochem Biophys Res Commun 217: 217-223.

Nishiyama, K., Suzuki, T. and Tokuda, H. (1996) Cell 85: 71-81.

Nishiyama, K.I., Fukuda, A., Morita, K. and Tokuda, H. (1999) EMBO J 18: 1049-1058.

Nohara, T., Nakai, M., Goto, A. and Endo, T. (1995) FEBS Lett 364: 305-308.

Nouwen, N., de Kruijff, B. and Tommassen, J. (1996) Proc Natl Acad Sci USA 93: 5953-5957.

Oliver, D., Norman, J. and Sarker, S. (1998) J Bacteriol 180: 5240-5242.

Oliver, D.B. and Beckwith, J. (1981) Cell 25: 765-772.

Oliver, D.B. and Beckwith, J. (1982) Cell 30: 311-319.

Oliver, D.B., Cabelli, R.J., Dolan, K.M. and Jarosik, G.P. (1990) Proc Natl Acad Sci USA 87: 8227-8231.

Packer, J.C. and Howe, C.J. (1996) Plant Mol Biol 31: 659-665.

Park, S.K., Kim, D.W., Choe, J. and Kim, H. (1997) Biochem Biophys Res Commun 235: 593-597.

Phoenix, D.A., Kusters, R., Hikita, C., Mizushima, S. and de Kruijff, B. (1993) J Biol Chem 268: 17069-17073.

Pogliano, K.J. and Beckwith, J. (1993) Genetics 133: 763-773.

Pogliano, K.J. and Beckwith, J. (1994) J Bacteriol 176: 804-814.

Pohlschroder, M., Murphy, C. and Beckwith, J. (1996) J Biol Chem 271: 19908-19914.

Pohlschroder, M., Prinz, W.A., Hartmann, E. and Beckwith, J. (1997) Cell 91: 563-566.

Prinz A., Behrens, C., Rapoport, T.A., Hartmann, E. and Kalies, K.U. (2000) EMBO J 19: 1900-1906.

Rajapandi, T., Dolan, K.M. and Oliver, D.B. (1991) J Bacteriol 173: 7092-7097.

Ramamurthy, V. and Oliver, D. (1997) J Biol Chem 272: 23239-23246.

Rand, R.P. and Sengupta, S. (1972) Biochim Biophys Acta 255: 484-492.

Rensing, S.A. and Maier, U.G. (1994) Mol Phylogenet Evol 3: 187-191.

Rietveld, A.G., Koorengevel, M.C. and de Kruijff, B. (1995) EMBO J 14: 5506-5513.

Riggs, P.D., Derman, A.I. and Beckwith, J. (1988) Genetics 118: 571-579.

Rollo, E.E. and Oliver, D.B. (1988) J Bacteriol 170: 3281-3282.

Saaf, A., Wallin, E. and von Heijne, G. (1998) Eur J Biochem 251: 821-829.

Sagara, K., Matsuyama, S. and Mizushima, S. (1994) J Bacteriol 176: 4111-4116.

Saier, M.H.J., Paulsen, I.T., Sliwinski, M.K., Pao, S.S., Skurray, R.A. and Nikaido, H. (1998) FASEB J 12: 265-274.

Sako, T. and Iino, T. (1988) J Bacteriol 170: 5389-5391.

Salavati, R. and Oliver, D. (1995) RNA 1: 745-753.

Salavati, R. and Oliver, D. (1997) J Mol Biol 265: 142-152.

Samuelson, J.C., Chen, M., Jiang, F., Moller, I., Wiedmann, M., Kuhn, A., Phillips, G.J. and Dalbey, R.E. (2000) Nature 406: 637-641.

Sato, K., Mori, H., Yoshida, M. and Mizushima, S. (1996) J. Biol Chem 271: 17439-17444.

Sato, K., Mori, H., Yoshida, M., Tagaya, M. and Mizushima, S. (1997b) J Biol Chem 272: 20082-20087.

Sato, K., Mori, H., Yoshida, M., Tagaya, M. and Mizushima, S. (1997a) J Biol Chem 272: 5880-5886.

Scaramuzzi, C.D., Hiller, R.G. and Stokes, H.W. (1992b) Curr Genet 22: 421-427.

Scaramuzzi, C.D., Stokes, H.W. and Hiller, R.G. (1992a) FEBS Lett 304: 119-123.

Schatz, G. and Dobberstein, B. (1996) Science 271: 1519-1526.

Schatz, P.J., Bieker, K.L., Ottemann, K.M., Silhavy, T.J. and Beckwith, J. (1991) EMBO J 10: 1749-1757.

48

Schatz, P.J., Riggs, P.D., Jacq, A., Fath, M.J. and Beckwith, J. (1989) Genes Dev. 3, 1035-1044.

Schiebel, E., Driessen, A.J.M., Hartl, F.U. and Wickner, W. (1991) Cell 64 ; 927-939.

Schmidt, M., Ding, H., Ramamurthy, V., Mukerji, I. and Oliver, D. (2000) J Biol Chem 275: 15440-15448.

Schmidt, M.G., Dolan, K.M. and Oliver, D.B. (1991) J. Bacteriol 173: 6605-6611.

Schmidt, M.G. and Oliver, D.B. (1989) J Bacteriol 171: 643-649.

Schuenemann, T.A., Delgado-Nixon, V.M. and Dalbey, R.E. (1999) J Biol Chem 274: 6855-6864.

Scotti, P.A., Urbanus, M.L., Brunner, J., de Gier, J.W., von Heijne, G., van der Does, C., Driessen, A.J.M., Oudega, B. and Luirink, J. (2000) EMBO J 19: 542-549.

Seoh, H.K. and Tai, P.C. (1997) J Bacteriol 179: 1077-1081.

Shilton, B., Svergun, D.I., Volkov, V.V., Koch, M.H., Cusack, S. and Economou, A. (1998) FEBS Lett 436: 277-282.

Shimizu, H., Nishiyama, K. and Tokuda, H. (1997) Mol Microbiol 26: 1013-1021.

Shimoike, T., Akiyama, Y., Baba, T., Taura, T. and Ito, K. (1992) Mol Microbiol 6: 1205-1210.

Shinkai, A., Mei, L.H., Tokuda, H. and Mizushima, S. (1991) J Biol Chem 266: 5827-5833.

Shiozuka, K., Tani, K., Mizushima, S. and Tokuda, H. (1990) J Biol Chem 265: 18843-18847.

Sianidis, G., Karamanou, S., Vrontou, E., Boulias, K., Repanas, K., Kyrpides, N., Politou, A.S. and Economou, A (2001) EMBO J 20: 961-970.

Stader, J., Gansheroff, L.J. and Silhavy, T.J. (1989) Genes Dev 3: 1045-1052.

Suzuki, H., Nishiyama, K. and Tokuda, H. (1999) J Biol Chem 274: 31020-31024.

Tani, K., Shiozuka, K., Tokuda, H. and Mizushima, S. (1989) J Biol Chem 264: 18582-18588.

Tani, K., Tokuda, H. and Mizushima, S. (1990) J Biol Chem 265: 17341-17347.

Taura, T., Baba, T., Akiyama, Y. and Ito, K. (1993) J Bacteriol 175: 7771-7775.

Tokuda, H., Akimaru, J., Matsuyama, S., Nishiyama, K. and Mizushima, S. (1991) FEBS Lett 279: 233-236.

Tokuda, H., Shiozuka, K. and Mizushima, S. (1990) Eur J Biochem 192: 583-589.

Tseng, T. T., Gratwick, K. S., Kollman, J., Park, D., Nies, D. H., Goffeau, A. and Saier, M. H. Jr. (1999) J Mol Microbiol Biotechnol 1: 107-125.

Uchida, K., Mori, H. and Mizushima, S. (1995) J Biol Chem 270: 30862-30868.

Ulbrandt, N.D., London, E. and Oliver, D.B. (1992) J Biol Chem 267: 15184-15192.

van de Vossenberg, J.L., Albers, S.V., van der Does, C., Driessen A.J.M. and van Klompenburg, W. (1998) Mol Microbiol 29: 1125-1127.

van der Does, C., den Blaauwen, T., de Wit, J.G., Manting, E.H., Groot, N.A., Fekkes, P., Driessen, A.J.M. (1996) Mol Microbiol 22: 619-629.

van der Does, C., Manting, E.H., Kaufmann, A., Lutz, M. and Driessen, A.J.M. (1998) Biochemistry 37: 201-210.

van der Does, C., Swaving, J., van Klompenburg, W. and Driessen, A.J.M. (2000) J Biol Chem 275: 2472-2478.

van der Wolk, J., Klose, M., Breukink, E., Demel, R.A., de Kruijff, B., Freudl, R. and Driessen, A.J.M. (1993) Mol Microbiol 8: 31-42.

van der Wolk, J.P., de Wit, J.G. and Driessen, A.J.M. (1997) EMBO J 16: 7297-7304.

van der Wolk, J.P., Fekkes, P., Boorsma, A., Huie, J.L., Silhavy, T.J. and Driessen, A.J.M. (1998) EMBO J 17: 3631-3639.

van Klompenburg, W. and de Kruijff, B. (1998) J Membr Biol 162: 1-7.

van Klompenburg, W., Ridder, A.N., van Raalte, A.L., Killian, A.J., von Heijne, G. and de Kruijff, B. (1997) FEBS Lett 413: 109-114.

van Voorst, F., van der Does, C., Brunner, J., Driessen, A.J.M. and de Kruijff, B. (1998) Biochemistry 37: 12261-12268.

van Wely, K.H.M., Swaving, J., Broekhuizen, C.P., Rose, M., Quax, W.J. and Driessen, A.J.M. (1999) J Bacteriol 181: 1786-1792.

Vogel, H., Fischer, S. and Valentin, K. (1996) Plant Mol Biol 32: 685-692.

von Heijne, G. (1989) Nature 341: 456-458.

von Heijne, G. (1997) Mol Microbiol 24: 249-253.

Yamada, H., Tokuda, H. and Mizushima, S. (1989) J Biol Chem 264: 1723-1728.

Yamane, K., Akiyama, Y., Ito, K. and Mizushima, S. (1990) J Biol Chem 265: 21166-21171.

Yamane, K., Ichihara, S. and Mizushima, S. (1987) J Biol Chem 262: 2358-2362.

Yang, Y.B., Yu, N. and Tai, P.C. (1997) J Biol Chem 272: 13660-13665.

Chapter 3

THE TAT PROTEIN EXPORT PATHWAY

Tracy Palmer[1,2] and Ben C. Berks[1]

[1]Centre for Metalloprotein Spectroscopy and Biology, School of Biological Sciences, University of East Anglia, Norwich NR4 7TJ, UK
[2]Department of Molecular Microbiology, John Innes Centre, Colney Lane, Norwich NR4 7UH, UK

1.   INTRODUCTION

The Tat (twin arginine translocation) system is a recently discovered protein transport pathway which is found in the cytoplasmic membranes of many bacteria, and in the energy-transducing membranes of plant organelles. Proteins are targeted to the Tat export machinery by N-terminal signal peptides harbouring a distinctive 'twin arginine' motif. The Tat system serves to export a subset of periplasmic and periplasmic-facing proteins that normally require redox cofactors for activity. Overwhelming evidence suggests that substrate proteins are translocated across the membrane in a folded conformation, with their redox cofactors already bound, and in some cases as hetero-oligomers. This remarkable translocase is the first example of a protein transport system transporting folded substrates across an energy-coupling membrane. Here, the salient features of the bacterial Tat protein export pathway are discussed.

2.   THE TAT SIGNAL PEPTIDE

Proteins destined for export by the Tat pathway, like those exported by the Sec machinery, are synthesized with N-terminal signal peptides. However, the signal peptide on each precursor is specific for the transport pathway and is therefore able to mediate mutually exclusive sorting between

B. Oudega (ed.), Protein Secretion Pathways in Bacteria, 51–64.
© 2003 Kluwer Academic Publishers. Printed in the Netherlands.

the Sec and Tat pathways. For example, the signal peptide of the Tat substrate trimethylamine-*N*-oxide (TMAO) reductase (TorA) when fused to the periplasmic P2 domain of *E. coli* leader peptidase (LEP) mediates exclusively Tat-dependent export, whilst the Sec-dependent signal peptide of pectate lyase (PelB) when fused to the small subunit of *E. coli* hydrogenase 2 directs the enzyme to the Sec translocon (Cristóbal *et al.*, 1999; Rodrigue *et al.*, 1999).

Despite sharing a similar tripartite structure (see chapters 1 and 2, this volume), Tat signal peptides differ from their Sec counterparts in several critical respects. Firstly, Tat signal peptides contain a conserved amino acid sequence motif at the n-region/h-region boundary. This motif can be defined as S-R-R-x-F-L-K, where the consecutive arginine residues are invariant, the frequency of the other motif residues is greater than 50% and x is normally a polar amino acid (Berks, 1996; Stanley *et al.*, 2000). A second distinguishing feature is that the h-regions of Tat signal peptides are significantly less hydrophobic than those of Sec targeting signals (Cristóbal *et al.*, 1999). Thirdly, Tat signal peptides frequently contain basic residues in the c-region, which are almost never found at the equivalent position of Sec signal peptides (Wexler *et al.*, 1998). A fourth striking feature is that Tat signal peptides are markedly longer than their Sec counterparts, in some cases reaching up to 58 amino acids in length. Much of this extra length can be attributed to a highly extended n-region. The possible significance of the n-region will be discussed in more detail in the section discussing 'substrates of the Tat pathway'. Some examples of Tat targeting signals are given in Table 1.

Table 1. Examples of typical Tat signal peptides. The signal sequences are all derived from *E. coli* proteins unless otherwise stated: HyaA – small subunit of hydrogenase 1; TorA – TMAO reductase; FdnG – α-subunit of formate dehydrogenase-N; SufI –SufI protein; DmsA – catalytic subunit of DMSO reductase; Gfo – glucose-fructose oxidoreductase from *Zymomonas mobilis*. The twin arginines of each signal peptide are shown embossed and other motif residues are in bold. The h-region is underlined. The n-region, which may be highly variable in length, is amino-terminal to the twin arginine motif.

Sequence	Protein
MNNEETFYQAMRRQGVT**RR**S**FLK**YCSLAATSLGLGAGMAPKIAWA	HyaA
MNNNDLFQAS**RR**R**F**LAQLGGLTVAGMLGPSLLTPRRATAAQA	TorA
MDVS**RR**QFF**K**ICAGGMAGTTVAALGFAPKQALA	FdnG
MSLS**RR**QFIQASGIALCAGAVPLKASA	SufI
MKTKIPDAVLAAEVS**RR**GLVKTTAIGGLAMASSALTLPFSRIAHA	DmsA
MTNKISSSDNLSNAVSATDDNASRTPNLT**RR**ALVGGGVGLAAAGALASGLQ	Gfo
S**RR**x**FLK**	consensus

2.1   The twin arginine motif

The most distinctive feature of Tat signal peptides is the 'twin arginine' motif. The paired arginines are almost universally conserved in the signal peptides of proteins exported by the Tat pathway and are critical for efficient transport. This has led to the designation of 'twin arginine' signal peptides, and to the translocation system as the Tat (twin arginine translocation) pathway. A number of site-directed mutagenesis experiments has confirmed the importance of the consecutive arginine residues. A complete export block was observed when the twin arginine residues of the *Zymomonas mobilis* glucose-fructose oxidoreductase (GFOR) were mutated singularly or in tandem to lysines (Halbig *et al.*, 1999b). Similarly, conservative substitution of both arginine residues of the TorA signal peptide of a TorA:LepP2 fusion also abolished Tat-dependent export (Cristóbal *et al.*, 1999). In the same vein, replacement of the twin arginine residues in the signal peptide of the *Wolinella succinogenes* hydrogenase with twin glutamines, and of the first arginine residue of the motif present in pre-nitrous oxide reductase of *Pseudomonas stutzeri* with glutamate were also sufficient to prevent export (Gross *et al.*, 1999; Dreusch *et al.*, 1997). However, in some circumstances export is still possible when a single arginine residue is conservatively substituted for lysine. The export of a *Desulfovibrio vulgaris* hydrogenase Tat signal peptide/β-lactamase fusion protein expressed in *E. coli* was only partially inhibited when the first arginine of the consensus motif was substituted with a range of amino acids, although it is not clear whether export was mediated by the Tat pathway. More recently replacement of the first arginine of the signal peptide of *E. coli* pre-SufI with lysine (but not alanine) still allowed slow but significant export (Stanley *et al.*, 2000). These results indicate that, in some signal peptide contexts, both consensus arginine residues are not obligatory for Tat targeting.

The role and importance of the non-arginine residues of the consensus motif is less clear. Using pre-SufI and pre-YacK as model substrates, the function of the other consensus residues has been addressed (Stanley *et al.*, 2000). After the arginines, the consensus phenylalanine is the most highly conserved amino acid, being found in approximately 80 % of twin arginine signal peptides. Conservative substitution of this residue with leucine was without effect, but replacement with alanine or tyrosine led to a dramatic decrease in the rate of export suggesting that the hydrophobicity of the amino acid at this position is important for the export process. The consensus leucine residue also contributes to efficient export since substitution with a non-hydrophobic side chain results in a small decrease in the export rate. The function of the conserved serine residue is not clear.

Originally suggested to provide a cap to the α–helical h-region, replacement with a non-capping residue did not significantly affect the export kinetics. Perhaps the most surprising observation was that substitution of the conserved lysine residue actually increased the rate of export. This has lead to speculation that the role of the lysine may be to mediate interactions with accessory proteins necessary to ensure that cofactor insertion has been achieved prior to export (Stanley *et al.*, 2000).

2.2 Hydrophobicity of the h-region and the c-region positive charge

The functional importance of the difference in hydrophobicity between Sec and Tat signal sequences has been addressed using a TorA:P2 fusion construct (Cristóbal *et al.*, 1999). The TorA signal peptide is able to confer Tat-dependent export on the water-soluble P2 fragment of LEP, which is normally a substrate of the Sec pathway. However, it is not able to redirect full length LEP, an integral membrane protein, onto the Tat pathway, indicating that the Sec targeting information present in the transmembrane helices of LEP (SRP-dependent) overrides the Tat targeting properties of the TorA signal. Increasing the h-region hydrophobicity of the TorA:P2 construct resulted in a re-routing of the export from the Tat to the Sec pathway, even though the twin arginine motif remained intact. Although the Sec-dependent export of P2 was at least an order of magnitude slower than authentic Sec substrates, subsequent substitution of the TorA signal peptide c-region positive charges resulted in very rapid and Sec-dependent export (Cristóbal *et al.*, 1999). Taken together with the observation that mutagenesis of the c-region positive charge of pre-SufI does not affect the export rate, this suggests that the positive charge acts as a 'Sec-avoidance' signal (Bogsch *et al.*, 1997; Cristóbal *et al.*, 1999; Stanley *et al.*, 2000).

2.3 Specificity in Tat signal sequences

There is some evidence that the Tat machinery from different bacteria may show specificity towards their cognate twin arginine signal peptides. Although the GFOR enzyme is efficiently exported by the *Z. mobilis* Tat system, when the enzyme is heterologously expressed in *E. coli* it is not exported but is cytoplasmically located and enzymatically active. However, when the native GFOR signal sequence is precisely replaced by that of *E. coli* TorA, the hybrid protein is exported by *E. coli* in a Tat-dependent manner (Blaudeck *et al.*, 2001). This observation suggests that

Tat signal peptides are not universally recognised by different Tat translocases. There is also some indication that Tat signal peptides may play specific roles in the assembly of their cognate redox enzymes. If the signal sequence of DmsA, the catalytic subunit of *E. coli* DMSO reductase is replaced by that of *E. coli* TorA, the membrane-bound DMSO reductase activity is reduced 5 fold, suggesting that the signal sequence influences assembly of its passenger protein (Sambasivarao *et al.*, 2000).

2.4   Export of heterologous proteins by the Tat pathway

There is a growing body of evidence that heterologous proteins, when fused to a twin arginine signal peptide, can be targeted for export by the Tat pathway. In addition to the export of β-lactamase and the P2 domain of LEP, which are both normally targeted to the periplasm by the Sec pathway, two recent reports have described the Tat-specific export of jellyfish green fluorescent protein (GFP) (Thomas *et al.*, 2001; Santini *et al.*, 2001). In both cases, fusion of GFP after the TorA signal conferred the export of the protein in an active form. Since GFP in incapable of folding in the periplasm (Feilmeier *et al.*, 2000), this strongly supports the contention that the Tat pathway is capable of transporting fully folded substrates (see section on 'substrates of the Tat pathway'). These observations suggest that the Tat system may prove to be the transport system of choice for the export of recombinantly expressed foreign proteins, particularly those which have proved refractory for secretion by the Sec system, and is potentially highly relevant for industrial production of proteins.

3.   TAT TRANSLOCON COMPONENTS

Two genetic loci encoding components of the Tat machinery have been identified in *E. coli* (Bogsch *et al.*, 1998; Sargent *et al.*, 1998; Weiner *et al.*, 1998). The *tatA* operon, which maps to 86 minutes on the *E. coli* chromosome, encodes the four genes *tatABCD*, while *tatE*, found at 14 minutes, is moncistronic. Both transcription units are expressed constitutively in *E. coli*, indicating a requirement for the Tat pathway under all growth conditions (Jack *et al.*, 2001). The genetic organisation of the *E. coli tat* genes is depicted schematically in Fig. 1.

56

E. coli 86.6 min

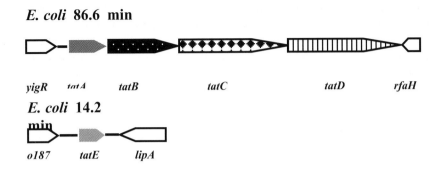

yigR   *tatA*   *tatB*   *tatC*   *tatD*   *rfaH*

E. coli 14.2 min

o187   *tatE*   *lipA*

Fig. 1. Genetic organisation of genes encoding Tat components in *E. coli*.

TatA, TatB and TatE are sequence-related proteins with a common structure. As outlined in Fig. 2, each is predicted to comprise a membrane-spanning α–helix at the N-terminus, immediately followed by a cytoplasmically-located amphipathic helix and then a C-terminal region of variable length. TatA and E display greater than 50 % sequence identity and share overlapping functions in Tat-dependent protein export. Thus single deletions in *tatA* or *tatE* lead to a decrease in the export of a range of Tat substrate proteins, but deletion of both is required to observe a complete export block (Sargent *et al.*, 1998). Generally the *tatA* mutant shows a more severe export defect than the *tatE* strain. This is consistent with genetic studies which indicate that *tatA* is transcribed and translated at a 50-100 fold higher level than *tatE* (Jack *et al.*, 2000). Despite sharing 20 % sequence identity with TatA/TatE, TatB has a distinct function in the export pathway. An in-frame deletion in *tatB* is sufficient to abolish the export of seven endogenous Tat substrates. The *tatB* mutant strain cannot be complemented in *trans* by supplying extra copies of *tatA* or *tatE*, and likewise the *tatA/tatE* mutant strain cannot be rescued by plasmid-borne *tatB* (Sargent *et al.*, 1998). It has been speculated that the TatABE proteins may serve as receptors for different classes of Tat substrates (Chanal *et al.*, 1998). However, the homology amongst this family of proteins is low (the only absolutely conserved residue is a glycine which forms an inter-domain hinge at the transmembrane helix/amphipathic helix boundary) and there is no conservation of polar amino acids which might be expected to interact with the twin arginines of the signal peptide. Genetic expression studies suggest that the ratio of TatA:TatB production is approximately 25:1, which is in good agreement with the levels of TatA and TatB found in membranes and in isolated TatAB complexes (Jack *et al.*, 2001; Sargent *et al.*, 2001).

The TatC protein is highly hydrophobic and is predicted to have six transmembrane helices, with the N- and C-termini located on the cytoplasmic face of the membrane. It is strictly required for protein export by the Tat pathway (Bogsch *et al.*, 1998). The *tatC* gene is transcribed and translated at a two fold lower rate than *tatB*, and 50 fold lower than *tatA*, suggesting that it may be the least abundant of the essential pathway components (Jack *et al.*, 2001). TatC proteins show the highest level of amino acid conservation. Fifteen amino acids, several of which are polar residues, are strictly conserved amongst the eubacterial TatC proteins and eight of these are also conserved amongst the eukaryotic homologues. Many of these conserved residues fall within a predicted cytoplasmic loop between helices two and three and in transmembrane segments 3 and 4. On the basis of this, it has been proposed that TatC may be the component of the transporter which recognises the signal peptide (Berks *et al.*, 2000). Further support for this contention comes from the study of Tat-dependent protein export in the Gram-positive bacterium *Bacillus subtilis*. This organism contains two copies of a *tatC*-like gene, each of which is linked to a gene encoding a TatA/B/E homologue. However, export of the Tat pathway substrate PhoD is only dependent upon one of the *tatC* genes, and independent of the second copy. These observations suggest that TatC acts as a specificity determinant for PhoD export, possibly at the level of signal peptide recognition (Jongbloed *et al.*, 2000).

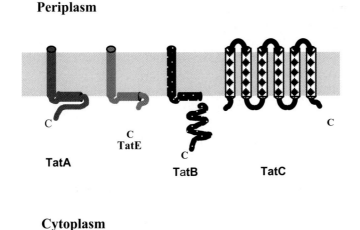

Fig. 2. Cellular location and predicted topologies of Tat components.

The *tatABC* genes in *E. coli* and other closely related enteric bacteria are co-transcribed with a fourth gene, *tatD*, encoding a soluble cytoplasmic protein with nuclease activity (Wexler *et al.*, 2000). Analysis of an *E. coli* strain carrying in-frame deletions in *tatD* and two further homologues, *ycfH* and *yjjV*, indicates that TatD family proteins play no obligate role in Tat-dependent protein export. Furthermore, Northern blot analysis demonstrates that although *tatD* is transcribed with *tatABC*, the presence of a putative stem-loop structure in the *tatC-tatD* intergenic region serves to greatly depress the level of TatD synthesis (Wexler *et al.*, 2000; Jack *et al.*, 2001).

4.   SUBSTRATES OF THE TAT PATHWAY

The vast majority of substrates that are exported by the Tat pathway are proteins containing redox cofactors, and are often required for respiratory growth. Indeed Tat substrates have been identified which bind any one or more of a number of different cofactor types including the molybdenum cofactor, iron sulphur clusters, copper and NADP (for a comprehensive list see Berks, 1996). Strong evidence suggests that such proteins acquire their cofactors, and therefore attain a folded conformation, in the cytosol. Strains of *E. coli* which are blocked for the synthesis of the molybdenum cofactor fail to export periplasmic molybdoenzymes (Santini *et al.*, 1998). Likewise, mutants of the *Z. mobilis* GFOR that are defective in the binding of NADP are drastically affected for export (Halbig *et al.*, 1999). These observations suggest that cofactor binding is a pre-requisite for export. Moreover in *tat* mutant strains, substrate proteins accumulate in the cytoplasm and in many cases have enzyme activity indicating that cofactor insertion has already occurred.

There is currently no evidence to indicate whether the bacterial Tat system is also capable of transporting unfolded proteins. However, work on the closely related ΔpH pathway from isolated plant thylakoids has indicated that this system is able to translocate both tightly folded proteins, and proteins which are malfolded (Hynds *et al.*, 1998). The development of an *in vitro* system to measure Tat transport in isolated bacterial membranes is required in order to address whether this observation also holds true for the prokaryotic transporter. However, regardless of this, it is likely that a proof-reading mechanism exists *in vivo* in order to prevent the interaction of immature or incorrectly assembled proteins with the translocase. This is particularly pertinent to Tat substrate proteins that are exported as hetero-oligomers, where the signal sequence resides on just one of the subunits. The best studied examples are the bacterial uptake hydrogenases, which minimally comprise of an iron-sulphur cluster-containing small subunit, and a large subunit which harbours the nickel-iron active site. A twin arginine signal peptide is found only on the small subunit, but this directs export of both the large and small subunit to the periplasmic side of the membrane. Export does not occur until the small subunit has bound to the large subunit and the maturation of the large subunit [Ni-Fe] cofactor is complete. Moreover, in the absence of the large subunit, the small subunit accumulates in the cytoplasm in its precursor form (Rodrigue *et al.,* 1999). A clue to how the cell co-ordinates the assembly of the enzyme and the export event is provided by the recent work of Oresnik *et al.,* (2001). The authors identified a protein that was specifically able to interact with the twin arginine signal sequence of the molybdenum cofactor-binding DmsA protein. The protein, designated DmsD, was shown to belong to a family of proteins often encoded in the structural operons of periplasmic molybdoenzymes. DmsD shows some sequence similarity to the product of the *torD* gene, which is co-expressed with the *torA* and *torC* genes required to respire TMAO. TorD is a cytoplasmically located chaperone protein that specifically interacts with the unfolded form of TorA prior to molybdenum cofactor insertion (Pommier *et al.,* 1998). These observations point to a mechanism whereby the signal sequence and unfolded mature region of the apo-molybdoenzyme is bound by a specific chaperone protein, which serves to maintain the apoenzyme in a form competent for cofactor insertion, and to shelter the signal peptide to prevent interaction with the Tat translocase. Presumably once the cofactor has bound to the enzyme, the chaperone is displaced, revealing the signal peptide and thus allowing export. It is not clear whether such a mechanism is also operational for other Tat substrate proteins. However, it is interesting to note that the structural operons for many redox enzymes exported by the Tat pathway contain extra accessory genes which might encode export/assembly chaperones. Moreover, the signal peptides for proteins containing the same cofactor from different bacteria exhibit striking sequence conservation in addition to the twin arginine motif. This marked sequence conservation is usually found in the signal peptide n-region, which often may be greatly extended (for an example see the hydrogenase signal sequences aligned in Berks *et al.,* 2000). The highly extended and conserved n-regions may mediate interactions with chaperones required for cofactor insertion and/or oligomerization. This may also provide an explanation for the inability of *E. coli* to recognise the extended *Z. mobilis* GFOR Tat signal sequence.

5.   MECHANISM OF PROTEIN TRANSLOCATION

The *E. coli* Tat translocase is comprised minimally of three proteins – TatA (or TatE), TatB and TatC. Phylogenetic analysis indicates that the encoding genes very frequently show genetic linkage but that they do not cluster with any further genes, suggesting that they probably form the core components of the Tat export system. Consistent with this proposal, overexpression of the *tatABC* genes results in a marked increase in flux of a protein with a mutant signal peptide through the Tat pathway (Sargent *et al.,* 2001). A recent study demonstrated the co-immunoprecipitation of TatA and TatB proteins from detergent solubilized membranes, suggesting a physiological interaction between these components of the Tat pathway (Bolhuis *et al.,* 2000). This has now been confirmed by isolation of a complex comprising the TatA and TatB proteins from the cytoplasmic membrane of cells overexpressing the *tatABC* genes. This complex, which has been purified to apparent homogeneity, has an apparent molecular mass by gel filtration of approximately 600 kDa and contains a molar excess of TatA over TatB of around 15. The mass of the complex is similar to that estimated for the size of the functional Tat transporter of plant thylakoid membranes (Berghöfer and Klösgen, 1999). Negative stain electron microscopy of the isolated TatAB complex results in the visualisation of annular structures with a central cavity of 65 Å in diameter. The diameter of this cavity is of an appropriate order of magnitude to accommodate a folded Tat substrate protein (which are not likely to exceed more than 70 Å, see Berks *et al.,* 2000 for a more detailed discussion). It should be noted that based on structural considerations alone, the minimal number of α–helices required to enclose an aqueous channel of 70 Å would be at least 20, which is consistent with a model where multiple copies of (at least) one of the Tat proteins would be present. Based on these observations, a model for how the Tat translocase may appear in the membrane is depicted in Fig. 3.

Although the purified high molecular weight material may be a Tat subcomplex, clearly it cannot represent a functional translocase since the essential TatC protein has been lost. In the model depicted in Fig. 3, TatC is proposed to be peripheral to the TatA/TatB ring, and to interact with TatB, since there is some evidence that TatC is unstable in a *tatB* mutant (Sargent *et al.,* 1999). According to the model, TatC is depicted to interact with the signal peptide, a prediction based on amino acid sequence conservation within the TatC protein family (see above).

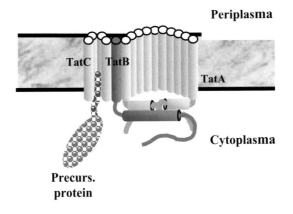

Fig. 3. A model for the structural organisation of the Tat preprotein translocase of *E. coli*. The model is shown in cross section. A folded substrate protein is depicted bound to TatC by its signal peptide.

A further striking feature of the Tat translocase is that it should be capable of exporting substrate proteins which vary greatly in cross-sectional area. The smallest known Tat substrate is the 9 kDa high-potential iron-sulphur protein (HiPIP) of *Chromatium vinosum*, which has a maximum diameter of approximately 30 Å (Carter *et al.*, 1974), whilst the largest, including the two subunits (HyaAB) of *E. coli* hydrogenase 1 and those of *E. coli* formate dehydrogenase-N (FdnGH) have a maximum diameter approaching 70 Å (Volbeda *et al.*, 1995). The Tat system must be capable of forming a tight seal around the substrate protein, regardless of its size, in order to prevent leakage of protons and other ions across the cytoplasmic membrane. One mechanism by which this might be achieved is if the pore-forming subunits actively assemble around the substrate protein. Alternatively, the individual subunits of the pore may slide against each other in an 'iris'-like mechanism as the size of the substrate protein varies. This seal might also be expected to extend above or below the width of the lipid bilayer, since some of the larger Tat substrates could probably not be accommodated within the thickness of the lipid bilayer (around 50 Å). This would require portions of the translocon, for example the amphipathic helices and C-terminal regions of TatA and TatB, to protrude above one or both membrane surfaces during the transport process. They may also control access to the pore at the cytoplasmic face of the membrane.

It has been proposed that the signal peptide inserts into the Tat translocase via a loop mechanism, in a manner analogous to signal peptide interaction in the Sec pathway (Fincher *et al.*, 1998). Here the N-terminus of the signal peptide and the bulk of the mature region of the protein remain at the cytoplasmic face of the membrane, whilst the tip of the loop, which contains the site of signal peptide cleavage by leader peptidase, is exposed at the periplasmic face of the membrane. Supportive evidence for this suggestion is provided by the Rieske iron-sulphur proteins of the cytochrome bc_1 and $b_6 f$ complexes. The Rieske proteins are synthesised with uncleaved twin arginine signal sequences which serve as membrane anchors, and consistent with the loop model, the twin arginine motif is located at the cytoplasmic side of the membrane (van Doren *et al.*, 1993). Tat signal peptides have identical cleavage sites to those of Sec signals and it has been proposed that they are also processed by leader peptidase (Berks, 1996).

It is likely that protein transport through the bacterial Tat apparatus is energised solely by the transmembrane proton gradient, as has been demonstrated for the plant thylakoid ΔpH import machinery (Mould and Robinson, 1991). Conclusive proof of this awaits the development of an *in vitro* system to measure Tat-dependent protein import into inverted vesicles derived from the bacterial cytoplasmic membrane. However, indirect evidence supports this contention. It has been demonstrated that export by the Tat pathway in whole cells of *E. coli* or *Z. mobilis* is sensitive to the addition of protonophores which collapse the protonmotive force (Wiegert *et al.*, 1996; Santini *et al.*, 1998; Cristóbal *et al.*, 1999; Blaudeck *et al.*, 2001). It should also be noted that the classical Sec pathway inhibitor, azide, has also been reported to partially (Santini *et al.*, 1998; Blaudeck *et al.*, 2001) or completely block export by the bacterial Tat pathway (Wiegert *et al.*, 1996), although this may be an indirect effect of, for example, inhibition of the respiratory chain.

6.   FUTURE WORK AND PROSPECTS

Clearly, we are at the early stages of understanding the structure, function and mechanism of the Tat protein export system, and many questions remain to be answered. The development of an *in vitro* system will allow us to further explore the mechanism of and requirements for protein translocation. For example, it would be pertinent to ask whether any soluble proteins are necessary for operation of the bacterial Tat pathway, particularly given the observation that water-soluble stromal factors are not required for protein import by the thylakoid ΔpH pathway (Mould *et al.*,

1991; Cline *et al.,* 1992). A further relevant question which could be addressed is whether folding of the substrate protein is a necessity for the transport process. An *in vitro* system should also permit the development of cross-linking strategies to identify which of the known Tat proteins (if any) is involved in recognition of the twin arginine signal peptide, and in forming subunit-subunit interactions. It is not clear whether there are any further components of the Tat translocase waiting to be discovered. These might reasonably be identified by co-purification with known Tat subunits. Isolation of complexes and sub-complexes of Tat components should help us to define the minimal subunit requirement for a functional translocase. This will require the reconstitution of purified translocase material into liposomes and development of a functional assay for Tat activity based on import of substrate proteins or signal peptide binding. It is also anticipated that structural studies on isolated domains through to entire complexes will become feasible. As our study of the redox enzymes which serve as substrate for the Tat systems continues, it is likely that further chaperone proteins involved in co-ordinating the assembly and export processes will be uncovered.

7.   ACKNOWLEDGEMENTS

We thank members of our laboratories, past and present for their contribution to our research on protein transport, and to our many colleagues with whom we have discussed the Tat system. Research in the authors' laboratories is supported by the Biotechnology and Biological Sciences Research Council, the Commission of the European Community through the programme ExporteRRs, the Wellcome Trust, the John Innes Centre, The University of East Anglia and The Royal Society.

8.   REFERENCES

Berghöfer, J. and Klösgen, R.B. (1999) FEBS Lett 460: 328-332.
Berks, B.C. (1996) Mol Microbiol 22: 393-404.
Berks, B.C., Sargent, F. and Palmer, T. (2000) Mol Microbiol 35: 260-274.
Blaudeck, N., Sprenger, G.A., Freudl, R. and Wiegert, T. (2001) J Bacteriol 183: 604-610.
Bogsch, E., Brink, S. and Robinson, C. (1997) EMBO J 16: 3851-3859.
Bogsch, E., Sargent, F., Stanley, N.R., Berks, B.C., Robinson, C. and Palmer, T. (1998) J Biol Chem 273: 18003-18006.
Bolhuis, A., Bogsch, E.G. and Robinson, C. (2000) FEBS Lett 472: 88-92.
Carter, C.W.jr., Kraut, J., Freer, S.T., N.-H.Xuong, Alden, R.A. and Bartsch, R.G. (1974) J Biol Chem 249: 4214.

64

Chanal, A., Santini, C.-L. and Wu, L.-F. (1998) Mol Microbiol 30: 674-676.

Cline, K., Ettinger, W.F. and Theg, S.M. (1992) J Biol Chem 267: 2688-2696.

Cristóbal, S., de Gier, J.-W., Nielsen, H. and von Heijne, G. (1999) EMBO J 18: 2982-2990.

Dreusch, A., Bürgisser, D.M., Heizmann, C.W. and Zumft, W.G. (1997) Biochim Biophys Acta 1319: 311-318.

Feilmeier, B.J., Iseminger, G., Schroeder, D., Webber, H. and Phillips, G.J. (2000) J Bacteriol 182: 4068-4076.

Fincher, V., McCaffery, M. and Cline, K. (1998) FEBS Lett 423: 66-70.

Gross, R., Simon, J. and Kröger, A. (1999) Arch Microbiol 172: 227-232.

Halbig, D., Wiegert, T., Blaudeck, N., Freudl, R. and Sprenger, G.A. (1999) Eur J Biochem 263: 543-551.

Hynds, P.J., Robinson, D. and Robinson, C. (1998) J Biol Chem 273: 34868-34874.

Jack, R.L., Sargent, F., Berks, B.C., Sawers, G. and Palmer, T. (2001) J Bacteriol 183: 1801-1804.

Jongbloed, J.D.H., Martin, U., Antelman, H., Hecker, M., Tjalsma, H., Venema, G., Bron, S., van Dijl, J.M. and Müller, J. (2000) J Biol Chem 275: 41350-41357.

Mould, R.M. and Robinson, C. (1991) J Biol Chem 266: 12189-12193.

Mould, R.M., Shackleton J.B. and Robinson, C. (1991) J Biol Chem 266: 17286-17289.

Oresnik, I.J., Ladner, C.L. and Turner, R.J. (2001) Mol Microbiol 40: 323-331.

Pommier, J., Mejean, V., Giordano, G. and Iobbi-Nivol, C. (1998) J Biol Chem 273: 16615-16620.

Rodrigue, A., Chanal, A., Beck, K., Müller, M. and Wu, L.-F. (1999) J Biol Chem 274: 13223-13228.

Sambasivarao, D., Turner, R.J., Simala-Grant, J.L., Shaw, G., Hu, J. and Weiner, J.H. (2000) J Biol Chem 275: 22526-22531.

Santini, C.-L., Ize, B., Chanal, A., Müller, M., Giordano, G. and Wu, L.-F. (1998) EMBO J 17: 101-112.

Santini, C-L., Bernadac, A., Zhang, M., Chanal, A., Ize, B., Blanco, C. and Wu, L-F. (2001) J Biol Chem 276: 8159-8164.

Sargent, F., Bogsch, E., Stanley, N.R., Wexler, M., Robinson, C., Berks, B.C. and Palmer, T. (1998) EMBO J 17: 3640-3650.

Sargent, F., Stanley, N.R., Berks, B.C. and Palmer, T. (1999) J Biol Chem 274: 36073-36083.

Sargent, F., Gohlke, U., de Leeuw, E., Stanley, N.R., Palmer, T., Saibil, H.R. and Berks, B.C. (2001) (submitted).

Stanley, N.R., Palmer, T. and Berks, B.C. (2000) J Biol Chem 275: 11591-11596.

Thomas, J.D., Daniel, R.A., Errington, J. and Robinson, C. (2001) Mol Microbiol 39: 47-53.

Van Doren, S.R., Yun, C.-H., Crofts, A.R. and Gennis, R.B. (1993) Biochemistry 32: 628-636.

Volbeda, A., Charon, M.-H., Piras, C., Hatchikian, E.C., Frey, M. and Fontecilla-Camps, J.C. (1995) Nature 373: 580-587.

Weiner, J.H., Bilous, P.T., Shaw, G.M., Lubitz, S.P., Frost, L., Thomas, G.H., Cole, J.A. and Turner, R.J. (1998) Cell 93: 93-101.

Wexler, M., Bogsch, E., Klösgen, R.B., Palmer, T., Robinson, C. and Berks, B.C. (1998) FEBS Lett 431: 339-342.

Wexler, M., Sargent, F., Jack, R.L., Stanley, N.R., Bogsch, E.G., Robinson, C., Berks, B.C. and Palmer, T. (2000) J Biol Chem 275: 16717-16722.

Wiegert, T., Sahm, H. and Sprenger, G.A. (1996) Arch Microbiol 166: 32-41.

Chapter 4

ASSEMBLY OF INNER MEMBRANE PROTEINS IN *ESCHERICHIA COLI*

David Drew[1], Linda Fröderberg[1], Louise Baars[1], Joen Luirink[2] and Jan-Willem de Gier[1]

[1]Department of Biochemistry and Biophysics, Stockholm University
106 91 Stockholm, Sweden
[2]Department of Molecular Microbiology, IMBW/BioCentrum Amsterdam
Faculty of Biology, Vrije Universiteit Amsterdam
De Boelelaan 1087
1081 HV Amsterdam, NL

1. INTRODUCTION

For a long time it was generally assumed that the assembly of inner membrane proteins in *Escherichia coli* occurs spontaneously, and that only the translocation of large periplasmic domains of inner membrane proteins requires the aid of a protein machinery, the Sec-translocase. However, evidence obtained in recent years indicates that most, if not all, inner membrane proteins require the assistance of protein factors at different stages during their assembly.

For a good understanding of the assembly of inner membrane proteins (IMPs) in *Escherichia coli*, some knowledge of the translocation of proteins across the inner membrane is required. Most of the components of the machinery involved in the translocation of proteins across the inner membrane of *E. coli*, the so-called Sec-machinery, have been identified in genetic screens (Danese and Silhavy, 1998). Biochemical studies have further characterized the Sec-machinery (see chapter 2; Manting and Driessen, 2000). Most secretory proteins are kept in a translocation competent state by the chaperone SecB. The SecB-preprotein complex is targeted at a late stage during translation or after translation to the inner membrane, where the pre-protein is delivered at the Sec-translocase. The Sec-translocase mediates the translocation of secretory proteins across the

65

B. Oudega (ed.), Protein Secretion Pathways in Bacteria, 65–82.

inner membrane. The core of the Sec-translocase consists of the integral membrane proteins SecY, SecE, and SecG, and the peripheral subunit SecA. SecA, which exists as a dimer, forms together with SecYEG a proton-motive force- and ATP-driven molecular machine that drives the stepwise translocation of secretory proteins across the inner membrane.

There is no consensus on how the Sec-components form a functional Sec-translocase. Electron microscopic examination of purified translocase indicated that under translocation conditions four SecYEG heterotrimers assemble, catalyzed by dimeric SecA, into one large protein complex, forming a protein-conducting channel (Manting *et al.,* 2000). In contrast, a monomeric translocase was proposed based on co-immuno-precipitation and cross-linking experiments (Yahr and Wickner, 2000). SecD, SecF, and YajC are additional translocase subunits that are not pertinent for the translocation reaction *per se*. They form a subcomplex that under certain conditions can be co-purified with the SecYEG complex. SecDFYajC has been implicated in diverse processes, like SecA functioning, release of translocated proteins and maintenance of the proton-motive force across the inner membrane.

Initially, the assembly of IMPs in *E. coli* was studied in Sec-mutant strains selected for protein secretion defects, and with the SecA inhibitor sodium azide. These studies indicated that IMPs can insert independent of the Sec-machinery into the inner membrane, and that only large domains of IMPs that have to cross the inner membrane require the aid of the Sec-translocase (von Heijne, 1994; von Heijne, 1997). However, it should be emphasized that the strains that were used in those studies had been selected for secretion defects rather than defects in IMP assembly. Furthermore, sodium azide only partially inhibits the activity of the Sec-translocase subunit SecA. A more prominent and general role for the Sec-translocase is suggested by recent *in vivo* studies using improved conditional Sec-strains and by *in vitro* studies in which the membrane insertion of IMPs has been partially reconstituted. These studies also identified YidC as a novel inner membrane protein factor that may play a key role in membrane insertion of IMPs both in conjunction with the Sec-translocase and as a separate entity.

In this chapter, the changing view on the insertion of IMPs will be discussed. In addition, the involvement of accessory factors in the steps that precede and follow membrane insertion will be briefly mentioned: targeting to the inner membrane and assembly into the final, lipid embedded and functional structure. At the end of this chapter, the use of *E. coli* as a vehicle to over-express membrane proteins will be discussed.

2. MEMBRANE TARGETING

In *E. coli,* most IMPs seem to be targeted to the inner membrane via the SRP-targeting pathway (see for a review (Herskovits *et al.*, 2000)). In principle, the SRP-targeting pathway in *E. coli* is homologous to the SRP-targeting pathway that targets both secretory and membrane proteins to the endoplasmic reticulum (ER) membrane in eukaryotes (Rapoport *et al.*, 1996). The eukaryotic SRP binds during biosynthesis to the N-terminal signal sequence of secretory and membrane proteins when it is just exposed outside the ribosome. Further translation is inhibited until the SRP contacts its receptor at the ER membrane, and subsequently dissociates from the nascent chain. The core of the eukaryotic SRP consists of a GTP-binding protein, the 54 kDa subunit (SRP54) and the 7SL RNA. They participate in a larger complex that binds to the signal sequence via SRP54. At the ER membrane, the SRP makes contact with its receptor subunit SRα, which is tethered to the membrane via the integral membrane subunit SRβ. Both SRα and SRβ are GTP-binding proteins.

In *E. coli*, Ffh (Fifty four homolog), 4.5S RNA and FtsY are homologous to the eukaryotic SRP54, 7SL RNA and SRα, respectively. Genetic screens employed to identify components involved in secretion of proteins across the inner membrane have never pointed to a role of Ffh, 4.5S RNA and FtsY in protein secretion. Accordingly, depletion of Ffh, 4.5S RNA and FtsY does not affect secretion of most proteins across the inner membrane. It should be emphasized that these findings do not exclude a role for the SRP in targeting of secretory proteins under all circumstances. However, the preferred substrates for SRP-mediated targeting appear to be IMPs. Assembly of most (of only a limited test set of) IMPs is strongly hampered upon the depletion of Ffh, 4.5S RNA and FtsY, which might explain the essential nature of these components. Furthermore, a recent genetic screen designed to identify components involved in the biogenesis of IMPs in *E. coli*, yielded several mutations in the *ffs* gene that encodes 4.5S RNA (Tian *et al.*, 2000). Finally, cross-linking studies have demonstrated that the *E. coli* SRP binds preferentially to particularly hydrophobic targeting signals present in the N-terminus of nascent IMPs (Valent *et al.*, 1997; Valent *et al.*, 1995). Binding of the SRP might be regulated by trigger factor, that also interacts with nascent polypeptides but without any obvious sequence specificity (Beck *et al.*, 2000). At present, the relationship between this function of trigger factor and its postulated roles as a chaperone and folding catalyst is not clear.

In contrast to its eukaryotic counterpart, FtsY is located both in the cytoplasm and at the inner membrane (Luirink *et al.*, 1994) and can already interact with ribosome/nascent chain/SRP complexes in the cytosol.

Consequently, FtsY might play a direct role in the targeting of these complexes. The mechanism and regulation of membrane association of FtsY is enigmatic but appears complex involving both protein and direct lipid contacts (a detailed discussion can be found in (Herskovits *et al.*, 2000)). Upon interaction with membrane lipids, the GTPase activity of FtsY (and probably also of the bound SRP) is stimulated, which results in dissociation of the targeting complex. The released nascent chain is then transferred to the translocation site in a process that is poorly defined in both pro- and eukaryotic cells.

In vitro studies using artificially truncated nascent chains suggested that the *E. coli* SRP can mediate co-translational targeting, but the *in vivo* relevance of these observations is uncertain (Houben *et al.*, 2000). To enable co-translational targeting, the mammalian and yeast SRPs slow down translation upon interaction with the nascent chain. However, the SRP subunits responsible for this feature have no homologous proteins in *E. coli*. Therefore, it is assumed that the *E. coli* SRP cannot arrest translation, although this has not been proven experimentally. Alternative mechanisms of translational pausing in IMPs might be considered. On the other hand, the short distance to the translocase and the relative speed of prokaryotic translocation may make such a mechanism less relevant. Even a post-translational role of the *E. coli* SRP cannot formally be excluded, and has a precedent in chloroplast SRP that interacts post-translationally with thylakoid membrane proteins that are imported from the cytosol. Notably, the *E. coli* SRP is essential for the targeting of a model IMP with only one transmembrane segment (signal anchor sequence) that is located at the very C-terminus of the protein, suggesting a post-translational mode of targeting (Cristobal *et al.*, 1999).

In *E. coli* cells completely depleted for SRP, assembly of SRP-dependent IMPs is not completely blocked (Newitt *et al.*, 1999). How can these IMPs be assembled in the absence of a functional SRP-targeting pathway? Possible alternative and rather speculative mechanisms to explain this include ribosome, mRNA and chaperone-mediated targeting. Recently, it has been shown that the large subunit of the mammalian ribosome can remain associated with the ER membrane after termination of translation (Potter and Nicchitta, 2000). This large subunit can, if it is programmed with mRNA encoding secretory or membrane proteins, recruit a small subunit and subsequently mediate the insertion and translocation of natural SRP substrates in an SRP-independent fashion.

Interestingly, a role for FtsY in the general targeting of ribosomes to the *E. coli* inner membrane has also been postulated, but the fate and function of these targeted ribosomes awaits further analysis (Herskovits and Bibi, 2000). In addition, the affinity of ribosomes for the translocase appears to be a conserved feature that also applies to *E. coli* (Prinz *et al.*, 2000). This affinity together with an exposed targeting signal might be sufficient to target a sub-population of IMPs to the translocase in the absence of SRP.

Alternatively, the mRNA might contain targeting information that functions independently from the SRP. There are some precedents for this mechanism in other systems. In rice, storage proteins can be targeted to specific locations in the ER membrane by signals in the mRNA (Choi *et al.*, 2000). In yeast, targeting signals appear to be present in both translated and nontranslated regions in the mRNA of a subset of mitochondrial proteins (Sanchirico *et al.*, 1998). In *Salmonella typhimurium*, secretion of a substrate of the type III secretion pathway was recently shown to be coupled to secretion via a complex mechanism that involves both amino acid and mRNA targeting signals (Karlinsey *et al.*, 2000).

Finally, It has been suggested, based on an *in vitro* study, that in *E. coli* the chaperone GroEL can mediate the post-translational targeting of the IMP lactose permease (Bochkareva *et al.*, 1996). In addition, evidence has been presented that GroEL can make a link with the inner membrane, by interacting with the Sec-translocase component SecA (Bochkareva *et al.*, 1998).

3.   MEMBRANE INSERTION

As mentioned in the introduction, until recently it was generally assumed that in *E. coli* the Sec-translocase was only involved in the translocation of large periplasmic domains of IMPs. However, detailed biochemical analysis of the insertion of membrane proteins into the ER membrane in eukaryotes inspired a re-evaluation of the role of the Sec-translocase in the assembly of IMPs in *E. coli*.

At the ER membrane, the SRP is released from the arrested nascent polypeptide in a GTP-dependent process that requires both subunits of the SR (Millman and Andrews, 1997) and the translocon, the Sec61 complex (Song *et al.*, 2000). The nascent chain, freed of SRP, enters the translocon that is sealed to the ribosome to form one continuous channel to allow a tightly coupled translation/translocation reaction. The core of the Sec61 complex consists of the Sec61α, Sec61β and Sec61γ components. The Sec61α and Sec61β components are homologous to the bacterial Sec-translocase components SecY and SecE, respectively (Rapoport *et al.*, 1996). There is no homolog of the Sec-translocase component SecA in the Sec61-translocon, which may be related to different mechanisms to energize insertion and translocation. *Vice versa*, the ER translocon contains an accessory component, TRAM, which has no structural homologue in *E. coli*.

Purified Sec61-heterotrimers form oligomers and each oligomer contains 3-4 Sec61-heterotrimers that constitute the functional translocation channel, which is approximately 50 Å wide during translocation. Oligomer formation is stimulated by the association of the Sec61-heterotrimer subunit Sec61α with ribosomes (Hanein *et al.*, 1996; Song *et al.*, 2000), which is mediated by the 28S rRNA of the large ribosomal subunit (Prinz *et al.*, 2000). During early stages of nascent chain insertion the Sec61-channel is blocked at the lumenal side by BiP, which both maintains the permeability barrier at rest and 'pulls' polypeptide chains across the ER membrane during translocation (Hamman *et al.*, 1998; Matlack *et al.*, 1999).

Hydrophilic polypeptides are translocated across the membrane. They can be glycosylated and partially folded at the lumenal side of the membrane while still attached to the ribosome at the cytosolic side. Hydrophobic transmembrane (TM) segments get trapped in the translocon, and at some stage move out into the lipid bilayer (Martoglio *et al.*, 1995). The exact mechanism by which this is achieved is not clear, but the mechanism appears to differ depending on the nature, number and orientation of TM segments. Using a cross-linking strategy, the TM region of a single spanning membrane protein with N_{cyt}-C_{lum} topology was shown to contact first Sec61α and secondly TRAM, before being released into the lipids upon termination of translation. It was suggested that TRAM functions to recognize and transfer TM segments from the core translocon into the lipid bilayer (Do *et al.*, 1996). In a similar approach, nascent opsin, a multi-spanning membrane protein was found to be close to Sec61α and Sec61β, respectively, during integration into the ER membrane (Laird and High, 1997). Recently, the nature of the first TM segment of a nascent N_{lum}-C_{cyt} protein was shown to have a profound influence on its interactions with membrane components. The hydrophobic first TM in the nascent chain rapidly contacted lipids, whereas a mutated, slightly less hydrophobic TM was cross-linked to TRAM and Sec61α. A model was proposed whereby, upon membrane integration, the TM equilibrates between aqueous and lipid compartments of the membrane depending on its hydrophobicity. TRAM was proposed to play a crucial role in retention of the mildly hydrophobic TM (and perhaps also the signal sequence of secretory proteins) at the aqueous translocon (Heinrich *et al.*, 2000).

While genetic studies have hinted at the involvement of the *E. coli* Sec-translocase components SecA, SecE and SecY in the membrane insertion of IMPs (de Gier *et al.*, 1998b; Newitt and Bernstein, 1998; Qi and Bernstein, 1999), it is fair to say that biochemical analysis of the membrane insertion process in *E. coli* still lags behind the studies described above that involve the ER membrane. Recent *in vitro* cross-linking studies indicate that IMPs that are targeted by the SRP/FtsY pathway insert at the SecAYEG translocase that is also used for the translocation of substrates of the SecB targeting route. Nascent model IMPs of different topology and complexity were released from the SRP after docking of FtsY at the membrane, and inserted into the membrane close to SecY, and, depending on the model protein, to SecA (Beck *et al.*, 2000; Scotti *et al.*, 2000; Valent *et al.*, 1998). The mechanism of transfer of the nascent chain from the SRP to the translocase is enigmatic (like in eukaryotic systems), but could involve direct interactions of FtsY and the SRP with the translocase that remain to be established. Interestingly, the *Bacillus subtilis* Ffh appears to have affinity for SecA in an *in vitro* binding study (Bunai *et al.*, 1999). Alternatively, the affinity of ribosomes for the translocase might contribute to the transfer reaction (Prinz *et al.*, 2000). It is also possible that the membrane surrounding of the translocase is enriched in acidic phospholipids that bind FtsY. Notably, like FtsY, SecA also binds preferentially to acidic phospholipids. Whatever the mechanism of transfer, it should occur fast since in *E. coli*, in contrast to eukaryotes, the presence of the Sec-translocase is not strictly required for the release of the nascent chain from the SRP (Scotti *et al.*, 1999).

Photo cross-linking studies using the N_{in}-C_{out} model IMP FtsQ have shown that the earliest interactions of the TM segment with the membrane involve both SecY and lipids (Urbanus *et al.*, 2001). These interactions are detected when the TM segment is not even fully exposed outside the ribosome, reminiscent of the situation in the ER described above, suggesting a conserved mechanism of insertion. It remains to be demonstrated whether the insertion takes place in the interior of the translocation channel (to which the lipids would have access) or between the interface of the translocase and the lipid surroundings. The hydrophilic regions of membrane targeted nascent FtsQ were shown to have extensive contact with both SecA and SecY. In contrast, a polytopic IMP, MtlA, has been found to contact only SecY in a similar approach which might be related to the presence of a large periplasmic loop in FtsQ and not in MtlA, that assembles independent of SecA *in vivo* (Beck *et al.*, 2000; Werner *et al.*, 1992). For SecA independent IMPs, like MtlA, bacterial ribosomes might play an important role as a trigger for the oligomerization of SecYEG trimers into a functional translocation unit. Furthermore, the physical link between the ribosome and the translocase may serve as a handle to push the nascent chain into the translocase obviating the need for SecA to energize IMP insertion (Prinz *et al.*, 2000).

Notably, SecA is not required for the initial membrane insertion of FtsQ (Scotti *et al.*, 2000), whereas it is essential for the complete assembly of the protein *in vivo* (Urbanus *et al.*, 2001). It is unclear how the need for

SecA recruitment at the translocation site is sensed at an early stage, whereas it is only necessary at a later stage (Neumann-Haefelin *et al.*, 2000).

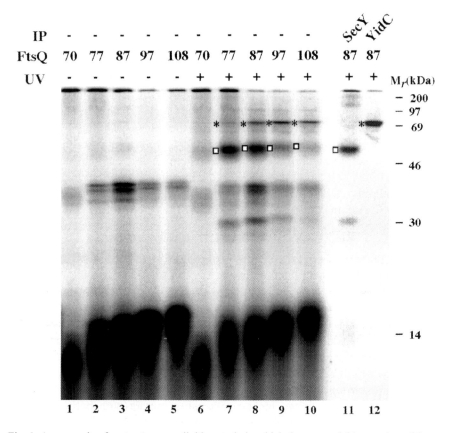

Fig. 1. An example of an *in vitro* crosslinking study in which the sequential interaction of the signal anchor sequence of nascent FtsQ with SecY and YidC is shown. FtsQ constructs of different lengths were used; the size (in number of amino acids) is indicated above the figure. The various constructs were prepared to contain a UV-inducible crosslinking group in the middle of the signal anchor sequence at position 40 (FtsQTAG40) as well as radio-labeled amino acids; details can be found in Scotti *et al.*, 2000 and Urbanus *et al.*, 2001. Following preparation of the constructs, the various nascent chains in complex with ribosomes and SRP were incubated with inverted membrane vesicles containing the translocase in the presence of the SRP receptor FtsY and ATP. Next, a UV treatment was applied and the samples were extracted with sodium carbonate to separate membrane bound from unbound constructs. Then SDS PAGE analysis was carried out and phosphorimaging. The first five lanes (from the left) show the non-crosslinked samples. The last five lanes show the sodium carbonate resistant pellets of the five crosslinked constructs. As controls SecY and YidC have been immunoprecipitated at the right of the figure, next to the markers. The shorter FtsQ constructs were found to crosslink to SecY, whereas the longer constructs were found to crosslink YidC in stead of SecY. Lipid crosslinks are not shown.

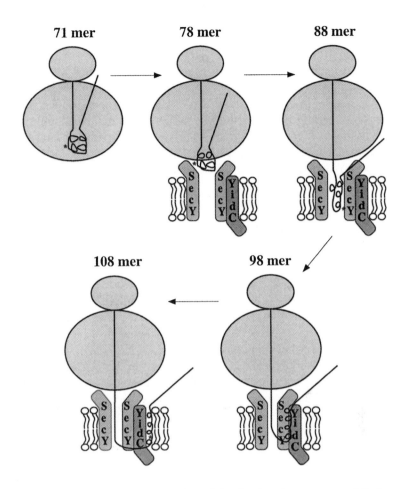

Fig. 2. Model of the sequential interaction of the signal anchor sequence of FtsQ nascent chains of different sizes with SecY and with YidC. The model is an interpretation of the data presented in Fig. 1 of this chapter.

When translocation intermediates of secretory proteins are stuck in the inner membrane they can be photo-cross-linked to SecA and SecY (Joly and Wickner, 1993). Again, these interaction studies show that a similar or identical channel is used for both secretory and membrane proteins. The translocase scans polypeptide chains for hydrophobic domains, and if amino acid stretches are sufficiently hydrophobic translocation stops (Duong and Wickner, 1998). How do transmembrane segments move from the translocase into the lipid bilayer? Recently, *in vitro* crosslinking (see Fig. 1

and Fig. 2) and *in vivo* depletion studies have identified a novel component that might be involved in this process: YidC (Houben *et al.*, 2000; Samuelson *et al.*, 2000; Scotti *et al.*, 2000). At a nascent chain length of approximately 100 amino acids, the signal anchor sequence of two nascent model inner membrane proteins of different topology, FtsQ and Lep were specifically photo-cross-linked to YidC and to lipids while their hydrophilic regions were still close to SecA and SecY, suggesting that YidC is close to the translocase during IMP insertion. Even more striking, YidC could be co-purified with the Sec-translocase and overexpression of translocase components positively affected YidC expression, suggesting a functional relationship. YidC has never been picked up in genetic screens employed to identify components involved in protein translocation. Accordingly, upon the depletion of YidC, protein translocation is hardly affected, suggesting that YidC is not involved in protein translocation (Samuelson *et al.*, 2000). Numerous questions remain to be answered. Is YidC a permanent component of the each translocase or is it specifically recruited depending on the substrate protein that enters the translocation channel? The latter possibility would imply that distinct translocases exist that might share the core (SecYEG) but differ in accessory components (SecA, SecDFYajC and YidC). Is YidC a stoichiometric component of the functional translocase? Does YidC also play a role in the lateral diffusion of cleaved signal sequences? At least, the sequential interaction of the TM in nascent FtsQ with SecY and YidC suggests that YidC acts after SecY has received the TM in the membrane (Scotti *et al.*, 2000; Urbanus *et al.*, submitted for publication). A similar order of interactions was detected in the mammalian ER with TRAM in the role of YidC. It remains to be established whether YidC and TRAM, which show no obvious similarity in primary sequence, are homologous in function.

YidC appears to be present in excess over SecYEG (M. Urbanus and J. Luirink, unpublished results) suggesting YidC might have an additional function distinct from the SecYEG complex. Indeed, depletion of YidC in a conditional strain not only affected the biogenesis of Sec-dependent IMPs but also of Sec-independent proteins like the small M13 procoat protein (Samuelson *et al.*, 2000). It seems possible that YidC somehow facilitates lipid-partitioning of the TMs of simple Sec-independent proteins without actually forming a translocation channel. It should be noted that a direct interaction between Sec-independent IMPs and YidC has not yet been reported. Therefore, it can not be formally excluded that the effect of YidC depletion on the biogenesis of these proteins is indirect.

YidC is homologous to the yeast mitochondrial IMP Oxa1p (Sääf *et al.*, 1998) (Fig. 3). It has been shown that Oxa1p is directly involved in the assembly of both nuclear and mitochondrially encoded IMPs (Hell *et al.*,

1998; Hell *et al.*, 2001). Notably, in yeast mitochondria there are no equivalents of the *E. coli* SRP, the SRP-receptor, or any of the Sec-components (Glick and von Heijne, 1996) again indicative of an independent role of Oxa1p and YidC in IMP insertion.

E. coli YidC

- polytopic 60 kDa inner membrane protein of unknown function

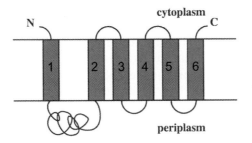

- homologous to yeast mitochondrial Oxa1p:

-polytopic 45 kDa inner membrane protein

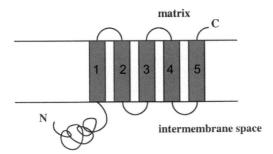

Fig. 3. Model of the membrane folding and topology of YidC, an *E. coli* IMP of unknown function but found to crosslink to nascent inner membrane proteins. In the lower part of the figure the homologous yeast mitochondrial protein Oxa1p is show. This latter protein is involved in assembly of respiratory chain protein complexes and it has been implicated in a novel mechanism of transport of so-called N-tail proteins.

4.   ASSEMBLY   AND   FOLDING   OF   INNER MEMBRANE PROTEINS

Like in the ER, it is unclear to what extent TM segments of polytopic IMPs that insert into the *E. coli* translocase move one by one into the lipid bilayer or *en bloc* after assembly and partial folding in the translocase. In this respect it should be mentioned that the translocase is sufficiently wide to accommodate more than one TM at a time (Hamman *et al.*, 1997). Whatever the mechanism, the lipid bilayer is the final destination of an IMP. Evidence is accumulating that lipids have a more specific role than merely providing a solvent for IMPs (Bogdanov and Dowhan, 1999). Phosphatidylethanolamine appears to assist the folding of the IMP lactose permease by interacting with non-native intermediates. In this respect, lipids can be considered to function as molecular chaperones for IMPs. IMPs also contain cytoplasmic and periplasmic loops that have to be folded correctly during or after synthesis. Cytoplasmic and periplasmic chaperones, that assist the folding of soluble proteins, could be involved in the folding of these loops. Also, membrane bound chaperones like e.g. DjlA that has a cytosolic, DnaJ-like, domain and FtsH might assist the folding of these loops. FtsH probably plays a prominent role in the quality control of IMPs acting both as a chaperone and protease (Akiyama and Ito, 2000; Akiyama *et al.*, 1994).

Besides the hydrophobic TM segments, IMPs contain more information that determines their final structure and topology in the membrane. The distribution of positively charged amino acid residues is a major determinant of the transmembrane topology of IMPs (von Heijne, 1989). The '**positive inside rule**' states that regions of IMPs facing the cytoplasm are generally enriched in arginyl and lysyl residues, whereas translocated regions are largely devoid of these residues. It is, despite several in depth studies, still not clear how positively charged amino acid residues determine the topology of IMPs (van de Vossenberg *et al.*, 1998). Recently, the structure of the *E. coli* SRP was solved (Batey *et al.*, 2000). It revealed that the 4.5S RNA directly participates in the formation of the signal sequence binding groove. It is tempting to speculate that the positively charged amino acid residues flanking the signal anchor sequence of IMPs are attracted by the negatively charged 4.5S RNA, thereby positioning the signal anchor sequence of the IMP in the binding pocket in such a way that the first transmembrane segment is delivered in the right orientation to the Sec-translocase.

Remarkably, the IMP lactose permease, which consists of 12 transmembrane segments, can be expressed in two separate entities, that still assemble into a functional protein (Bibi and Kaback, 1990). It has been shown that between these two fragments, there is an amino acid stretch that delays translation so that the first half of the protein gets the opportunity to fold properly, before the second half of the protein enters the stage (Weinglass and Kaback, 2000). The IICB(Glc) subunit of the *E. coli* glucose transporter can also be expressed in two halves and still forms an active protein. However, the two halves can only form an active protein if they are encoded by a dicistronic operon but not if encoded by two genes on two different replicons. This suggests that spatial proximity of the nascent polypeptide chains is important for folding and membrane assembly (Beutler *et al.*, 2000).

5. ALTERNATIVE INNER MEMBRANE TARGET-ING AND INSERTION PATHWAYS

Our knowledge of the assembly of IMPs in *E. coli* is based on monitoring the analysis of only a very limited number of (model) IMPs. This may have biased our ideas on IMP assembly. Recently, evidence has emerged that the SRP/Sec-translocase pathway is not the only IMP targeting/insertion pathway in *E. coli*. M13 procoat protein is the best known example of an IMP that requires neither the SRP-targeting pathway nor any of the Sec-components for its assembly (de Gier *et al.*, 1998). M13 procoat protein consists of two transmembrane segments connected by a short periplasmic loop. It has a mildly hydrophobic signal sequence that is below the SRP binding threshold (de Gier *et al.*, 1998). Interestingly, if the hydrophobicity of the signal sequence of M13 procoat protein is beefed up, M13 procoat protein is funneled into the SRP/Sec-translocase pathway (de Gier *et al.*, 1998). As mentioned above, upon depletion of YidC, insertion of wild-type M13 procoat protein into the inner membrane is almost completely blocked (Samuelson *et al.*, 2000). This indicates that YidC is involved in both Sec-dependent and Sec-independent membrane protein insertion (Samuelson *et al.*, 2000).

In vivo depletion studies have provided evidence for a third IMP targeting/insertion pathway in *E. coli*: the SRP/YidC pathway (Fröderberg *et al.*, submitted for publication). It seems that in *E. coli* the SRP-targeting pathway can deliver IMPs directly to YidC, without involvement of any of the Sec-components. It has been suggested, based on antibody inhibition experiments, that in chloroplasts which have a prokaryotic ancestor, a similar pathway is operational (Moore *et al.*, 2000). The chloroplast SRP can deliver membrane proteins, without involvement of any of the cpSec-components, to the thylakoidal membrane protein Albino3, which is the chloroplast homologous protein of YidC.

In *E. coli*, besides the Sec protein secretion pathway, another protein secretion pathway is operational, the TAT-pathway (see chapter 3; Berks *et al.*, 2000). Pre-proteins transported by the TAT-pathway usually bind redox cofactors and fold or even oligomerize before translocation across the membrane, whereas the Sec-machinery can only accommodate non-folded polypeptide chains. TAT substrates are characterized by an N-terminal signal peptide that contains a consensus "twin arginine" (RR) motif. So far, no clear example of an IMP targeted/inserted through the TAT-pathway has been found. However, in chloroplasts, where a similar TAT-pathway is operational, insertion of at least one thylakoidal membrane protein, that does not contain a twin arginine motif, is mediated by the TAT-pathway (Summer *et al.*, 2000). Thus, it cannot be ruled out that in *E. coli* the TAT pathway assists the biogenesis of certain IMPs.

As mentioned before, it is not known if in *E. coli* one and the same Sec-translocase is involved in the assembly of all Sec-dependent IMPs, or that there are different 'specialized' Sec-translocases. It is also not known if Sec-translocases are 'ready to use', or that Sec-translocases are assembled during the IMP insertion process, depending on the needs of the IMP that is to be inserted into the inner membrane. It has been shown that secretory and membrane proteins engage different domains/components of the Sec-translocase (Koch and Muller, 2000; Newitt and Bernstein, 1998). It indicates that one and the same subunit functions differently in Sec-dependent protein translocation and IMP insertion. This may explain why the assembly of IMPs, except the ones with large periplasmic domains, is not affected in *E. coli* mutant strains that are selected for protein secretion defects.

It appears that in *E. coli* at least three different IMP targeting and insertion pathways exist: the SRP/Sec-translocase pathway, the YidC pathway, and the SRP/YidC pathway. The picture is emerging that in *E. coli*, IMP targeting/insertion pathways consist of modules (the SRP, SecYEG, SecAYEG and YidC modules), and that different combinations of these modules (SRP/YidC, SRP/SecYEG/YidC, SRP/SecAYEG/YidC and YidC by itself) can make up targeting/insertion pathways that mediate the insertion of defined sets of IMPs. In addition, the SecAYEG module can co-operate with the chaperone SecB to constitute a pathway devoted to the translocation of proteins across the inner membrane (Fig. 4). The

SecDFYajC complex probably constitutes an extra module, which can interact with the SecYEG and SecAYEG modules, and possibly also with the YidC module (Duong and Wickner, 1997).

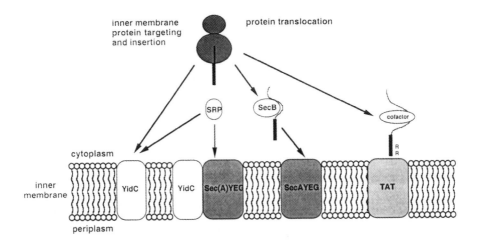

Fig. 4. Inner membrane protein targeting and insertion pathway modules in *E. coli*. In *E. coli*, membrane protein targeting/insertion pathways consist of modules: the SRP, SecYEG, SecAYEG and YidC modules. Different combinations of these modules (SRP/YidC, SRP/SecYEGYidC, SRP/SecAYEGYidC and YidC by itself) can make up membrane protein insertion pathways that mediate the insertion of defined sets of inner membrane proteins into the inner membrane. In addition, the SecAYEG module can cooperate with the chaperone SecB to constitute a pathway devoted to the translocation of proteins across the inner membrane. For the sake of completeness the TAT protein translocation pathway is also depicted.

6. CONCLUSIONS, OVER-EXPRESSION OF MEMBRANE PROTEINS AND FUTURE PERSPECTIVES

So far, studies on the biogenesis of IMPs in *E. coli* have mainly focussed on identifying the components involved, and efforts have been made to reconstruct targeting and insertion pathways. It is very likely that not all components involved in IMP assembly have been identified yet, making it also possible that there are still unidentified membrane protein assembly pathways in *E. coli*. Many important topics concerning the biogenesis of IMPs, such as the role of the proton-motive force, lipids, the quality control of IMP biogenesis, the location of IMP biogenesis in the cell, and the formation of membrane protein complexes have barely been touched yet.

The over-expression of integral membrane proteins is an important bottleneck in studies of membrane protein structure and function (see e.g. the Structural Genomics Supplement of Nature Structural Biology, November 2000). Over-expression of membrane proteins in a membrane is favored to over-expression in inclusion bodies, since it is relatively easy to isolate membrane proteins from membranes and usually impossible to isolate them from inclusion bodies. *E. coli* is one of the most widely used vehicles to over-express both pro- and eukaryotic membrane proteins. It would already be a big step forward if one could determine quickly and accurately whether a membrane protein can be over-expressed in the inner membrane or not. Therefore, in our lab a membrane protein over-expression screen has been developed, based on the method described by Waldo *cum suis* (Waldo *et al.,* 1999), with which the (over)expression status of large numbers of membrane proteins can be tested rapidly (Drew *et al.,* submitted for publication). In this screen Green Fluorescent Protein (GFP) is fused to the C-terminus of the membrane protein that has to be over-expressed. If the over-expressed membrane protein/GFP-fusion ends up in inclusion bodies, the GFP is misfolded and therefore not fluorescent. However, if the membrane protein/GFP-fusion is expressed in the inner membrane GFP folds properly and is fluorescent. The amount of fluoresence is an indication as to the (over)expression levels of the membrane protein in the inner membrane. This GFP-fusion approach is an extremely powerful tool to rapidly screen large numbers of membrane proteins for over-expression in the inner membrane.

Finally, it is very well conceivable that the rapidly extending knowledge of IMP assembly and the physiological consequences of membrane protein over-expression can be very helpful to tailor *E. coli* strains for the over-expression of membrane proteins in the inner membrane.

7.   REFERENCES

Akiyama, Y. and Ito, K. (2000) EMBO J 19: 3888-3895.
Akiyama, Y., Shirai, Y. and Ito, K. (1994) J Biol Chem 269: 5225-5229.
Batey, R.T., Rambo, R.P., Lucast, L., Rha, B. and Doudna, J.A. (2000) Science 287: 1232-1239.
Beck, K., Wu, L.F., Brunner, J. and Muller, M. (2000) EMBO J 19: 134-143.
Berks, B.C., Sargent, F. and Palmer, T. (2000) Mol Microbiol 35: 260-274.
Beutler, R., Kaufmann, M., Ruggiero, F. and Erni, B. (2000) Biochemistry 39: 3745-3750.
Bibi, E. and Kaback, H. (1990) Proc Natl Acad Sci USA 87: 4325-4329.
Bochkareva, E., Seluanov, A., Bibi, E. and Girshovich, A. (1996) J Biol Chem 271: 22256-22261.
Bochkareva, E.S., Solovieva, M.E. and Girshovich, A.S. (1998) Proc Natl Acad Sci USA 95: 478-483.

Bogdanov, M. and Dowhan, W. (1999) J Biol Chem 274: 36827-36830.

Bunai, K., Yamada, K., Hayashi, K., Nakamura, K. and Yamane, K. (1999) J Biochem (Tokyo) 125: 151-159.

Choi, S.B., Wang, C., Muench, D.G., Ozawa, K., Franceschi, V.R., Wu, Y. and Okita, T.W. (2000) Nature 407: 765-767.

Cristobal, S., Scotti, P., Luirink, J., von Heijne, G. and de Gier, J.W. (1999) J Biol Chem 274: 20068-20070.

Danese, P.N. and Silhavy, T.J. (1998) Annu Rev Genet 32: 59-94.

De Gier, J., Scotti, P., Sääf, A., Valent, Q., Kuhn, A., Luirink, J. and von Heijne, G. (1998a) Proc Natl Acad Sci USA 95: 14646-14651.

Do, H., Falcone, D., Lin, J., Andrews, D.W. and Johnson, A.E. (1996) Cell 85: 369-378.

Duong, F. and Wickner, W. (1997) EMBO J 16: 2756-2768.

Duong, F. and Wickner, W. (1998) EMBO J17: 696-705.

Glick, B.S. and von Heijne, G. (1996) Prot Sci 5: 2651-2652.

Hamman, B.D., Hendershot, L.M. and Johnson, A.E. (1998) Cell 92: 747-758.

Hanein, D., Matlack, K.E.S., Jungnickel, B., Plath, K., Kalies, K.-U., Miller, K.R., Rapoport, T. and Akey, C.W. (1996) Cell 87: 721-732.

Heinrich, S.U., Mothes, W., Brunner, J. and Rapoport, T.A. (2000) Cell 102: 233-244.

Hell, K., Herrmann, J.M., Pratje, E., Neupert, W. and Stuart, R.A. (1998) Proc Natl Acad Sci USA 95: 2250-2255.

Hell, K., Neupert, W. and Stuart, R.A. (2001) EMBO J 20: 1281-1288.

Herskovits, A.A. and Bibi, E. (2000) Proc Natl Acad Sci USA 97: 4621-4626.

Herskovits, A.A., Bochkareva, E.S. and Bibi, E. (2000) Mol Microbiol 38: 927-937.

Houben, E.N., Scotti, P.A., Valent, Q.A., Brunner, J., de Gier, J.L., Oudega, B. and Luirink, J. (2000) FEBS Lett: 476: 229-233.

Joly, J.C. and Wickner, W. (1993) EMBO J 12: 255-263.

Karlinsey, J.E., Lonner, J., Brown, K.L. and Hughes, K.T. (2000) Cell 102: 487-497.

Koch, H.G. and Muller, M. (2000) J Cell Biol 150: 689-694.

Laird, V. and High, S. (1997) J Biol Chem 272: 1983-1989.

Luirink, J., ten Hagen-Jongman, C.M., van der Weijden, C.C., Oudega, B., High, S., Dobberstein, B. and Kusters, R. (1994) EMBO J 13: 2289-2296.

Manting, E.H. and Driessen, A.J. (2000) Mol Microbiol 37: 226-238.

Manting, E.H., van Der Does, C., Remigy, H., Engel, A. and Driessen, A.J. (2000) EMBO J 19: 852-861.

Martoglio, B., Hofmann, M.W., Brunner, J. and Dobberstein, B. (1995) Cell 81: 207-214.

Matlack, K.E., Misselwitz, B., Plath, K. and Rapoport, T.A. (1999) Cell 97: 553-564.

Millman, J. and Andrews, D. (1997) Cell 89: 673-676.

Moore, M., Harrison, M.S., Peterson, E.C. and Henry, R. (2000) J Biol Chem 275: 1529-1532.

Neumann-Haefelin, C., Schafer, U., Muller, M. and Koch, H.G. (2000) EMBO J 19: 6419-6426.

Newitt, J.A. and Bernstein, H.D. (1998) J Biol Chem 273: 12451-12456.

Newitt, J.A., Ulbrandt, N.D. and Bernstein, H.D. (1999) J Bacteriol 181: 4561-4567.

Potter, M.D. and Nicchitta, C.V. (2000) J Biol Chem. 275: 33828-33835.

Prinz, A., Behrens, C., Rapoport, T.A., Hartmann, E. and Kalies, K.U. (2000) EMBO J 19: 1900-6.

Qi, H.Y. and Bernstein, H.D. (1999) J Biol Chem 274: 8993-8997.

Rapoport, T.A., Jungnickel, B. and Kutay, U. (1996) Annu Rev Biochem 65: 271-303.

Samuelson, J.C., Chen, M., Jiang, F., Moller, I., Wiedmann, M., Kuhn, A., Phillips, G.J. and Dalbey, R.E. (2000) Nature 406: 637-641.

Sanchirico, M.E., Fox, T.D. and Mason, T.L. (1998) EMBO J 17: 5796-5804.

82

Scotti, P.A., Urbanus, M.L., Brunner, J., de Gier, J.W., von Heijne, G., van der Does, C., Driessen, A.J., Oudega, B. and Luirink, J. (2000) EMBO J 19: 542-549.

Scotti, P.A., Valent, Q.A., Manting, E.H., Urbanus, M.L., Driessen, A.J., Oudega, B. and Luirink, J. (1999) J Biol Chem 274: 29883-29888.

Song, W., Raden, D., Mandon, E.C. and Gilmore, R. (2000) Cell 100: 333-343.

Summer, E.J., Mori, H., Settles, A.M. and Cline, K. (2000) J Biol Chem 275: 23483-23490.

Sääf, A., Monne, M., de Gier, J.W. and von Heijne, G. (1998) J Biol Chem 46: 30415-30418.

Tian, H., Boyd, D. and Beckwith, J. (2000) Proc Natl Acad Sci USA 97: 4730-4735.

Urbanus, M.L., Scotti, P.A., Fröderberg, L., Sääf, A., de Gier, J.-W., Brunner, J., Samuelson, J.C., Dalbey, R.E., Oudega, B. and Luirink, J. (2001) EMBO Reports, in press.

Valent, Q.A., de Gier, J.W.L., von Heijne, G., Kendall, D.A., ten Hagen-Jongman, C.M., Oudega, B. and Luirink, J. (1997) Mol Microbiol 25: 53-64.

Valent, Q.A., Kendall, D.A., High, S., Kusters, R., Oudega, B. and Luirink, J. (1995) EMBO J 14: 5494-5505.

Valent, Q.A., Scotti, P.A., High, S., de Gier, J.W., von Heijne, G., Lentzen, G., Wintermeyer, W., Oudega, B. and Luirink, J. (1998) EMBO J 17: 2504-2512.

Van de Vossenberg, J.L., Albers, S.V., van der Does, C., Driessen, A.J. and van Klompenburg, W. (1998) Mol Microbiol 29: 1125-1157.

Von Heijne, G. (1989) Nature 341: 456-458.

Von Heijne, G. (1994) FEBS Lett 346: 69-72.

Von Heijne, G. (ed.) (1997) Membrane Protein Assembly. R. G. Landes Press, Austin, TX.

Waldo, G.S., Standish, B.M., Berendzen, J. and Terwilliger, T.C. (1999) Nat Biotechnol 17: 691-5.

Weinglass, A.B. and Kaback, H.R. (2000) Proc Natl Acad Sci USA 97: 8938-43.

Werner, P.K., Saier, M.H. and Müller, M. (1992) J Biol Chem 267: 24523-24532.

Chapter 5

BIOGENESIS OF OUTER MEMBRANE PROTEINS

Jan Tommassen and Romé Voulhoux

Department of Molecular Microbiology and Institute of Biomembranes
Utrecht University
Padualaan 8
3584 CH Utrecht, NL

1.   INTRODUCTION

The cell envelope of Gram-negative bacteria consists of two membranes, the inner membrane, which is a phospholipid bilayer, and the outer membrane, which is an asymmetrical bilayer with phospholipids and lipopolysaccharides (LPS) in the inner and outer monolayers, respectively. The membranes are separated by the peptidoglycan-containing periplasm. Both membranes contain proteins. Whereas integral inner-membrane proteins span the membrane by hydrophobic α-helical segments, outer-membrane proteins (OMPs) present an entirely different structure, the β-barrel. These proteins consist mainly of antiparallel amphipathic β-strands, exposing their hydrophobic residues to the membrane and their hydrophilic residues to the interior of the barrel (Koebnik et al., 2000). Most integral OMPs are involved in transport processes, such as the import of nutrients or the export of proteins. Examples are the porins, which form hydrophilic pores in the outer membrane to allow for the passage of small hydrophilic molecules by passive diffusion, the receptors, which present a binding site for specific nutrients and mediate their uptake in an energy-consuming process, and the secretins, which form large oligomers and are involved in protein secretion (Bitter and Tommassen, 1999; Koebnik et al., 2000). Other OMPs, for example OmpA protein of Escherichia coli, have a structural role in maintaining the integrity of the outer membrane, or function in adhesion to or invasion of eukaryotic target cells, e.g. the Opa proteins of pathogenic

B. Oudega (ed.), Protein Secretion Pathways in Bacteria, 83–97.

Neisseria (Billker *et al.*, 2000). The outer membrane contains only few enzymes. Examples are the OmpT protein of *E. coli* (Vandeputte-Rutten *et al.*, 2001), which is a protease, and the outer membrane phospholipase A, which is produced by many different Gram-negative bacteria (Brok *et al.*, 1998). In addition to integral OMPs, the outer membrane contains a number of lipoproteins, which are attached to the membrane via their N-terminal lipid moiety. The proteinaceous part protrudes either into the periplasm, as is the case for the major lipoprotein of *E. coli* also known as Braun's lipoprotein or Lpp (Braun and Rehn, 1969), or into the extracellular medium, as is the case, for example, for the neisserial LbpB and TbpB proteins, which function as parts of the lactoferrin and transferrin receptor, respectively (Cornelissen and Sparling, 1994; Pettersson *et al.*, 1997).

OMPs are synthesized in the cytoplasm with an N-terminal signal sequence, which directs them to the Sec machinery in the inner membrane (see chapter 2). After transport across the inner membrane via the Sec machinery, the signal sequence is cleaved off by either one of two signal peptidases, one of which, i.e. signal peptidase II, is specific for lipoproteins (Dev and Ray, 1984). Subsequently, the proteins have to fold into their native conformation, and to be targeted to and inserted into the outer membrane. In this chapter, the current knowledge of these late events in OMP biogenesis will be reviewed.

2. BIOGENESIS OF INTEGRAL OUTER MEBRANE PROTEINS

2.1 Folding and insertion of OMPs *in vitro*

Many denatured OMPs, including the outer membrane phospholipase A (Dekker *et al.*, 1995) and the trimeric porins OmpF (Eisele and Rosenbusch, 1990) and PhoE (Jansen *et al.*, 2000) of *E. coli*, have been demonstrated to fold spontaneously *in vitro* in the presence of detergents, suggesting that they fold *in vivo* within the membrane environment. Furthermore, spontaneous folding and insertion of OmpA (Surrey and Jähnig, 1992; Radionova *et al.*, 1995; Kleinschmidt and Tamm, 1996; 1999) and of the porin OmpF (Surrey *et al.*, 1996) into liposomes of 1,2-dioleoyl-*sn*-glycero-3-phosphocholine (DOPC) has been reported (for a review, see Tamm *et al.*, 2001), suggesting that the folding and insertion of OMPs are spontaneous processes. However, the insertion of OMPs into such liposomes fails to explain why OMPs insert exclusively into the outer membrane and not into the inner membrane. The specific lipid composition of the outer membrane, containing LPS as a major component, could provide an explanation for the exclusive insertion of OMPs into this membrane. Indeed, the folding of *in vitro* synthesized outer membrane porin PhoE in low concentrations of the detergent Triton X-100 is strongly stimulated by LPS (de Cock and Tommassen, 1996). Furthermore, LPS has been reported to induce the folding *in vitro* of OmpF (Sen and Nikaido, 1991) and OmpA (Freudl *et al.*, 1986) as well. A crucial role for LPS in the biogenesis of OMPs is supported by the observation that deep-rough mutants of *E. coli* and *Salmonella typhimurium*, which lack a substantial portion of the core region of the LPS, have reduced amounts of OMPs (Koplow and Golfine, 1974; Ames *et al.*, 1974). However, whereas LPS is essential for the viability of *E. coli* and *S. typhimurium*, this is not the case in *Neisseria meningitidis*, where the *lpxA* gene, encoding the enzyme catalyzing the first step of LPS biosynthesis could be inactivated (Steeghs *et al.*, 1998). This mutant is viable and produces an outer membrane devoid of LPS, in which all OMPs examined are normally assembled (Steeghs *et al.*, 2001). Thus, an obligatory role for LPS in OMP folding and targeting *in vivo* can be questioned.

The folded forms of PhoE obtained after *in vitro* folding in the presence of LPS are not trimers, but rather native-like monomers, which can be discriminated from denatured monomers by their faster migration during SDS-polyacrylamide gel electrophoresis (SDS-PAGE). Since these folded monomers can trimerize and insert into added outer membranes, when the Triton X-100 concentration is increased, this form of the protein is considered to represent a *bona fide* intermediate in the biogenesis of PhoE (de Cock and Tommassen, 1996). This view has been supported by the detection of folded monomers of PhoE *in vivo* in pulse-chase experiments (Van Gelder and Tommassen, 1996). However, the conversion of these folded monomers into trimers is rather slow. Probably, these folded monomers detected *in vivo*, in contrast to those obtained in the *in vitro* folding assay, are derived from metastable trimers, which dissociate during sample preparation for SDS-PAGE (Jansen *et al.*, 2000). Also in the case of other OMPs, such as LamB and OmpF, the occurrence of metastable trimers as assembly intermediates has been described (Misra *et al.*, 1991; Laird *et al.*, 1994). The existence of a folded monomeric form of PhoE as an assembly intermediate has been questioned in studies of a missense mutant of PhoE, which could trimerize *in vivo* and in a detergent system *in vitro*, but which failed to form a folded monomer with LPS in the *in vitro* assay (Jansen *et al.*, 2000). Therefore, the possibility that the folded monomers represent an *in vitro* folding artifact, and that folding is initiated at the subunit interface (Schulz, 1995), rather than within the individual subunits, which subsequently trimerize, is still open.

2.2   Folding *in vivo* and passage through the periplasm

The observation that denatured OMPs fold *in vitro* in the presence of detergents, suggests that they may fold *in vivo* within the membrane, and, hence, that membrane insertion might precede folding. However, the membrane-spanning segments of OMPs are amphipathic β-strands, and a hydrophobic surface compatible with membrane insertion is only formed after folding of the proteins into a β-barrel structure. Hence, folding must be concomitant with, or even precede membrane insertion. Consistently, deletions that disrupt the β-strands strongly inhibit OMP folding and insertion (Bosch *et al.*, 1986), whereas deletions that affect only the cell surface-exposed loops affect neither folding nor outer membrane insertion (Koebnik, 1999). The fact that disulfide-bond formation is catalyzed *in vivo* by the periplasmic enzyme DsbA (Bardwell *et al.*, 1991) was used to verify the hypothesis that OMP folding precedes membrane insertion (Eppens *et al.*, 1997). Making use of the crystal structure of PhoE (Cowan *et al.*, 1992), cysteines were introduced into the protein by site-directed mutagenesis at positions that are far apart in the primary structure, but are in sufficiently close proximity to allow disulfide bond formation when the protein is correctly folded. It has been demonstrated that the disulfide bond is formed *in vivo* in a DsbA-dependent manner (Eppens *et al.* 1997). Since the cysteines are present in positions that are inaccessible to a periplasmic enzyme after insertion of the protein into the membrane, the disulfide bond must have been introduced into the protein prior to its insertion into the membrane. Hence, these experiments demonstrate that the protein is at least partially folded before its insertion into the outer membrane.

Generally, unfolded proteins can reach their native state spontaneously *in vitro* (Anfinsen, 1973). For a long time, protein self-assembly has been thought to occur *in vivo* as well. However, nowadays it is well accepted that correct folding of proteins in the cytoplasm depends on the function of other proteins, called molecular chaperones and folding catalysts. Similarly, the observation that OMPs fold spontaneously *in vitro* in detergent systems or in liposomes does not implicate that their folding is also unassisted *in vivo*. Many chaperones and folding catalysts are known to exist in the periplasm (see chapter 6), and since the folding of OMPs takes place, at least partially, in this compartment, these proteins are implicated in OMP folding. Two periplasmic chaperones with a role in OMP biogenesis have been identified so far. One of them, SurA, has initially been identified as a protein required for survival during the stationary phase (Tormo *et al.*, 1990). Subsequently, the protein has shown to be a peptidyl-prolyl *cis/trans* isomerase (PPIase) (Lazar and Kolker, 1996; Missiakas *et al.*, 1996: Rouvière and Gross, 1996) and *surA* mutants have found to be defective in the assembly of various OMPs (Lazar and Kolter, 1996). More specifically, the conversion of unfolded monomers of an OMP into folded monomers (which should probably be interpreted as metastable trimers) has been found to be affected in *surA* mutant cells (Rouvière and Gross, 1996). The strong effect of the *surA* mutation on OMP biogenesis is somewhat surprising, because of the presence in the periplasm of many other PPIases that can potentially take over the SurA function. Moreover, the inactivation of *rotA*, which reduces the total PPIase activity in the periplasm to barely detectable levels, does not result in any phenotype (Kleerebezem *et al.*, 1995). Hence, the role of SurA in OMP biogenesis may not be related to its PPIase activity. Indeed, the kinetics of the assembly of a mutant form of PhoE, in which all prolines have been substituted by alanines, appeared as strongly affected by the *surA* mutation as was that of the wild-type PhoE (Eppens *et al.*, unpublished observation). Therefore, besides being a PPIase, SurA appears to have chaperone activity. Consistently, a mutant form of SurA lacking the PPIase domain has shown to retain its chaperone activity (Behrens *et al.*, 2001).

Another periplasmic protein implicated in OMP biogenesis is Skp (*s*eventeen *k*ilodalton *p*rotein). Like *surA* mutants, *skp* mutants are viable, but they contain reduced amounts of OMPs (Chen and Henning, 1996). Skp has been shown to bind selectively to OMPs *in vitro*. However, it seems unlikely that Skp is involved in the folding of OMPs. It has been shown that Skp bound to PhoE synthesized *de novo* in an *in vitro* system, but it was released when the folding of PhoE was induced with LPS and Triton X-110 (de Cock *et al.*, 1999). These results suggest an early role of Skp in OMP biogenesis. Consistently, Skp has been shown to interact with PhoE still in contact with the Sec translocon (Harms *et al.*, 2001). Possibly, by its affinity for non-native OMPs and phospholipids, Skp may play a role in the targeting of OMPs to their assembly sites in the membrane (de Cock *et al.*, 1999).

2.3   Insertion into the outer membrane

It has been reported that OmpA can spontaneously fold and insert into small unilamellar vesicles of DOPC *in vitro* (Tamm *et al.*, 2001). However, this does not exclude the involvement of a proteinaceous machinery in the insertion of OMPs into the outer membrane *in vivo*. Similarly, whereas some small inner-membrane proteins, such as the M13 procoat protein, have been reported to insert spontaneously into liposomes *in vivo* (Geller and Wickner, 1985), their insertion into the inner membrane is assisted *in vivo* (Samuelson *et al.*, 2000). The insertion of inner-membrane proteins *in vivo* generally requires the Sec translocon and the integral inner-membrane protein YidC (see chapter 2). Since the insertion of inner-membrane proteins requires a proteinaceous machinery, the involvement of such a machinery in OMP insertion can be expected. This expectation is underscored by the observation that the kinetics of folding and insertion of OmpA into the liposomes *in vitro* is very slow, taking 20-30 minutes to go to completion at 37°C (Tamm *et al.*, 2001).

Recently, a component of the putative OMP-insertion machinery has been identified in *N. meningitidis* (Voulhoux *et al.*, 2002) This component is an integral OMP, designated Omp85, and it has shown to be essential for the viability of the bacteria. Furthermore, homologues of the *omp85* gene have been detected in all sequenced Gram-negative bacterial genomes examined (Fig. 1). In many of these genomes, this gene is located adjacent to the *skp* gene, encoding a periplasmic chaperone involved in OMP biogenesis. To study the role of Omp85 in OMP biogenesis, the chromosomal copy of the gene has been inactivated in a strain carrying a copy of *omp85* under *lac*-promoter control on a plasmid. By omission of the inducer of the *lac* promoter from the growth medium, the cells are depleted of Omp85. Under those conditions, unfolded forms of all OMPs examined, including porins, receptors, a secretin and outer membrane phospholipase A, have been found to accumulate, probably as aggregates. The unfolded forms could be extracted from the cell envelopes with urea, demonstrating that they were properly not assembled into the outer membrane. A direct role for Omp85 in OMP biogenesis was further supported by demonstrating an interaction of Omp85, immobilized on a nitrocellulose membrane, with a non-native form of the porin PorA in an overlay assay (Voulhoux *et al.*, 2002). An attractive possibility is that Omp85 recognizes the OMP signature that is present at the C-terminus of the vast majority of bacterial OMPs (Struyvé *et al.*, 1992). It is of interest to note that a homologue of Omp85, designated Toc75, is also present in chloroplasts, where it functions as a component of the chloroplast protein-import machinery (Reumann *et al.*, 1999). The presence of Omp85 homologues in Gram-negative bacteria and chloroplasts suggests a common evolutionary origin. Probably, after the transfer of the ancestral gene to the plant nucleus early in endosymbiosis, the transporter might have reversed orientation and recruited additional proteins to mediate import of proteins into the stroma of the chloroplast.

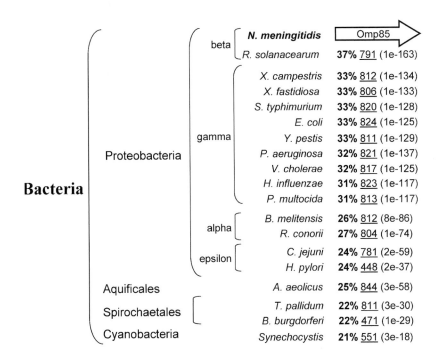

Fig. 1. Conservation of Omp85 in Gram-negative bacteria. Data base searches revealed the presence of homologues of the Neisserial *omp85* gene in all completely finished genomes examined. For each homologue, the percentage of identity over a certain number of amino acid residues (underligned) and the corresponding E-values (in brackets) are indicated at the right. The E value reports the number of hits expected to be found by chance.

Interestingly, some additional components of the chloroplast protein-importmachinery appear to have prokaryotic homologues as well (Reumann and Keegstra, 1999). Cross-linking experiments (Manning *et al.*, 1998) and semi-native SDS-PAGE (Voulhoux *et al.*, 2002) have indicated that Omp85 is part of a multisubunit complex. It will be interesting to identify the other components of this complex and to determine whether they are involved in OMP assembly as well.

In an independent study, another OMP involved in outer-membrane biogenesis was identified in *E. coli* (Braun and Silhavy, 2002). Also this OMP, designated Imp or OstA, is essential for the viability of the bacteria. The gene is located in an operon, which includes the *surA* gene encoding a periplasmic chaperone involved in OMP biogenesis. Homologues of this protein are present in many Gram-negative bacteria. OMPs assemble correctly into their native conformation upon Imp depletion, but they appear in a membrane fraction with a higher density in sucrose gradient centrifugation than that of the outer membrane, suggesting an increased protein to lipid ratio. Similarly, OMPs have shown to appear in a high-density membrane fraction in a conditional *msbA* mutant at the restrictive temperature. In such an *msbA* mutant, the transport of LPS and phospholipids across the inner membrane is disturbed, leading to an increased protein to lipid ratio in the outer membrane (Doerrler *et al.*, 2001). Thus, it seems plausible that Imp/OstA plays a direct role in the insertion of these lipid components into the outer membrane.

3. BIOGENESIS OF LIPOPROTEINS

3.1 Lipidation and maturation of lipoproteins

The majority of bacterial lipoproteins shares the unique structural feature of an *N*-terminal N-acyl-diacylglycerylcysteine, suggesting a common biosynthetic pathway for the lipidation. This pathway has been studied extensively for Lpp (or Braun's lipoprotein), the major lipoprotein of *E. coli*. Lpp and other lipoproteins are synthesized in the cytoplasm in a precursor form, the prolipoprotein, with an N-terminal signal sequence. A lipoprotein-specific consensus sequence was found at the C-terminal end of the signal sequence, i.e. Leu-(Ala/Ser)-(Gly/Ala)-Cys, where the cysteine represents the amino acid residue at position +1 relative to the processing site (Lai *et al.*, 1981). Lipidation and maturation of the lipoproteins takes place after translocation over the inner membrane via the Sec machinery. Treatment of the cells with globomycin, a specific inhibitor of the lipoprotein-specific signal peptidase II, resulted in the accumulation of lipidated prolipoprotein (Hussain *et al.*, 1980), strongly suggesting that lipidation occurs before the processing of the precursor. Consistently, lipidation of the prolipoprotein has found to be a prerequisite for processing by signal peptidase II (Tokunaga *et al.*, 1982).

The complete biosynthetic pathway for the formation of mature lipoproteins requires three enzymes (Fig. 2A). First, a diacylglycerol group is transferred from phosphatidylglycerol to the sulfydryl group of the cysteine at position +1 of the precursor. This step has found to be catalyzed by the phosphatidylglycerol:prolipoprotein diacylglyceryl-transferase (Lgt) (Sankaran and Wu, 1994). The diacylglycerylprolipoprotein is subsequently matured by the signal peptidase II (Dev and Ray, 1984), after which the free

A

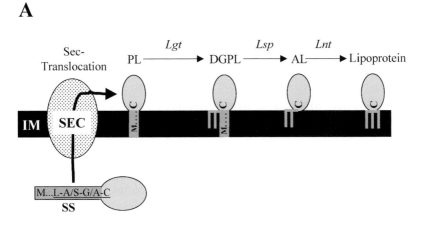

Cytoplasm

B

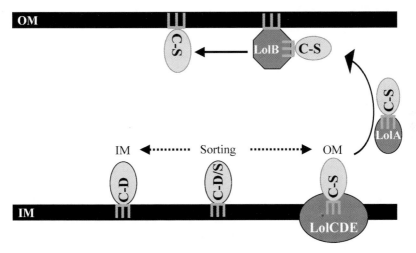

Fig. 2. Biogenesis of lipoproteins. Panel A: Lipidation and maturation of lipoproteins. After transport via the Sec machinery, the prolipoproteins are converted into mature lipoproteins in three successive steps at the periplasmic side of the inner membrane. PL, prolipoprotein; DGPL, diacylglyceryl-prolipoprotein; AL, apolipoprotein; SS, signal sequence. Panel B: Depending on the sorting signal, the mature lipoproteins remain in the inner membrane, or they are transported to the outer membrane via the Lol system. Further details are described in the text.

α-amino group of the N-terminal residue is acylated by the phospholipid:apolipoprotein transacetylase (Lnt) (Gupta *et al.*, 1993) leading to the mature lipoprotein.

3.2   Sorting signals in lipoproteins

Lipoproteins are present in *E. coli* at the periplasmic sides of both the inner and the outer membrane. The amino acid residue at position +2 in the mature lipoprotein, i.e. just after the lipidated cysteine, has been proposed to constitute the sorting signal. Lipoproteins with an aspartate at this position are retained in the inner membrane, whereas those with a serine or other residues at this position appear to localize to the outer membrane (Yamaguchi *et al.*, 1988). This "+2 rule" is strongly supported by the observation that the substitution of an aspartate for the serine at position +2 of Lpp results in the relocalization of the lipoprotein to the inner membrane. This initial +2 rule appears to be an oversimplification and it has recently been refined. First, the nature of the amino acid at position +3 can influence the potency of the inner membrane retention signal (Gennity and Inouye, 1991). Thus, for example, when histidine or lysine is present at position +3, some lipoproteins are localized in the outer membrane even if aspartate is present at position +2. Further studies have demonstrated that the strongest inner membrane retention signals are the couples Asp-Asp, Asp-Glu and Asp-Gln at positions +2 and +3, respectively (Terada *et al.*, 2001). Secondly, it has been found in a genetic selection procedure that also the presence of a phenylalanine at position +2 leads to at least a partial relocalizaton of an outer membrane lipoprotein to the inner membrane (Seydel *et al.*, 1999). Furthermore, it has been demonstrated in the same study by systematic substitutions of the amino acid at position +2 that also tryptophan, tyrosine, proline, and glycine function partially as inner membrane retention signals. Again, the strength of these alternative retention signals appears to be determined by the amino acid present at position +3. Thus, for example the couples Trp-Asn and Phe-Asn at position +2 and +3 appears to function as inner membrane retention signals, whereas the couples Trp-Ser and Phe-Ser do not (Terada *et al.*, 2001).

3.3   Transport of lipoproteins to the outer membrane

The transport of lipoproteins devoid of an inner-membrane retention signal is mediated by a set of five proteins, the Lol proteins (Fig. 2B). Three proteins, LolC, LolD, and LolE, constitute a transporter in the inner membrane of the ATP-binding cassette family (Yakushi *et al.*, 2000). This transporter releases lipoproteins destined for the outer membrane from the inner membrane in an ATP-dependent manner, leading to the formation of a complex between the lipoprotein and the periplasmic chaperone LolA (Matsuyama *et al.*, 1995). By the interaction with LolA, the lipid moiety of the lipoprotein may be shielded from the aqueous environment of the periplasm. The LolA-lipoprotein complex then crosses the periplasm and interacts with a receptor in the outer membrane. This receptor, LolB, mediates the anchoring of the lipoprotein in the outer membrane and releases LolA from the complex (Matsuyama *et al.*, 1997). LolB itself is also a lipoprotein and is, like the other Lol proteins, essential for the viability of the bacteria. Whereas outer-membrane specific lipoproteins have shown to stimulate ATP hydrolysis by the LolCDE transporter, even in the absence of LolA, inner-membrane specific lipoproteins do not (Masuda *et al.*, 2002). Thus, lipoproteins with an inner-membrane retention signal are apparently not recognized by the LolCDE transporter, thereby remaining in the inner membrane.

3.4   Transport of lipoproteins to the cell surface

The Lol system for the transport of lipoproteins to the outer membrane has been extensively studied in *E. coli*. The outer-membrane lipoproteins characterized in *E. coli* are all exposed on the periplasmic side of the membrane. Other bacteria are known to expose lipoproteins at the cell surface, for example the LbpB and TbpB components of the lactoferrin and transferrin receptors, respectively, of *N. meningitidis* (Cornelissen and Sparling, 1994; Pettersson *et al.*, 1997). Such lipoproteins may be transported to the outer leaflet of the outer membrane via an extension of the Lol system. The existence of such an extension of the Lol pathway would implicate the requirement of a sorting signal to discriminate between periplasmically- and surface-exposed lipoproteins. Alternatively, sorting could already take place at the inner-membrane level. The only surface-exposed lipoprotein that has been studied in this respect is the starch-debranching enzyme pullulanase of *Klebsiella oxytoca* (Pugsley, 1993). It is interesting to note that pullulanase contains an aspartate at position +2, indicating that it is not a substrate for the Lol system. Indeed, the transport of pullulanase to the cell surface requires a dedicated machinery, a so-called type II secretion system. Type II secretion systems are widely disseminated among Gram-negative bacteria and are generally used for the secretion of non-lipidated hydrolytic enzymes and toxins into the extracellular environment (see chapter 8). In the absence of a functional type II secretion system, pullulanase is retained in the inner membrane (Pugsley, 1993). Thus, in *K. oxytoca*, sorting between periplasmically and surface-exposed outer-membrane lipoproteins takes place at the inner-membrane level. However, this is not necessarily always the case. For example, no genes for a type II secretion system can be distinguished in the sequenced genomes of *N. meningitidis* (Parkhill *et al.*, 2000; Tettelin *et al.*, 2000). Furthermore, the amino acid residues at positions +2 and +3 of LbpB and TbpB are Ile-Gly and Leu-Gly, respectively (Pettersson *et al.*, 1998), suggesting that these proteins are substrates for the Lol system. Indeed, *lol* genes have been identified in the *N. meningitidis* genomes (Parkhill *et al.*, 2000; Tettelin *et al.*, 2000). Thus, LbpB and TbpB might be transported to the cell surface via an extension of the Lol system, which remains to be identified. However, it cannot be excluded that other mechanisms for the transport of lipoproteins to the outer membrane do exist. Homologues of the *lol* genes have been found in most of the other sequenced Gram-negative bacterial genomes, including for example *Yersinia pestis, Vibrio cholerae, Haemophilus influenzae, Pseudomonas aeruginosa*, and *S. typhimurium*. However, they seem to be absent in the genomes of *Campylobacter jejuni, Helicobacter pylori*, and, more particularly, in that of *Borrelia burgdorferi*. This latter bacterium exposes many lipoproteins at the cell surface, suggesting that these lipoproteins reach the outer membrane in a Lol-independent manner.

4.   CONCLUSIONS AND PERSPECTIVES

In spite of much research and results in the past decades, there is still a lot to be learned about the late steps of OMP biogenesis. With respect to integral OMPs, two periplasmic chaperones, Skp and SurA, have been identified, but their exact role in OMP biogenesis is far from clear. The exact place and timing of OMP folding (associated with the inner or outer membrane or free in the periplasm: prior to or concomitant with outer membrane insertion) remain to be identified. The recent identification of an outer membrane component of the OMP insertion machinery, Omp85, is expected to boost studies on OMP insertion both *in vivo* and *in vitro*. For lipoproteins, the targeting pathway to the outer membrane is now well established. However, the interactions between the Lol components and their substrates, and the transfer of the lipoproteins from the outer membrane receptor LolB into the membrane need to be studied in further detail. Furthermore, nothing is known with respect to the transport of lipoproteins across the outer membrane to the cell surface. This pathway remains to be identified. The same is true for the transport pathway of phospholipids and LPS from the inner membrane to the outer membrane. An intriguing and testable hypothesis is that Lol-like systems are involved in their transport. Possibly, the OMP designated Imp or OstA is involved in this pathway. Finally, the mechanism of transport of macromolecules, such as OMPs, across the peptidoglycan layer has not even been touched. In conclusion, much research remains to be done before a global view of the biogenesis of the bacterial outer membrane can be developed.

5. REFERENCES

Ames, G.F.L., Spudich, E.N. and Nikaido, H. (1974) J Bacteriol 117: 406-416.

Anfinsen, C.B. (1973) Science 181: 223-230.

Bardwell, J.C.A., McGovern, K. and Beckwith, J. (1991) Cell 67: 85-91.

Behrens, S., Maier, R., de Cock, H., Schmid, F.X. and Gross C.A. (2001) EMBO J 20: 285-294.

Billker, O., Popp, A., Gray-Owen, S.D. and Meyer, T.F. (2000) Trends Microbiol 8: 258-260.

Bitter, W. and Tommassen, J. (1999) Trends Microbiol 7: 4-6.

Bosch, D., Leunissen, J., Verbakel, J., de Jong, M., van Erp, H. and Tommassen, J. (1986) J Mol Biol 189: 449-455.

Braun, V. and Rehn, K. (1969) Eur J Biochem 10: 426-438.

Braun, M. and Silhavy T. J. (2002) Mol Microbiol 45: 1289-1302.

Brok, R.G.P.M., Boots, A.P., Dekker, N., Verheij, H.M. and Tommassen, J. (1998) Res Microbiol 149: 703-710.

Chen, R and Henning, U. (1996) Mol Microbiol 19: 1287-1294.

Cornelissen, C.N. and Sparling, P.F. (1994) Mol Microbiol 14: 843-850.

Cowan, S.W., Schirmer, T., Rummel, G., Steiert, M., Ghosh, R., Pauptit, R.A., Jansonius, J.N. and Rosenbusch, J.P. (1992) Nature 358: 727-733.

De Cock, H., Schäfer, U., Potgeter, M., Demel, R., Müller, M. and Tommassen, J. (1999) Eur J Biochem 259: 96-103.

De Cock, H. and Tommassen, J. (1996) EMBO J 15: 5567-5578.

Dekker, N., Merck, K., Tommassen, J. and Verheij, H.M. 1995. Eur J Biochem 232: 214-219.

Dev, I.K. and Ray, P.H. (1984) J Biol Chem 259: 11114-11120.

Doerrler, W.T., Reedy, M.C. and Raetz C.R.H. (2001) J Biol Chem 276: 11461-11464.

Eisele, J.L. and Rosenbusch, J.P. (1990) J Biol Chem 265: 10217-10220.

Eppens, E.F., Nouwen, N. and Tommassen, J. (1997) EMBO J 16: 4295-4301.

Freudl, R., Schwarz, H., Stierhof, Y.-D., Gamon, K., Hindennach, I. and Henning, U. (1986). J Biol Chem 261: 11355-11361.

Geller, B.L. and Wickner, W. (1985) J Biol Chem 260: 13281-13285.

Gennity, J.M. and Inouye, M. (1991) J Biol Chem 266: 16458-16464.

Gupta, S.D., Gan, K., Schmid, M.B. and Wu, H.C. (1993) J Biol Chem 268: 16551-16556.

Hussain, M., Ichihara, S. and Mizushima, S. (1980) J Biol Chem 255: 3707-3712.

Jansen, C., Heutink, M., Tommassen, J. and de Cock, H. (2000) Eur J Biochem 267: 3792-3800.

Kleinschmidt, J.H. and Tamm, L.K. (1996) Biochemistry 35: 12993-13000.

Kleinschmidt, J.H. and Tamm, L.K. (1999) Biochemistry 38: 4994-5005.

Kleerebezem, M., Heutink, M. and Tommassen, J. (1995) Mol Microbiol 18: 313-320.

Koebnik, R. (1999) J Bacteriol 181: 3688-3694.

Koebnik, R., Locher K.P. and Van Gelder P. (2000) Mol Microbiol 37: 239-253.

Koplow, J. and Goldfine, H. (1974) J Bacteriol 117: 527-543.

Lai, J.S., Sarvas, M., Brammer, W.J., Neugebauer, K. and Wu, H.C. (1981) Proc Natl Acad Sci USA 78: 3506-3510.

Laird, M.W., Kloser, A.W. and Misra, R. (1994) J Bacteriol 176: 2259-2264.

Lazard, S.W. and Kolter, R. (1996) J Bacteriol 178: 1770-1773.

Manning, D.S., Reschke, D.K. and Judd, R.C. (1998) Microbial Pathogen 25: 11-21.

Masuda, K., Matsuyama, S. and Tokuda H. (2002) Proc Natl Acad Sci USA 99: 7390-7395.

Matsuyama, S., Tajima, T. and Tokuda H. (1995) EMBO J 14: 3365-3372.

Matsuyama, S., Yokota, N. and Tokuda, H. (1997) EMBO J 16: 6947-6955.

Misra, R., Peterson, A., Ferenci, T. and Silhavy, T.J. (1991) J Biol Chem 266: 13592-13597.

Missiakas, D., Betton, J.-M. and Raina, S. (1996) Mol Microbiol 21: 871-884.

Parkhill, J., Achtman, M., James, K.D., Bentley, S.D., Churcher, C., Klee, S.D., Morelli, G., Basham, D., Brown, D., Chillingworth, T., Davies, R.M., Davis, P., Devlin, K., Feltwell, T., Hamlin, N., Holroyd, S., Jagels, K., Leather, S., Moule, S., Mungall, K., Quail, M.A., Rajandream, M.A., Rutherford, K.M., Simmonds, M., Sankaran, K. and Wu, H.C. (1994) J Biol Chem 269: 19701-19706.

Pettersson, A., Poolman, J.T., van der Ley, P. and Tommassen, J. (1997) Antonie van Leeuwenhoek 71: 129-136.

Pettersson, A., Prinz, T., Umar, A., van der Biezen, J. and Tommassen, J. (1998) Mol Microbiol 27: 599-610.

Pugsley, A.P. (1993) Microbiol Rev 57: 50-108.

Radionova, N.A., Tatulian, S.A., Surrey, T., Jähnig, F. and Tamm, L.K. (1995) Biochemistry 34: 1921-1929.

Reumann, S., Davila-Aponte, J. and Keegstra, K. (1999) Proc Natl Acad Sci USA 96: 784-789.

Reumann, S. and Keegstra, K. (1999) Trends Plant Sci 4: 302-307.

Rouvière, P.E. and Gross, C.A. (1996) Genes Dev 10: 3170-3182.

Samuelson, J.C., Chen, M., Jiang, F., Moller, I., Wiedmann, M., Kuhn, A., Phillips G.J. and Dalbey, R.E. (2000) Nature 406: 637-641.

Schulz, G.E. (1995) In: Advances in Cell and Molecular Biology of Membranes and Organelles (Tartahoff, A.M. and Dalbey, R.E., eds), Vol. 4, pp.175-184, JAI Press Inc., Amsterdam.

Sen, K. and Nikaido, H. (1991) J Bacteriol 173: 926-928.

Seydel, A., Gounon, P. and Pugsley, A.P. (1999) Mol Microbiol 34: 810-821.

Skelton, J., Whitehead, S., Spratt, B.G. and Barrell, B. (2000) Nature 404: 502-506.

Steeghs, L., de Cock, H., Evers, E., Zomers, B., Tommassen, J. and van der Ley, P. (2001) EMBO J 20: 6937-6945.

Steeghs, L., den Hartog, R., den Boer, A., Zomer, B., Roholl, P. and van der Ley, P. (1998) Nature 392: 449-450.

Struyvé, M., Moons, M. and Tommassen, J. (1992) J Mol Biol 218 :141-148.

Surrey, T. and Jähnig, F. (1992) Proc Natl Acad Sci USA 89: 7475-7461.

Surrey, T., Schmid, A. and Jähnig, F. (1996) Biochemistry 35: 2283-2288.

Tamm, L.K., Arora, A. and Kleinschmidt J.H. (2001) J Biol Chem 276: 32399-32402.

Terada, M., Kuroda, T., Matsuyama, S. and Tokuda, H. (2001) J Biol Chem 276: 47690-47694.

Tettelin, H., Saunders, N.J., Heidelberg, J., Jeffries, A.C., Nelson, K.E., Eisen, J.A., Ketchum, K.A., Hood, D.W., Peden, J.F., Dodson, R.J., Nelson, W.C., Gwinn, M.L., DeBoy, R., Peterson, J.D., Hickey, E.K., Haft, D.H., Salzberg, S.I., White, O., Fleischmann, R.D., Dougherty, B.A., Mason, T., Ciecko, A., Parksey, D.S., Blair, E., Cittone, H., Clark, E.B.,

Cotton, M.D., Utterback, T.R., Khouri, H., Qin, H., Vamathevan, J., Gill, J., Scarlato, V., Masignani, V., Pizza, M., Grandi, G., Sun, L., Smith, H.O., Fraser, C.M., Moxon, E.R., Rappuoli, R. and Venter, J.C., (2000) Science 287: 1809-1815.

Tokunaga, M., Tokunaga, H. and Wu, H.C. (1982) Proc Natl Acad Sci USA 79: 2255-2259.

Tormo., A., Almirón, M. and Kolter, R. (1990) J Bacteriol 172: 4339-4347.

Vandeputte-Rutten, L., Kramer, R.A., Kroon, J., Dekker, N., Egmond, M.R. and Gros, P. (2001) EMBO J 20: 5033-5039.

Van Gelder, P. and Tommassen, J. (1996) J Bacteriol 178: 5320-5322.

Voulhoux, R., Bos, M.P., Geurtsen, J., Mols, M. and Tommassen, J. (2002) Science, in press.

Yakushi, T., Masuda, K., Narita, S., Matsuyama, S. and Tokuda, H. (2000) Nat Cell Biol 2: 212-218.

Yamaguchi, K., Yu, F. and Inouye, M. (1988) Cell 53: 423-432.

Chapter 6

CHAPERONES AND FOLDING CATALYSTS INVOLVED IN THE GENERAL PROTEIN SECRETION PATHWAY OF *ESCHERICHIA COLI*

Nellie Harms[1] and Hans de Cock[2]

[1]Department of Molecular Microbiology, IMBW/BioCentrum Amsterdam
Faculty of Biology, Vrije Universiteit Amsterdam, De Boelelaan 1087
1081 HV Amsterdam, NL
[2]Molecular Microbiology and Institute of Biomembranes
Utrecht University, Padualaan 8,
3584 CH, Utrecht, NL

1. INTRODUCTION

Cytoplasmic proteins leaving the exit site of the ribosome fold into their native conformation, whereas most proteins destined for the inner membrane or the extracytoplasmic environment are targeted to the inner membrane (IM). These latter proteins are protected against premature folding, misfolding, aggregation and degradation. These proteins are inserted into or translocated across the inner membrane in a non-native state. Folding and assembly into their native conformation occurs in or at the membrane but outside the cytoplasm. Proteins that are translocated across the IM either remain in the periplasm or are incorporated into the outer membrane. In addition, they can be translocated across the outer membrane where they can end up at the outer cell surface or in the extracellular environment. The precision and speed of these processes is ensured by targeting factors, molecular chaperones and folding catalysts. Molecular chaperones are molecules that prevent unfavorable interactions of a substrate and guides the substrate (a protein) into a productive export and folding pathway. In contrast, folding catalysts speed up the kinetics of folding by catalysing intrinsic rate limiting steps in protein folding, like the

B. Oudega (ed.), Protein Secretion Pathways in Bacteria, 99–119.
© 2003 *Kluwer Academic Publishers. Printed in the Netherlands.*

cis/trans isomerization of X-Pro peptide bonds and disulfide bond formation.

The list of targeting factors, chaperones and folding catalysts that are involved in protein secretion in *Escherichia coli* has increased strongly over the last decade (Table 1). Detailed knowledge is present on the functioning of a number of components that have a role in the initial steps in the targeting and translocation of secretory proteins across the IM. Much less is known of the components that have a role in processes following IM translocation. This chapter focuses on the targeting and folding processes that function in the general secretion pathway in *Escherichia coli*.

2. CYTOPLASMIC CHAPERONES

2.1 Overview

Many proteins that are synthesized in the cytoplasm are ultimately found in extracytoplasmic environments. The proteins that have their function in the IM or the extracytoplasmic environment have to be targeted to the IM and to be prevented from premature folding, misfolding, aggregation and degradation in the cytoplasm. An exception on this rule are the proteins that are translocated via the Tat-pathway that will not be discussed here (see chapter 3) and possibly a number of proteins that are secreted by specific systems (see other chapters). Integration into and translocation across the IM occurs in a non-native state. Decision-making and sorting of the proteins start at the exit site of the ribosome where folding catalysts, chaperones and targeting factors interact with the nascent polypeptide. Information of a protein's destiny and the pathway that is used to reach its final location resides on the primary structure of the protein itself. Presecretory proteins are equipped with an amino-terminal signal sequence. The signal sequence of *E. coli* presecretory proteins contains a positively charged amino terminus followed by a hydrophobic core and a signal sequence cleavage site. Most inner membrane proteins (IMPs) lack a distinct signal cleavage site. However, IMPs are anchored in the IM by hydrophobic α–helical trans-membrane (TM) domains. These TM segments act as targeting signals for the IMPs. The targeting signals are recognized by targeting factors that direct the proteins to the translocon (translocase) at the IM. This translocase consists of the subunits SecY, SecE and SecG, which form the core of the translocon complex. Several other modules (such as SecA, SecDSecFYajC, YidC) function in the context of the type of protein that is delivered at the translocation site (reviewed in Manting *et al.* (2000) and de Gier and Luirink (2001), see also chapter 2 and 4).

Table 1. Targeting factors, chaperones and folding catalysts involved in the general secretion pathway of *Escherichia coli*.

Location	Mr kDa	Form	Functions
Cytoplasm			
TF	60		PPIase, chaperone
SRP	4.5S RNA 48		targeting, IM proteins
SecB	16	tetramer	targeting; chaperone secreted proteins
DnaK	70		chaperone
DnaJ	40		chaperone
GroEL	60	two rings of 7 subunits	chaperonin
Periplasm			
Skp (also in IM)	17	tetramer	chaperone; IM release factor of OM proteins
LolA	20		chaperone; IM release factor of lipoproteins
RotA/PpiA/CypA	19		PPIase, cyclophilin
FkpA	27	dimer	PPIase, FK506 binding protein; chaperone
SurA	44		PPIase, parvulin; chaperone OM proteins
PpiD (IM)	69		PPIase, parvulin
DsbA	21		disulfide oxidoreductase
DsbB (IM)	20		disulfide oxidoreductase
DsbC	23	dimer	disulfide isomerase; chaperone
DsbD (IM)	53		disulfide oxidoreductase
DsbE/ CcmG	18		disulfide oxidoreductase
DsbG	26	dimer	disulfide isomerase; chaperone

IM, inner membrane; OM, outer membrane; PPIase, peptidyl proplyl isomerase,

Proteins that are translocated across the IM have in principle three destinations in Gram-negative bacteria. They can either fold directly into their native conformation and remain in the periplasm, they can be targeted to and inserted into the outer membrane (OM) or they can be translocated across the OM and function at the cell surface or in the extracellular environment. A subgroup of proteins is directly translocated across both IM and OM thereby bypassing the periplasm. This group is not further discussed in this chapter. The secretion processes are assisted by molecular chaperones and folding catalysts. In the periplasm, chaperones that depend on ATP are not found, simply because ATP is not present in the periplasm (Rosen, 1987). A number of periplasmic chaperones and folding catalysts have been described in the last decade. The chaperones are often specific for their protein substrate and are referred to as dedicated chaperones. These will be described in other chapters. This section will deal with the various cytoplasmic chaperones and folding catalysts that function in the context of the general secretion pathway. Folding and assembly of IMPs and OMPs will be described in detail in other chapters.

2.2 The signal recognition particle

In *E. coli*, distinct targeting pathways exist of which the SRP pathway and the chaperone based pathway are very well described. Both *in vivo* as well as *in vitro* data indicate that the majority of IMPs follow the SRP pathway. The SRP has a high affinity for hydrophobic regions of TM segments. Binding to these regions ensures co-translational targeting of nascent polypeptides to the translocation site in the IM (Valent *et al.*, 1998; Beck *et al.*, 2000; Houben *et al.*, 2000). This co-translational targeting minimizes the exposure of hydrophobic regions of IMPs in the cytoplasm. It has been suggested recently that the SRP is essential in *E. coli* partly because efficient IMP targeting prevents accumulation of misfolded and aggregated IMPs in the cytoplasm (Bernstein and Hyndman, 2001). It has been observed that depletion of the SRP causes induction of the heat shock response and of the heat shock regulated proteases ClpQ and Lon. The activity of these proteases suppresses the otherwise lethal effect of SRP depletion (Bernstein and Hyndman, 2001). See chapter 1 for more details.

The early genetic screens that were developed to identify gene products that are involved in the targeting of presecretory proteins did not result in the identification of the SRP pathway, most likely due to the choice of the protein substrate that was used in the assay (periplasmic or OM proteins). *In vitro* data indicated however that the SRP could be cross-linked to signal sequences of presecretory proteins under special conditions. Valent *et al.*, (1997) studied a series of PhoA signal sequences with varying hydrophobicity. They demonstrated that the SRP could bind the signal sequence only when the hydrophobic core reached a certain hydrophobicity. Beck *et al.* (2000) showed that the SRP binds the OmpA signal sequence in the absence of the cytoplasmic folding catalyst Trigger Factor (TF), suggesting that TF influences the interaction of the SRP with the signal sequence. On basis of these *in vitro* data they hypothesized that TF has an regulatory role in directing presecretory proteins into the chaperone based pathway. However, recent *in vivo* experiments showed that presecretory proteins can be rerouted from the chaperone based pathway to the SRP pathway by exchanging the signal sequence for a TM segment of an IMP or by increasing the hydrophobicity of the signal sequence (Lee and Bernstein, 2001). These data indicated an important role for the SRP, instead of TF, in sorting proteins at the ribosomal exit site. The data of Beck and Lee fit, however, in the following sorting model. Hydrophobic signal sequences and transmembrane segments of IMPs interact with the SRP and are directed into the co-translational SRP pathway. Signal sequences with a hydrophobic core below a certain threshold bind the SRP with lower affinity and translation proceeds. TF than binds to the nascent polypeptide thereby weakening the binding with the SRP even further. Upon translation elongation binding sites for other chaperones become accessible and the presecretory protein is directed in the chaperone based post-translational targeting pathway.

2.3 Trigger factor

The first chaperone or folding catalyst that interacts with nascent presecretory proteins is the peptidyl prolyl isomerase Trigger Factor (TF) (Valent *et al.*, 1995; Hesterkamp *et al.*, 1996). TF is an abundant cytoplasmic protein with a high efficiency in catalyzing trans to cis prolyl bond isomerization during the refolding of denatured proteins. TF is associated with the 50S subunit in the *E. coli* ribosome (Stoller *et al.*, 1995) and was found to interact with a large variety of nascent proteins both *in vitro* as well as *in vivo* (Valent *et al.*, 1995; Hesterkamp *et al.*, 1996; Valent *et al.*, 1997; Deuerling *et al.*, 1999; Teter *et al.*, 1999). PrePhoE nascent chains as short as 57 amino acid residues have been shown to cross-link to TF. Recent *in vivo* studies present evidence that TF has a role in the cytosolic folding pathway together with the cytoplasmic chaperone DnaK (Deuerling *et al.*, 1999; Teter *et al.*, 1999). DnaK was found to interact under non-stress conditions with nascent chains with a length of 30 kDa or longer. Deletion of the non-essential gene encoding TF resulted in the doubling of the number of nascent chains interacting with DnaK. Under these conditions DnaK interacts with much shorter nascent chains, suggesting that TF is the first protein that interacts with nascent chains as soon as they emerge from the ribosome. Synthetic lethality has been found in strains that were mutated for both TF and DnaK, indicating that the function of TF and DnaK overlap. Several recent studies have indeed revealed that in addition to its isomerase activity, TF has chaperone activity. TF prevents the aggregation of recombinant proteins either in combination with the chaperonin GroEL-GroES or alone (Nishihara *et al.*, 2000), TF is necessary for the breakdown of abnormal proteins (Kandror *et al.*, 1999). Refolding of GAPDH is assisted by the chaperone function of TF (Huang *et al.*, 2000).

Does TF function in the general protein secretion pathway? A role for TF in secretion was originally demonstrated in *in vitro* experiments with OmpA. TF stimulated the *in vitro* translocation of proOmpA into inverted IM vesicles leading to the hypothesis that TF maintains the translocation competent state of proOmpA (Crooke and Wickner, 1987; Crooke *et al.*, 1988; Crooke *et al.*, 1988). Recently a role in secretion was demonstrated again by Beck *et al.* (2000) who suggested that TF has a role as a decision maker and directs presecretory proteins into the chaperone based targeting pathway (see also above). *In vivo* studies, however, did not reveal an effect of TF depletion in proOmpA secretion (Guthrie and Wickner, 1990). In the same study it was shown that TF could not overcome the export defects present in a SecB deletion strain. It is conceivable that TF, being positioned at the nascent chain exit site of the ribosomal tunnel, just plays a role as a general chaperone and protects the nascent chain from aggregation and improper interactions without making a distinction between cytosolic or secreted proteins. TF also binds to nascent chains of IMPs *in vitro* (Valent *et al.*, 1995; Valent *et al.*, 1997), while an interaction with the polytopic IMP MtlA could not be found (Beck *et al.*, 2000). It has been suggested that IMPs used in the studies of Valent *et al.* (1995, 1997) all contain a large periplasmic loop that needs the SecA protein for translocation. TF might play a role in directing these periplasmic loops to SecA. In contrast, MtlA does not contain such loops and therefore does not need SecA for proper insertion into the membrane. Another explanation might be that the constructs that were used in our laboratory are rather long and segments downstream of the TM domains are exposed to the cytoplasmic environment. In vivo these segments are probably never exposed to the cytosol since they are co-translationally inserted in the membrane and directly translocated over this membrane. Therefore, interaction of TF with IMPs should be studied with short nascent chains that only expose just the TM domain into the cytoplasm.

2.4   SecB

Presecretory proteins that do not bind the SRP or that bind the SRP with low affinity are not targeted by the SRP pathway. They do bind however TF that probably protects the protein against misfolding and aggregation in the initial stages of protein biosynthesis. Upon elongation, other molecular chaperones are required to prevent them from premature folding, aggregation and degradation, and to target the proteins to the IM.

Genetic screens developed to identify gene products involved in secretion of presecretory proteins resulted among others in the identification of the *secB* gene (Kumamoto and Beckwith, 1983), encoding a small protein of 17 kDa (Kumamoto and Nault, 1989). In contrast to the mutations in genes involved in the SRP pathway, null mutations in *secB* are viable provided that reduced growth speed are maintained to prevent generation of suppressor mutants. The export of a subset of presecretory proteins is influenced and are denoted traditionally as SecB-dependent proteins (Kumamoto and Gannon, 1988). Many proteins, however, are exported equally well in *secB* mutants.

SecB is a cytoplasmic homo-tetrameric protein with a molecular mass of 64 kDa (Watanabe and Blobel, 1989; Smith *et al.*, 1996). At physiological ionic strength and pH the SecB tetramer is stable (Smith *et al.*, 1996). Both biochemical and genetic data indicate that SecB is a dimer of dimers (Smith *et al.*, 1996; Muren *et al.*, 1999). The recently solved crystal structure of SecB from *Haemophilus influenza* confirmed this dimer of dimer organization (Xu *et al.*, 2000) (also see chapter 1 for a figure). The protein contains a dual function in preprotein translocation (reviewed in Randall and Hardy, 1995; Kim and Kendall, 2000). Upon binding with SecB preproteins are kept in a non-native state and consequently in a translocation competent state. In addition SecB functions as a targeting factor since it contains a SecA binding site. The latter is the peripheral membrane ATPase of the export apparatus that delivers the energy for translocation.

Since no SecB-preprotein co-crystals have been studied sofar the peptide-binding site of SecB was predicted based on several criteria. At first the site must be located on the surface of SecB. Since the SecB tetramer is very stable over a long range of pH values (Smith *et al.*, 1996), it is unlikely that binding of the peptide needs a change in the oligomeric state of the protein. Furthermore, the binding site must be hydrophobic and flexible enough to bind a wide variety of proteins. In the crystal structure a groove at the surface of SecB was found that meets these criteria. The tetramer contains four of these grooves, two on each side. The grooves at the same side of the molecule fuse together forming a large binding channel. Each groove can be subdivided into two subsites. The amino acid residues in the first subsite are aromatic, conserved and structurally flexible. They provide the necessary plasticity for binding peptides. This subsite recognizes hydrophobic/aromatic amino acid regions of the polypeptide and provide the recognition site for the SecB binding motif as described by Knoblauch et al (1999). These authors suggested that this motif is a nine-residue long sequence that is enriched in aromatic and basic residues. The second subsite contains amino acid residues that are hydrophobic but not aromatic. This subsite binds to more extended regions of the polypeptide ligand by forming regular main chain hydrogen bonds with them. Probably the long stretch of peptide ligands wraps around the binding grooves at both sides of the chaperone.

Genetic and biochemical studies revealed several residues in SecB that are important for the binding to SecA (Kimsey *et al.*, 1995; Fekkes *et al.*, 1998). These residues are conserved within the SecB family and can be found in the crystal structure clustered on the large β-sheet, forming a large acidic patch on the surface of the molecule.

SecB associates with nascent chains as they emerge from the ribosome either co-translationally at a late stage during translation or post-translationally (Kumamoto and Francetic, 1993; Randall *et al.* ,1997). The molecular basis for the specificity of the SecB-preprotein interaction was for a long time unclear. It has been suggested that SecB recognizes preproteins by binding their signal sequence domain (Watanabe and Blobel, 1995). However, signal sequence recognition is not a prerequisite for the binding *per se* (Lecker *et al.*, 1989; Liu, 1989; Randall *et al.*, 1990; Hardy and Randall, 1991). What is necessary, however, is that the protein is in a non-native conformation (Collier *et al.*, 1988; Randall *et al.*, 1990; de Cock and Tommassen, 1992). It is now generally believed that SecB binds the preprotein at its mature domain. Recently, the SecB-binding frames of several periplasmic proteins have been identified biochemically and all the regions are poised towards the centre of the proteins (Topping and Randall, 1994; Khisty *et al.*, 1995; Khisty and Randall, 1995; Smith *et al.*, 1997). Although SecB can bind to a spectrum of non-native polypeptides *in vitro*, *in vivo* it seems to be selective for peptides that are destined for translocation.

Why do some precursors require SecB for translocation, whereas others do not and how does SecB discriminate between preproteins and cytoplasmic proteins? Knoblauch *et al.* (1999) identified a SecB peptide binding motif by screening immobilized peptides for SecB binding. The motif is 9 amino acids long and contains aromatic and basic residues. However, such motifs could be found in cytosolic as well as in preproteins and consequently SecB cannot differentiate between secretory and non-secretory proteins solely on basis of binding specificity. It has been proposed that the formation of a complex between SecB and a protein that possesses a non-native structure will depend on the rate of folding of the polypeptide relative to its rate of association with SecB (Hardy and Randall, 1991; Randall and Hardy, 1995). Proteins containing a signal sequence presumably fold slowly (Hardy and Randall, 1991; Park and Randall, 1988) and are favored therefore to bind SecB and to enter the export pathway. This model has been questioned, however, because SecB can also bind fully folded proteins (Zahn *et al.*, 1996) and the rate of SecB-ligand association can be much faster than the rate of folding of SecB-independent proteins (Fekkes *et al.*, 1995). In addition this model does not explain why some proteins can be translocated in the absence of SecB. Kim showed that the SecB-independent protein PhoA can be converted to SecB-dependence by small well-defined mutations in two regions of the preprotein (Kim *et al.*, 2000). They suggested that prePhoA is inherently able to utilize SecB and that it does so when its ability to interact with the translocase is compromised. They hypothesize that all preproteins, which use the Sec transport machinery, are able to utilize SecB. However, if the preprotein interacts very well with the transport machinery, loss of SecB will simply not have an effect on transport. Although proteins have been defined as SecB-independent because they are transported rather efficiently in the absence of SecB, this does not necessarily mean that SecB plays no role in the transport of these proteins under normal conditions. It does, however, also not exclude the possibility that some preproteins are chaperoned by other proteins than SecB.

2.5　General cytoplasmic chaperones, the heat shock proteins

In eukaryotes, members of the Hsp70 chaperones and Hsp 60 chaperonin families play an important role in targeting polypeptides to different cellular compartments. In *E. coli*, little evidence is available that indicate a major role of the heat shock proteins in protein secretion. GroEL, the Hsp60 isologue in bacteria, was found to be involved in translocation of the SecB-independent precursor of ß-lactamase (Bochkareva *et al.*, 1988; Kusukawa and Ito, 1989). The DnaK, DnaJ and GrpE chaperone team was found to be involved in the export of alkaline phosphatase, ß-lactamase and ribose binding protein (Wild *et al.*, 1992; Wild *et al.*, 1996). In addition, a recent paper describes the interaction of GroEL with SecA and it has been suggested that GroEL is involved in the release of SecA from the membrane (Bochkareva *et al.*, 1998). Although a role in protein translocation of the heat shock proteins is apparent they clearly function as general chaperones that do not make a distinction between cytosolic and secreted proteins. The latter proteins are kept in a non-native, translocation competent state until they reach the translocation site at the IM, but the chaperones lack a specialized targeting function.

3. PERIPLASMIC CHAPERONES AND FOLDING CATALYSTS

3.1 Skp

Skp is a basic soluble protein with a molecular mass of 17 kDa. The proteins is located in the periplasm, but it is present in the IM fraction as well (Thome and Müller, 1991). The protein exists in two different states, as was demonstrated by their different sensitivity towards proteases (de Cock *et al.*, 1999). The conversion between these states can be modulated *in vitro* by phospholipids, LPS and bivalent cations (de Cock *et al.*, 1999). Skp appears to be involved in the folding of OMPs during their passage through the periplasm. It has been demonstrated that Skp binds selectively to OmpA and proteins of the bacterial porin family (Chen and Henning, 1996; de Cock *et al.*, 1999) and LPS (Geyer *et al.*, 1979). In addition, also non-OMPs seem to be substrate for binding (Bothmann and Plückthun, 1998; Hayhurst and Harris, 1999). A *skp* mutant strain was found to contain reduced concentrations of OMPs (Chen and Henning, 1996). Missiakas *et al.* (1996) proposed a role of Skp late in the assembly process, based on the observation that Skp overexpression, unlike that of SurA, did not suppress the phenotypic defects conferred by the expression of altered LPS. Skp may help to remove or exchange the original LPS molecule associated with folded monomers or pro-trimers of OMPs, thus facilitating its assembly. Recent data indicate however that Skp has an early role in the biogenesis of OMPs. De Cock *et al* (1999) has shown that Skp interacts specifically with non-native OMPs and has a high affinity for phospholipids. Schäfer *et al.* (1999) showed that OmpA interacts with Skp in close vicinity of the IM. Harms *et al.* (2001) demonstrated that Skp interacts with the N-terminus of the OMP PhoE already during translocation when this protein is still in a transmembrane orientation in the translocase. They suggested that the initial interaction occurs with the membrane bound, protease resistant form of Skp.

Skp is not essential for growth of *E. coli*, however cells can not tolerate the simultaneous loss of Skp and another periplasmic chaperone FkpA (Dartigalongue *et al.*, 2001). A mutation in the *skp* gene combined with one in the gene encoding the periplasmic protease DegP displays a synthetic conditional-lethal phenotype. Therefore, at elevated temperatures,

E. coli requires the periplasmic protease DegP to remove misfolded proteins from the periplasm (Schäfer *et al.*, 1999; Dartigalongue *et al.*, 2001). This indicates that a possible function of Skp is to maintain the solubility of early folding intermediates in the periplasm. In addition, Skp is involved in release of OmpA from the IM (Schäfer *et al.*, 1999), but release of the porin-like PhoE requires probably additional factors (Harms *et al.*, 2001).

3.2 LolA

Bacterial lipoproteins are a special class of secreted proteins, since they are processed by a specific signal peptidase (lipoprotein signal peptidase) and they contain an N-terminal cysteine residue that is lipid modified. The fatty acids can anchor the protein in the membrane. Lipoproteins are localized in the outer or in the inner membrane of *E. coli*, depending on the nature of the amino acid residue located next to the N-terminal fatty-acylated cysteine. An aspartic acid residue at this position (+2) causes localization in the IM, whereas lipoproteins with other amino acid residues at this position are actively displaced to the outer membrane. When expressed in spheroplasts, the major OM lipoprotein (Lpp) is retained in the IM in its mature form. A periplasmic chaperone, LolA, has been found to mediate the release of Lpp (Matsuyama *et al.*, 1995) or other OM-lipoproteins (Yokota *et al.*, 1999) from the IM by forming a soluble lipoprotein-LolA complex. The LolA dependent release is dependent on the OM-specific sorting signal, since no release can be found when an aspartic acid residue was present on the N-terminal second position of Lpp (Matsuyama *et al.*, 1995). In addition, the LolA dependent release requires ATP (Yakushi *et al.*, 1998). The ATPase involved in this reaction is the LolCDE complex that belongs to the ATP-binding cassette (ABC) transporter family (Yakushi *et al.*, 2000). The soluble lipoprotein-LolA complex is targeted to the OM. The OM receptor LolB mediates the outer membrane incorporation of the lipoprotein via the transient formation of a LolB-lipoprotein complex (Matsuyama *et al.*, 1997; Yokota *et al.*, 1999).

3.3 Peptidyl prolyl *cis/trans* isomerases

Peptidyl-proline residues can occur in either the *cis* or *trans* conformation, yet usually only one conformation is suitable to yield a correctly folded protein (Wulfing and Pluckthun, 1994). Peptidyl prolyl *cis/trans* isomerases (PPIases) catalyze this *cis/trans* isomerization of peptidyl-prolyl bonds (Göthel and Marahiel, 1999). PPIases have been identified both in the cytoplasm and in the periplasm, and they have been divided into three subfamilies on basis of their binding capacity to different drugs (Göthel and Marahiel, 1999): (i) the cyclophilins with RotA (also known as CypA or PpiA) as a periplasmic representative (Liu and Walsh, 1990; Hayano *et al.*, 1991); (ii) FK506-binding proteins with FkpA as periplasmic member (Horne and Young, 1995; Missiakas *et al.*, 1996); and (iii) the parvulins with SurA in the periplasm (Lazar and Kolter, 1996; Missiakas *et al.*, 1996; Rouviere and Gross, 1996) and PpiD in the IM with its catalytic domain in the periplasm (Dartigalongue and Raina, 1998).

The PPIase RotA is not essential in *E. coli. In vitro* this enzyme is a potent isomerase (Liu and Walsh, 1990; Hayano *et al.*, 1991), but it seems not to play an important role in protein folding and export since the kinetics of periplasmic protein export and folding and outer membrane assembly is not affected in a *rotA* mutant (Kleerebezem *et al.*, 1995).

Both SurA and PpiD have been implicated in OMP folding. Initially, SurA has been identified as a protein essential for stationary phase survival (Tormo *et al.*, 1990). By using a *surA* mutant strain, a role of SurA in the assembly of OMPs but not of periplasmic proteins was demonstrated in pulse-chase experiments (Lazar and Kolter, 1996; Rouviere and Gross, 1996). In the absence of SurA, the periplasmic folding of LamB appeared to be affected. SurA consists of a substantial N-terminal region, two iterative parvulin-like domains in the C–terminal half of the protein and a C-terminal tail. SurA exhibits only weak PPIase activity (Rouviere and Gross, 1996), but has chaperone activity (Behrens *et al.*, 2001). This chaperone like activity is independent of the PPIase activity and resides in the N-terminal domain coupled to the C-terminal tail. Interestingly, SurA has shown to possess a strong preference for OMPs over cytosolic or periplasmic proteins in co-immunoprecipitation experiments (Behrens *et al.*, 2001). The IMP PpiD has been isolated as a multicopy suppressor of a *surA* mutant (Dartigalongue and Raina, 1998). Its role in folding has been demonstrated by the fact that the levels of OMPs were reduced in a *ppiD* mutant, although less severe than in a *surA* null mutant. A *ppiD/surA* double mutant is lethal suggesting an overlapping function. PpiD showed significantly higher PPIase activity *in vitro* than SurA, in spite of the less pronounced phenotype of the *ppiD* mutant. Whether PpiD has a similar preference for OMPs as SurA has not been determined.

FkpA is a homodimer with each monomer consisting of 245 amino acid residues that are organized in two domains, an N-terminal domain of unknown function and the C- terminal FK506-binding domain (Arie *et al.*, 2001). FkpA exhibits PPIase activity although to a lesser extent than RotA. Since FkpA is able to suppress the aggregation of a defective folding variant of the maltose-binding protein, MalE31, at stoichiometric amounts, it has been proposed that FkpA functions as a chaperone for envelope proteins in the periplasm (Arie *et al.*, 2001). As found for SurA this chaperone-like activity did not depend on its PPIase activity, but requires the presence of the N-domain. The gene encoding the *fkpA* is not essential. Mutants defective in *fkpA* do not show a significant phenotype, but double mutants in *fkpA* and *degP* display a synthetic conditional-lethal phenotype (Arie *et al.*, 2001; Dartigalongue *et al.*, 2001), just as has been found for Skp (Schäfer *et al.*, 1999; Dartigalongue *et al.*, 2001). Interestingly, a double mutant in both *skp* and *fkpA* is lethal. In addition, a *fkpA/surA* double mutant displays a synthetic conditional-lethal phenotype and a *fkpA/surA/degP* triple mutant is non-viable (Dartigalongue *et al.*, 2001). All these results suggest that the functions of the periplasmic chaperones, Skp and the PPIases, are overlapping to a large extent and indicate that just as trigger factor in the cytoplasm, the PPIases have a chaperone-like function.

3.4   Disulfide oxidoreductases

Disulfide bonds are required for the stability and function of a large number of proteins. For many proteins, disulfide bonds are permanent features of the final folded structure, which enhance the stability of the protein. In Gram-negative bacteria the formation of stable disulfide bonds takes place in the periplasm. Six disulfide oxidoreductase (Dsb) proteins, DsbA, B, C, D, E and G, involved in disulfide bond formation, reduction and isomerization, have been identified and characterized. The Dsb proteins contain a conserved active site motif Cys-X-X-Cys and, with the exception of DsbB, they have the thioredoxin fold (Martin, 1995).

The periplasmic DsbA protein is the first discovered Dsb protein (Bardwell *et al.*, 1991). DsbA catalyzes disulfide bond formation in unfolded substrates (Joly and Schwartz, 1994; Frech *et al.*, 1996). The catalyzing capacity of the protein is caused by the destabilizing effect of the active site disulfide (Zapun *et al.*, 1993) and by the low pK$_a$ of Cys30 (Guddat *et al.*, 1998), which is the essential residue of the Cys-X-X-Cys motif of DsbA (Wunderlich *et al.*, 1995). DsbA's active site disulfide is the most oxidizing disulfide known, which explains why DsbA functions as an oxidant of thiols rather than a disulfide isomerase or reductant (Zapun *et al.*, 1993; Wunderlich and Glockshuber, 1993). During catalysis, DsbA is reduced and reoxidation is caused by the integral membrane protein DsbB (Missiakas *et al.*, 1993; Kishigami *et al.*, 1995). DsbB has two pairs of essential cysteines (Jander *et al.*, 1994). The disulfide bond between Cys104 and Cys130 in the carboxy-terminal periplasmic domain acts directly to oxidize DsbA (Guilhot *et al.*, 1995). The second pair of cysteines, Cys41 and Cys44, which are located in the N-terminal periplasmic loop, reoxidizes the first pair. Reoxidation of the second pair requires the presence of a functional respiratory chain (Kobayashi and Ito, 1999). Bader *et al.* (1999) demonstrated that DsbB uses quinones as electron acceptors, allowing various choices for electron transport to support disulfide bond formation. Under aerobic conditions the cytochrome *bo* oxidase is used, while under partially anaerobic conditions the electrons flow via cytochrome *bd* oxidase to oxygen. Menaquinone shuttles electrons to final electron acceptors during fermentation.

When a protein contains only two cysteines, DsbA will inevitably form the correct disulfide bond. However, since the formation of disulfide bonds by DsbA is very fast (Zapun *et al.*, 1993), DsbA may promote incorrect disulfide bonds in substrate proteins with multiple cysteines. Incorrect pairing has to be corrected for the protein to assume its native conformation. Two periplasmic redox proteins, DsbC (Missiakas *et al.*, 1994; Rietsch *et al.*, 1996; Sone *et al.*, 1998) and DsbG (Andersen *et al.*, 1997; Besette *et al.*, 1999) are involved in the isomerization of undesired disulfide bonds. DsbC is a homodimer with four thiol groups in each 23 kDa subunit, two in the active site Cys98 and Cys 101 and two at positions Cys141 and Cys 163 (Missiakas *et al.*, 1994; Shevchik *et al.*, 1994). The active site cysteines of the dimeric DsbC are normally in the reduced state (Zapun *et al.*, 1995). In the isomerization reaction a mixed disulfide between the isomerase and the substrate is formed initially. The resolution of this mixed substrate can follow two paths. Either DsbC reduces the disulfide bond, leaving the substrate to be re-oxidized by DsbA in a second and maybe third round. Alternatively, the mixed substrate is resolved by a nucleophilic attack of an alternate cysteine of the substrate, resulting in a disulfide bond rearrangement in the substrate protein. Reduction of oxidized DsbC relies on the IM protein DsbD (Sambongi and Ferguson, 1994; Crooke and Cole, 1995; Missiakas *et al.*, 1995). In the absence of DsbD, DsbC is completely oxidized. The reduction of DsbD depends on the cytoplasmic proteins thioredoxin and thioredoxin reductase (Rietsch *et al.*, 1997). DsbC contains apart from its isomerase activity also chaperone activity, since it is able to assist in the refolding of, among others, lysozyme (Chen *et al.*, 1999).

DsbG has been identified recently, but it remains under debate whether it functions as an oxidant or displays isomerase activity (Andersen *et al.*, 1997; Besette *et al.*, 1999). DsbD is also involved in reducing the DsbG protein (Besette *et al.*, 1999). DsbG is a dimeric protein that in stoichiometric amounts prevents the thermal aggregation of two classical chaperone substrate proteins. This indicates that alike DsbC and the PPIases, also DsbG has chaperone like activities.

Two distinct pathways assist the proper formation of disulfide bonds within folding proteins in the periplasm. The DsbA-DsbB pathway introduces disulfide bonds *de novo*, while the DsbC-DsbD pathway functions to isomerize disulfides. One of the questions in disulfide biology is how the isomerase pathway is kept separate from the oxidase pathway. Data from Bader *et al.* (2001) suggest that the dimerization of DsbC acts to protect this protein from oxidation by DsbB. This explains how DsbA and DsbC exhibit rather different redox activities in the same cellular compartment without interfering each other. In addition, these data might support the function of DsbG as isomerase, since this protein is also present as a dimer in the periplasm.

Finally, the DsbE protein is required to maintain the heme-binding site of apo-cytochrome *c* reduced and, therefore, the active site cysteines are in the reduced state (Fabianek *et al.*, 1998). The covalent bonds between the vinyl side chains of heme and the thiol group in the apo-cytochrome *c* are only established if the cysteines in the apo-cytochrome are kept reduced (Fabianek *et al.*, 2000). DsbD has been proposed to have a role in recycling the oxidized form of DsbE (Fabianek *et al.*, 1999). DsbE has been localized in the periplasm (Raina and Missiakas, 1997) and in the IM with its active site exposed to the periplasm (Fabianek *et al.*, 1998).

3.5 Other periplasmic heat shock proteins

Skp, the PPIases and at least DsbA and DsbC are part of the periplasmic heat shock regulon. Heat shock and environmental stress can lead to the misfolding of polypeptides in the cytoplasm as well as in the periplasm or membranes. *E. coli* responds to extracytoplasmic stress by activating the transcription of several genes. Two different signal transduction pathways are known that controls the transcription activation. Misfolding of proteins in the cell envelope and imbalanced synthesis of OMP uniquely induces the σ^E regulon, controlled by the $\alpha^2\beta\beta\sigma^E$ RNA polymerase. IM associated aggregates of misfolded proteins activate the other pathway that comprises the two-component regulatory system CpxA and CpxR (see for a recent review Raivio and Silhavy (1999)).

The genes encoding DsbA, RotA, PpiD and the DegP protease are under control of the Cpx pathway (Pogliano *et al.*, 1997; Dartigalongue and Raina, 1998). Those encoding Skp, FkpA, DsbC, and SurA are under control of the σ^E regulon (Danese and Silhavy, 1997; Dartigalongue *et al.*, 2001). The *degP* gene is under control of both regulons. Mutations in above mentioned genes lead to induction of the stress response indicating again that all these periplasmic folding catalysts are important for maintaining proper folding pathways in the periplasm. Recent data indicate that the $E\_^{E}$ RNA polymerase transcribes some fourty-three genes indicating that more proteins or non-protein compounds are probably involved in folding pathways in the periplasm. Several of them have already been identified and have their function either in the periplasm or the IM. Most of the gene products have unknown functions but several proteins were found to be involved in LPS biogenesis and one is a protease (Dartigalongue et al., 2001).

4.   NON-PROTEINACEOUS CHAPERONES

Studies on the assembly of E. coli lactose permease have revealed a role for phospholipids as molecular chaperones, now designated lipochaperones (Bogdanov and Dowhan, 1999). LacY appears to fold in vivo with the assistance of phosphatidyl ethanol amine (PE). A LacY folding defect caused by either in vivo or in vitro assembly in PE deficient membranes can be corrected by PE in vitro (Bogdanov et al., 1996). Once properly folded in vivo, PE is no longer required to maintain the native conformation. Membrane insertion of LacY and early steps in its conformational maturation appear to be independent of membrane phospholipid composition. However, late steps in the final folding of LacY requires a specific phospholipid-assisted process (Bogdanov and Dowhan, 1998).

Lipopolysaccharide and phospholipids play a role in the folding and assembly process of OMPs. In an early folding step the non-native mature OMP PhoE is converted into a folded-monomeric intermediate with native-like properties. In this step a specific requirement for glycolipid, LPS, low amounts of Triton X-100 and divalent cations has been found (de Cock et al., 1996). A non-lamellar structure of the lipidA of LPS and negative charges in the inner core region were found to be important (de Cock et al., 1999). Consistently, a much less efficient folding of PhoE was found with LPS isolated from deep rough mutants (de Cock and Tommassen, 1996). Similarly, the unfolded, processed form of OmpA, which accumulated in cells overproducing this protein, could be converted into a heat-modifiable form after addition of LPS (Freudl et al., 1986). The role of LPS in the folding of PhoE in vitro is consistent with several in vivo data. A severe defect in the biogenesis of porins was observed in deep-rough mutants (Tommassen and Lugtenberg, 1981; Ried et al., 1990) and an assembly defect of an OmpF mutant protein can be suppressed in vivo by mutations that affect the LPS/phospholipid ratio (Danese and Silhavy, 1998). Furthermore, the drug cerulenin, which inhibits fatty acid synthesis, strongly affects the biogenesis of the porins *in vivo*, consistent with an important role of newly synthesized LPS and/or phospholipids in porin biogenesis (Bocquet-Pages *et al.*, 1981; Pages *et al.*, 1982; Bolla *et al.*, 1988). It has been proposed that LPS stabilizes the trimerization-competent state of a monomeric intermediate by shielding the hydrophobic outer surface of the protein, including the subunit interface that comprises approximately 30% of the monomers surface. Removal of the LPS from the subunit interface can be the trigger to initiate trimer formation at the correct moment. Phospholipids may be involved in this uncapping of the interface, as has been recently proposed (de Cock *et al.*, 2001). Importantly, no LPS has been detected at the subunit interface of a trimer (Cowan *et al.*, 1992) and LPS is not required to maintain the stable structure of a trimer. In this respect, LPS can be regarded as a genuine lipochaperone.

Trimerization and OM insertion of the folded monomers of PhoE generated *in vitro* can be achieved by increasing the Triton X-100 concentration to 0.08% and by the addition of OM (de Cock and Tommassen, 1996). The role of Triton X-100 in this *in vitro* folding and assembly process is not clear, but it may mimic the role of a periplasmic chaperone, since Triton X-100 affects the assembly-competent state of PhoE (de Cock *et al.*, 1996). However, purified LPS, but no Triton X-100 is necessary to convert *in vitro* synthesized OmpF into trimers (Sen and Nikaido, 1991).

5.   CONCLUSIONS AND FUTURE PERSPECTIVE

The list of targeting factors, chaperones and folding catalysts that function in the general secretory pathway has increased strongly over the last decade. Detailed knowledge is present on the functioning of a number of these components that have a role in the initial steps in the targeting and translocation of secretory proteins across the IM. However, much less is known of the players that make up the pathways following IM translocation. Major questions are how these players function, how their timing of binding and release is regulated and in which order they work. Furthermore, it needs to be determined if these interactions take place at the IM or OM membrane or in the periplasm. In addition, it is still not clear whether the pathways for periplasmic and OM proteins strictly diverge at the IM translocon or whether a subset of the chaperones and folding catalysts function in both pathways after translocation.

Research on these items focuses on the biogenesis of OMPs. The current knowledge indicates that Skp and SurA have high preferences for OMPs and suggests that they create a specific OMP pathway and that they are not active in the pathway for periplasmic proteins. Furthermore, the connection between this molecular chaperone pathway and the lipochaperone pathway seems to be somewhere at the level of SurA. Overexpression of SurA has been shown to suppress the deep rough phenotype of a *htrM* mutant (Missiakas *et al.,* 1996). Skp has been shown to interact already with nascent OMPs in the translocon. This very early interaction might than be followed by an interaction with SurA and subsequently with lipochaperones.

It can be anticipated that more players will be identified in time. It is not unlikely that all these components form a machinery, in close proximity of the IM translocon, in order to efficiently guide the substrates into productive export and folding pathways in the cell envelope. Unraveling these pathways and understanding of the functioning of these pathways might in future be useful in light of developments of new drugs that prevent proper localization and folding of cell envelope components that play a critical role in the pathogenesis of microorganisms.

6.   REFERENCES

Andersen, C.L., Matthey-Dupraz, A., Missiakas, D. and Raina, S. (1997) Mol Microbiol 26: 121-132.

Arie, J.P., Sassoon, N. and Betton, J.M. (2001) Mol Microbiol 39: 199-210.

Bader, M., Muse, W., Ballou, D.P., Gassner, C. and Bardwell, J.C. (1999) Cell 98: 217-227.

Bader, M.W., Hiniker,A., Regeimbal, J., Goldstone, D., Haebel, P.W. Riemer, J., Metcalf, P., and Bardwell, J.C.A. (2001) EMBO J 20: 1555-1562.

Bardwell, J.C.A., McGovern, K. and Beckwith, J. (1991) Cell 65: 581-589.

Beck, K., Wu, L.F., Brunner, J. and Muller, M. (2000) EMBO J 19: 134-143.

Behrens, S., Maier, R., de Cock, H., Schmid, F.X. and Gross, C.A. (2001) EMBO J 20: 285-294.

Bernstein, H.D. and Hyndman, J.B. (2001) J Bacteriol 183: 2187-2197.

Besette, P.H., Cotto, J.J., Gilbert, H.F. and Georgiou, G. (1999) J Biol Chem 274: 7774-7782.

Bochkareva, E.S., Lissin, N.M. and Gishovich, A.S. (1988) Nature (London) 336: 254-257.

Bochkareva, E.S., Solovieva, M.E. and Girshovich, A.S. (1998) Proc Natl Acad Sci USA 95: 478-483.

Bocquet-Pages, C., Lazdunski, C. and Lazdunski, A. (1981) Eur J Biochem 118: 105-111.

Bogdanov, M. and Dowhan, W. (1998) EMBO J 17: 5255-5264.

Bogdanov, M. and Dowhan, W. (1999) J Biol Chem 274: 36827-36830.

Bogdanov, M., Sun, J., Kaback, H.R. and Dowhan, W. (1996) J Biol Chem 271: 11615-11618.

Bolla, J.-M., Lazdunski, C. and Pages, J.-M. (1988) EMBO J 7: 3595-3599.

Bothmann, H. and Plückthun, A. (1998) Nat Biotechnol 16: 376-380.

Chen, J., Song, J.L., Zhang, S., Wang, Y., Cui, D.F. and Wang, C.C. (1999) J Biol Chem 274: 19601-19605.

Chen, R. and Henning, U. (1996) Mol Microbiol 19: 1287-1294.

Collier, D.N., Bankaitis, V.A., Weiss, J.B. and Jr, P.J.B. (1988) Cell 53: 273-283.

Cowan, S.W., Schrimer, T., Rummel, G., Steiert, M., Ghosh, R., Pauptit, R.A., Jansonius, J.N. and Rosenbusch, J.P. (1992) Nature (London) 358: 727-733.

Crooke, E., Brundage, L., Rice, M. and Wickner, W. (1988) EMBO J 7: 1831-1835.

Crooke, E., Guthrie, B., Lecker, S., Lill, R. and Wickner, W. (1988) Cell 54: 1003-1011.

Crooke, E. and Wickner, W. (1987) Proc Natl Acad Sci USA 84: 5216-5220.

Crooke, H. and Cole, J. (1995) Mol Microbiol 15: 1139-1150.

Danese, P.N. and Silhavy, T.J. (1997) Gene Develop 11: 1183-1193.

Danese, P.N. and Silhavy, T.J. (1998) Ann Rev Genet 32: 59-94.

Dartigalongue, C., Missiakas, D. and Raina, S. (2001) J Biol Chem 23: 23.

Dartigalongue, C. and Raina, S. (1998) EMBO J 17: 3968-3980.

De Cock, H., Brandenburg, K., Wiese, A., Holst, O. and Seydel, U. (1999) J Biol Chem 274: 5114-5119.

De Cock, H., Pasveer, M., Tommassen, J. and Bouveret, E. (2001) Eur J Biochem 268: 865-875.

De Cock, H., Schafer, U., Potgeter, M., Demel, R., Muller, M. and Tommassen, J. (1999) Eur J Biochem 259: 96-103.

De Cock, H. and Tommassen, J. (1992) Mol Microbiol 6: 599-604.

De Cock, H. and Tommassen, J. (1996) EMBO J 15: 5567-5573.

De Cock, H., van Blokland, S. and Tommassen, J. (1996) J Biol Chem 271: 12885-12890.

De Gier, J.W. and Luirink, J. (2001) Mol Microbiol 40: 314-322.

Deuerling, E., Schulze-Specking, A., Tomoyasu, T., Mogk, A. and Bukau, B. (1999) Nature 400: 693-696.

Fabianek, R.A., Hennecke, H. and Thony-Meyer, L. (1998) J Bacteriol 180: 1947-1950.

Fabianek, R.A., Hennecke, H. and Thony-Meyer, L. (2000) FEMS Microbiol Rev 24: 303-316.

Fabianek, R.A., Hofer, T. and Thony-Meyer, L. (1999) Arch Microbiol 171: 92-100.

Fekkes, P., de Wit, J.G., van der Wolk, J.P., Kimsey, H., Kumamoto, C.A. and Driessen, A.J.M. (1998) Mol Microbiol 29: 1179-1190.

Fekkes, P., Denblaauwen, T. and Driessen, A.J.M. (1995) Biochem 34: 10078-10085.

Frech, C., Wunderlich, M., Glockshuber, R. and Schmid, F.X. (1996) EMBO J 15: 392-398.

Freudl, R., Schwarz, H., Stierhof, Y.-D.G., K., Hindennach, I. and Henning, U. (1986) J Biol Chem 261: 11355-11361.

Geyer, R., Galanos, C., Westphal, O. and Golecki, J.R. (1979) Eur J Biochem 98: 27-38.

Göthel, S.F. and Marahiel, M.A. (1999) CMLS Cell and Mol Life Sci 55: 423-436.

Guddat, L.W., Bardwell, J.C.A. and Martin, J.L. (1998) Struct 6: 757-767.

Guilhot, R., Jander, G., Martin, N.L. and Bechwith, J. (1995) Proc Natl Acad Sci USA 92: 9895-9899.

Guthrie, B. and Wickner, W. (1990) J Bacteriol 172: 5555-5562.

Hardy, S.J.S. and Randall, L.L. (1991) Science 251: 439-443.

Harms, N., Koningstein, G., Dontje, W., Muller, M., Oudega, B., Luirink, J. and de Cock, H. (2001) J Biol Chem 276: 18804-18811.

Hayano, T., Takahashi, N., Kato, S., Maki, N. and Suzuki, M. (1991) Biochem 30: 3041-3048.

Hayhurst, A. and Harris, W.J. (1999) Protein Expr Purif 15: 336-343.

Hesterkamp, T., Hauser, S., Lütcke, H. and Bukau, B. (1996) Proc Natl Acad Sci USA 93: 4437-4441.

Horne, S.M. and Young, K.D. (1995) Arch Microbiol 163: 357-365.

Houben, E.N., Scotti, P.A., Valent, Q.A., Brunner, J., de Gier, J.L., Oudega, B. and Luirink, J. (2000) FEBS Lett 476: 229-233.

Huang, G.C., Li, Z.Y., Zhou, J.M. and Fischer, G. (2000) Protein Sci 9: 1254-1261.

Jander, G., Martin, N.L. and Beckwith, J. (1994) EMBO J 13: 5121-5127.

Joly, J.C. and Schwartz, J.R. (1994) Biochem 33: 4231-4236.

118

Kandror, O., Sherman, M. and Goldberg, A. (1999) J Biol Chem 274: 37743-37749.

Khisty, V.J., Munske, G.R. and Randall, L.L. (1995) J Biol Chem 270: 25920-25927.

Khisty, V.J. and Randall, L.L. (1995) J Bacteriol 177: 3277-3282.

Kim, J. and Kendall, D.A. (2000) Cell Stress Chaperones 5: 267-275.

Kim, J., Luirink, J. and Kendall, D.A. (2000) J Bacteriol 182: 4108-4112.

Kimsey, H.H., Dagarag, M.D. and Kumamoto, C.A. (1995) J Biol Chem 270: 22831-22835.

Kishigami, S., Akiyama, Y. and Ito, K. (1995) FEBS Lett 364: 55-58.

Kleerebezem, M., Heutink, M. and Tommassen, J. (1995) Mol Microbiol 18: 313-320.

Knoblauch, N.T., Rudiger, S., Schonfeld, H.J., Driessen, A.J., Schneider-Mergener, J. and Bukau, B. (1999) J Biol Chem 274: 34219-34225.

Kobayashi, T. and Ito, K. (1999) EMBO J 18: 1192-1198.

Kumamoto, C.A. and Beckwith, J. (1983) J Bacteriol 154: 253-260.

Kumamoto, C.A. and Francetic, O. (1993) J Bacteriol 175: 2184-2188.

Kumamoto, C.A. and Gannon, P.M. (1988) J Biol Chem 263: 11554-11558.

Kumamoto, C.A. and Nault, A.K. (1989) Gene 75: 167-175.

Kusukawa, N., Yura, T., Ueguchi, C., Akiyama, Y. and Ito, K. (1989) EMBO J 8: 3517-3521.

Lazar, S.W. and Kolter, R. (1996) J Bacteriol 178: 1770-1773.

Lecker, S., Lill, R., Ziegelhoffer, T., Georgopoulos, C., Bassford, P.J., Kumamoto, C.A. and Wickner, W. (1989) EMBO J 8: 2703-2709.

Lee, H.C. and Bernstein, H.D. (2001) Proc Natl Acad Sci USA 98: 3471-3476.

Liu, G., Topping, T.B. and Randall, L.L. (1989) Proc Natl Acad Sci USA 86: 9213-9217.

Liu, J. and Walsh, C.T. (1990) Proc Natl Acad Sci USA 87: 4028-4032.

Manting, E.H., van der Does, C., Remigy, H., Engel, A. and Driessen, A.J. (2000) EMBO J 19: 852-861.

Martin, J.L. (1995) Struct 3: 245-250.

Matsuyama, S., Tajima, T. and Tokuda, H. (1995) EMBO J 14: 3365-3372.

Matsuyama, S., Yokota, N. and Tokuda, H. (1997) EMBO J 16: 6947-6955.

Missiakas, D., Betton, J.M. and Raina, S. (1996) Mol Microbiol 21: 871-884.

Missiakas, D., Georgopoulos, C. and Raina, S. (1993) Proc Natl Acad Sci USA 90: 7084-7088.

Missiakas, D., Georgopoulos, C. and Raina, S. (1994) EMBO J 13: 2013-2020.

Missiakas, D., Schwager, F. and Raina, S. (1995) EMBO J 14: 3415-3424.

Muren, E.M., Suciu, D., Topping, T., Kumamoto, C.A. and Randall, L.L. (1999) J Biol Chem 274:

Nishihara, K., Kanemori, M., Yanagi, H. and Yura, T. (2000) Appl Environ Microbiol 66: 884-889.

Pages, C., Lazdunsk, C. and Lazdunski, A. (1982) Eur J Biochem 122: 381-386.

Park, S., Liu, G., Topping, T.B., Cover, W. H. and Randall, L.L. (1988) Science 239: 1033-1035.

Pogliano, J., Lynch, A.S., Belin, D., Lin, E.C.C. and Beckwith, J. (1997) Gene Develop 11: 1169-1182.

Raina, S. and Missiakas, D. (1997) Annu Rev Microbiol 51: 179-202.

Raivio, T.L. and Silhavy, T.J. (1999) Curr Opin Microbiol 2: 159-165.

Randall, L.L. and Hardy, S.J.S. (1995) Trends in Biochemical Science 29: 65-69.

Randall, L.L., Topping, T.B. and Hardy, S.J.S. (1990) Science 248: 860-863.

Randall, L.L., Topping, T.B., Hardy, S.J.S., Pavlov, M.Y., Freistroffer, D.V. and Ehrenberg, M. (1997) Proc. Natl Acad Sci USA 94: 802-807.

Ried, G., Hindennach, I. and Henning, U. (1990) J Bacteriol 172: 6048-6053.

Rietsch, A., Belin, D., Martin, N. and Beckwith, J. (1996) Proc Natl Acad Sci USA 93: 13048-13053.

Rietsch, A., Bessette, P., Georgiou, G. and Beckwith, J. (1997) J Bacteriol 179: 6602-6608.

Rosen, B.P. (1987) In: *Escherichia coli* and *Salmonella typhimurium*: Cellular and Molecular Biology (Neidhardt, F.C., ed), pp. 760-767, American Society for Microbiology, Wahington D.C.

Rouviere, P.E. and Gross, C.A. (1996) Genes Dev 10: 3170-3182.

Sambongi, Y. and Ferguson, S.J. (1994) FEBS Lett 353: 235-238.

Schäfer, U., Beck, K. and Müller, M. (1999) J Biol Chem 274: 24567-24574.

Sen, K. and Nikaido, H. (1991) J Bacteriol 173: 926-928.

Shevchik, V.E., Condemine, G. and Robert-Baudouy, J. (1994) EMBO J 13: 2007-2012.

Smith, V.F., Hardy, S.J.S. and Randall, L.L. (1997) Protein Sci 6: 1746-1755.

Smith, V.F., Schwartz, B.L., Randall, L.L. and Smith, R.D. (1996) Protein Sci 5: 488-494.

Sone, M., Akiyama, Y. and Ito, K. (1998) J Biol Chem 273: 27756-27756.

Stoller, G., Rucknagel, K.P., Nierhaus, K.H., Schmid, F.X., Fischer, G. and Rahfeld, J.U. (1995) EMBO J 14: 4939-4948.

Teter, S.A., Houry, W.A., Ang, D., Tradler, T., Rockabrand, D., Fischer, G., Blum, P., Georgopoulos, C. and Hartl, F.U. (1999) Cell 97: 755-765.

Thome, B.M. and Müller, M. (1991) Mol Microbiol 5: 2815-2821.

Tommassen, J. and Lugtenberg, B. (1981) J Bacteriol 147: 118-123.

Topping, T.B. and Randall, L.L. (1994) Protein Sci 3: 730-736.

Tormo, A., Almiron, M. and Kolter, R. (1990) J Bacteriol 172: 4339-4347.

Valent, Q.A., de Gier, J.W.L., von Heijne, G., Kendall, D.A., ten Hagen-Jongman, C.M., Oudega, B. and Luirink, J. (1997) Mol Microbiol 25: 53-64.

Valent, Q.A., Kendall, D.A., High, S., Kusters, R., Oudega, B. and Luirink, J. (1995) EMBO J 14: 5494-5505.

Valent, Q.A., Scotti, P.A., High, S., de Gier, J.-W.L., von Heijne, G., Lentzen, G., Wintermeyer, W., Oudega, B. and Luirink, J. (1998) EMBO J 17: 2504-2512.

Watanabe, M. and Blobel, G. (1989) Proc Natl Acad Sci USA 86: 2728-2732.

Watanabe, M. and Blobel, G. (1995) Proc Natl Acad Sci USA 92: 10133-10136.

Wild, J., Altman, E., Yura, T. and Gross, C.A. (1992) Genes Dev 6: 1165-1172.

Wild, J., Rossmeissl, P., Walter, W.A. and Gross, C.A. (1996) J Bacteriol 178: 360-3613.

Wulfing, C. and Pluckthun, A. (1994) J Mol Biol 242: 655-669.

Wunderlich, M. and Glockshuber, R. (1993) J Biol Chem 268: 24547-24550.

Wunderlich, M., Otto, A., Maskos, K., Mucke, Seckler, R. and Glockshuber, R. (1995) J Mol Biol 247: 28-33.

Xu, Z., Knafels, J.D. and Yoshino, K. (2000) Nat Struct Biol 7: 1172-1177.

Yakushi, T., Masuda, K., Narita, S., Matsuyama, S. and Tokuda, H. (2000) Nat Cell Biol 2: 212-218.

Yakushi, T., Yokota, N., Matsuyama, S. and Tokuda, H. (1998) J Biol Chem 273: 32576-32581.

Yokota, N., Kuroda, T., Matsuyama, A. and Tokuda, H. (1999) J Biol Chem 274: 30995-30999.

Zahn, R., Perrett, S. and Fersht, A.R. (1996) J Mol Biol 261: 43-61.

Zapun, A., Missiakas, D., Raina, S. and Creighton, T.E. (1993) Biochem 32: 5983-5089.

Zapun, A., Missiakas, D., Raina, S. and Creighton, T.E. (1995) Biochem 34: 5075-5089.

Chapter 7

TYPE I PROTEIN SECRETION SYSTEMS IN GRAM-NEGATIVE BACTERIA: *ESCHERICHIA COLI* α-HEMOLYSIN SECRETION

Ivaylo Gentschev and Werner Goebel

Department of Microbiology, University of Wuerzburg,
D-97074 Wuerzburg, Germany

1.   INTRODUCTION

To date, different pathways, denominated type I, type II, type III, etc have been described in Gram-negative bacteria for the translocation of proteins across the two membranes of the cell envelope (Pugsley *et al.,* 1997; Hueck, 1998; Thanassi and Hultgren, 2000; Christie and Vogel, 2000). In addition these pathways, other mechanisms of protein translocation across the cell envelope have been described (see other chapters), which have not been given a type number or for which the type number is still on debate. Some secretion pathways, for instance the type II pathway, are dependent of the so-called general secretory pathway (GSP), whereas other pathways, for instance the type I and type III secretion systems, are independent of the general secretory pathway (GSP, Pugsley, 1993; Pugsley *et al.,* 1997). These latter pathways do not involve amino-terminal processing of the secreted proteins. Furthermore, protein secretion via the latter pathways occurs by a one-step mechanism through a transenvelope channel without the occurrence of periplasmic intermediates. In this chapter the general features of type I protein secretion pathways will be discussed and the mechanism of secretion of α-hemolysin by *E. coli* will be described in detail.

B. Oudega (ed.), Protein Secretion Pathways in Bacteria, 121–139.

2.   GENERAL FEATURES OF TYPE I PROTEIN SECRETION

Type I secretion (reviewed in Wandersman, 1996; Binet *et al.,* 1997; Young and Holland, 1999; Holland and Blight, 1999; Ludwig and Goebel, 1999) is exemplified by the *E. coli* α-hemolysin (Wagner *et al.,* 1983, Holland *et al.,* 1990) and *Erwinia chrysanthemi* protease (Letoffe *et al.,* 1990) secretion systems. A type I secretion apparatus always consists of three cell envelope proteins. The first component is an inner membrane (IM) protein with ATPase activity, which belongs to the ABC (ATP-binding cassette) superfamily of eukaryotic and prokaryotic protein exporters and which couples ATP hydrolysis to the substrate export (Saurin *et al.,* 1999; Holland and Blight,1999; Young and Holland, 1999). The second component is an IM-anchored protein that spans the periplasm; such proteins are termed membrane fusion proteins (MFP) (Dinh *et al.,* 1994). The third component is a general outer membrane protein (OMF for outer membrane factor) (Paulsen *et al.,* 1997).

The genes for the first two exporter proteins are usually linked to the structural gene of the secreted protein. These gene clusters are located either on chromosome-bound pathogenicity islands (Dobrindt *et al.,* 2000) or on transmissible plasmids (Ludwig and Goebel, 1999), suggesting that the genetic determinants for the type I secretion systems can be transmitted by horizontal gene transfer among gram-negative bacteria.

The secretion signal is located within the C-terminal 30-80 amino acid residues (Wandersman, 1996; Holland and Blight, 1999; Ludwig and Goebel, 1999) of the exported proteins and is, in most cases, not subject to proteolytic cleavage during translocation. However, the HasA and HasAp haemoproteins, secreted by *S. marcescens* and *P. aeruginosa*, respectively, are shorter than the same proteins secreted by a reconstituted system in *E. coli*. This size difference is due to C-terminal cleavage by proteases secreted by the original host bacteria (Letoffe *et al.,* 1994, 1998). Colicin V, another type 1 secreted protein, contains an N-terminal transport signal of the double-glycine type, which is cut-off during secretion (van Belkum *et al.,* 1997). There is no significant amino acid sequence similarity between the C-terminal secretion signals of different proteins. Nevertheless, recognition of a given secretion signal by export components of different secretion machineries was demonstrated in several cases (e.g. Masure *et al.,* 1990; Fath *et al.,* 1991; Sebo and Ladant, 1993; Duong *et al.,* 1994; Welch, 1995). Structural data obtained by CD and NMR spectroscopy for several type I C-terminal signal peptides show that these signals peptides are unstructured in aqueous solution, but consist of two or three non-interacting α–helices separated by short loops in membrane mimetic environments (Wolft *et al.,*

1994, 1997; Zhang *et al.,* 1995; Yin *et al.,* 1995; Izadi-Pruneyre *et al.,* 1999). It thus appears that the secretion signal may be defined by a higher-order structure and type I transporters may recognize a much wider diversity of primary sequences than previously anticipated. However, the precise structure of these C-terminal secretion signals remains unclear.

The proteins secreted via the type I secretion pathway can be subdivided into several protein families (Table 1): RTX (Repeats in Toxin)-proteins (e.g. α-hemolysin (HlyA) of *E. coli*), proteases (e.g. four different proteases of *Erwinia chrysanthemi*), lipases (e.g. *Serratia marcescens* lipase), S-layer proteins (e.g. SlaA-protein of *S. marcescens*), haemophores (the two haemoproteins of *S. marcescens* and *Pseudomonas aeruginosa*), bacteriocins and proteins of unknown function. The RTX-toxin family and the protease family are the two best-characterized groups of type I proteins. The RTX-toxins represent the largest family of proteins transported by the type I secretion pathway. These toxins are widespread among gram-negative pathogenic bacteria and represent pore-forming proteins (hemolysins and leukotoxins). RTX-toxins are synthesized as functionally inactive precursor proteins, which are post-translationally activated by covalent acylation of specific internal lysine during secretion (Issartel *et al.,* 1991; Ludwig and Goebel, 1999). All RTX proteins carry in addition to the C-terminal secretion signal a domain consisting of tandemly arranged glycine- and aspartate-rich repeats (UXGGXG(N/D)DX) which is located close to the C-terminal half of these proteins. The number of repeat units in this domain varies in the RTX-toxins between about 10 and 40 (Ludwig and Goebel, 1999) This domain is responsible for Ca^{2+} binding which is essential for biological activity of the RTX-toxins. Most proteases transported by the type I secretion pathway are involved in utilization of protein substrates from the environment as nutrients. The secretion signal sequences of these proteases are very similar, allowing the efficient transport by one secretion apparatus (Guzzo *et al.,* 1991). The secretion signals of RTX-toxins and proteases are different and secretion by exporters belonging to the two protein families occurs only with low efficiency.

Efficient secretion of heterologous proteins fused to type I secretion signal peptides has been also reported (Gentschev *et al.,* 1992b; 1996a, b; Leffoffe and Wandersman, 1992; Blight and Holland, 1994; Blight *et al.,* 1994a; Duong *et al.,* 1996; Palacios *et al.,* 2001). This possibility has been employed for the development of novel live vaccines (see below).

Table 1. Members of protein families secreted via the type I secretion pathway

Family	Secreted Protein	Organism	References
RTX-Toxins:	HlyA (Hemolysin)	*Escherichia coli*	Goebel and Hedgpeth, 1982
-II-	Ehly, Ehx (EHEC-Hemolysin)	*Escherichia coli*	Schmidt *et al.,* 1995 and 1996
-II-	EaggEC exotoxin	*Escherichia coli*	Baldwin *et al,* 1992
-II-	HlyA (Hemolysin)	*Proteus* ssp.	Koronakis *et al.,* 1987
-II-	HlyA (Hemolysin)	*Morganella morganii*	Welch, 1987
-II-	Hemolysin	*Moraxella bovis*	Billson *et al.,* 2000
-II-	ApxI-III (Hemolysins)	*Actinobacillus pleuropneumoniae*	Frey *et al.,* 1993
-II-	AshA (Hemolysin)	*Actinobacillus suis*	Burrows and Lo, 1992
-II-	Hemolysin	*Enterobacter cloacae*	Prada and Beutin, 1991
II-	LktA (Leukotoxin)	*Actinobacillus acti-nomycetemcomitans*	Kraig *et al.,* 1990
-II-	PllktA (Leukotoxin)	*Pasteurella haemolytica-like*	Chang *et al.,* 1993
-II-	LktA (Leukotoxin)	*Pasteurella haemolytica*	Lo *et al.,* 1987
-II-	CyaA (Adenylat cyclase toxin)	*Bordetella pertussis*	Weiss and Hewlett, 1986; Glaser *et al.,* 1988
Proteases:	AprA (alkaline protease)	*Pseudomonas aeruginosa*	Guzzo *et al.,* 1990
-II-	AprX protease	*Pseudomonas fluorescens* CY091	Liao and McCallus, 1998
-II-	PrtA,B,C,G (Proteases A, B, C and G)	*Erwinia chrysanthemi*	Wadersman *et al.,* 1987; Ghigo and Wandersman, 1992, 1994
-II-	ZapA(Metalloprotease)	*Proteus mirabilis*	Wassif *et al.,* 1995
-II-	PrtSM, PrtA (Zinc Metalloprotease)	*Serratia marcescens*	Létoffé *et al.,* 1991
Lipases:	Lip (Lipase)	*Pseudomonas fluorescens*	Duong *et al.,* 1994
-II-	LipA (Lipase)	*Serratia marcescens*	Akatsuka *et al.,* 1997
S-Layer Proteins:	SlaA	*Serratia marcescens*	Kawai *et al.,* 1998
-II-	RsaA	*Caulobacter crescentus*	Awram and Smit, 1998
Haemo-phores:	HasAp haemoproteins	*Pseudomonas aeruginosa*	Letoffe *et al.,* 1998

125

-II-	HasA (Haem-binding protein)	*Serratia marcescens*	Létoffé *et al.*, 1994
Bacteriocins:	CvaC (Colicin V)	*Eschericha coli*	Fath *et al.*, 1991
Nodulation:	NodO (nodulation-signalling protein)	*Rhizobium leguminosarum*	Economou *et al.*, 1990
Other Proteins:	FrpA and FrpC	*Neisseria meningitidis*	Thompson *et al.*, 1993
-II-	AprX (?)	*Pseudomonas aeruginosa*	Duong *et al.*, 2001

3. ESCHERICHIA COLI α-HEMOLYSIN SECRETION SYSTEM: THE PROTOTYPE OF TYPE I SECRETION SYSTEMS

E. coli α-hemolysin (HlyA) is the best-characterized protein secreted by a type I secretion machinery. This toxin acts as virulence factor in *E. coli* strains (UPEC) causing mainly urinary tract infections (Welch *et al.*, 1981; Hacker *et al.*, 1983). HlyA exhibits cytolytic and cytotoxic activity against a wide range of mammalian cell types (e.g. erythrocytes, granulocytes, monocytes and endothelial cells). Synthesis, activation and secretion of *E. coli* HlyA are determined by the *hlyCABD* operon (Wagner *et al.*, 1983; Felmlee *et al.*, 1985; Hess *et al.*, 1986). The hemolysin operon has been found in *E. coli* on large transmissible plasmids or on chromosome-bound pathogenicity islands (Knapp *et al.*, 1986; Ludwig and Goebel, 1999). The genes are transcribed from a single promoter which is situated in front of *hlyC* (Welch and Pellett, 1988; Vogel *et al.*, 1988). The transcription is strongly polar due to the presence of a rho-independent terminator in the *hlyA-hlyB* intergenic region (Welch and Pellett, 1988). This termination is suppressed by elongation protein RfaH and a cis-acting 5′ DNA sequence termed JUMP start element or ops element (operon polarity suppressor), which must act together to allow the efficient transcription of the entire *hly* operon (Bailey *et al.*, 1992). The same mechanism of transcriptional elongation/ antitermination is found for *rfa* and *tra* operons, which encode synthesis and secretion of components important for virulence and fertility of gram-negative bacteria (Bailey *et al.*, 1992, 1997).

The pro-HlyA is activated in the cytoplasm of the bacterial cell to the hemolytically active form by the HlyC protein, which acts as an fatty acid acyltransferase (Issartel *et al.*, 1991). This acylation of pro-HlyA by HlyC is not required for secretion (Ludwig *et al.*, 1987).

The HlyA export machinery is the best-studied type I secretion system and consists typically of the three components HlyB, HlyD and TolC (Wagner *et al.,* 1983; Wandersman and Delepelaire, 1990). Whereas the inner membrane proteins HlyB and HlyD are specific components of the transport apparatus of α-hemolysin, the third component, TolC, is a multifunctional protein located in the outer membrane of *E. coli* and it was shown to be part of at least four different export systems (Zgurskaya and Nikaido, 2000). HlyA carries at its C-terminus a secretion signal of about 50-60 amino acids in length (termed in the following HlyAs), which is recognized by the HlyB/HlyD/TolC-translocator, promoting direct secretion of the entire protein into the extracellular medium without the formation of periplasmic intermediates. In addition, the secretion of HlyA has shown to be *secA* and *secY*-independent (Gray *et al.,* 1989; Gentschev *et al.,* 1990).

4. COMPONENTS OF THE HEMOLYSIN SECRE-TION SYSTEM

4.1 HlyB: an ABC transporter

The transport component HlyB belongs to the ABC transportes which represent the largest superfamily of prokaryotic and eukaryotic proteins detected in living organisms (Higgins, 1992; Saurin *et al.,* 1999). These proteins are involved in ATP-driven unidirectional membrane transport (import or export) of molecules such as ions, carbohydrates, amino acids, antibiotics, polysaccharides and proteins. Members of this superfamily always possess a large, conserved protein domain bearing the ATP binding cassette (ABC), essential for coupling transport to ATP hydrolysis, and at least one cognate, but much less conserved membrane domain (MD). Most ABC transporters are composed of four-units, two ABC subunits and two MDs (Holland and Blight, 1999).

Surprisingly, a recent phylogenetic analysis of ABC transporters involved in type I protein secretion showed that these proteins belong to different clusters of the ABC exporter subfamily (Saurin *et al.,* 1999). RTX-ABC-exporters (prototype HlyB of *E. coli*) are closely related to TAP1 and TAP2 (peptide transporters associated with CTL-antigen processing; Elliott, 1997). All these proteins possess an N-terminal membrane domain and a C–terminal ABC domain (exhibiting a similar MD-ABC module) and transport similar substrates.

The topology of HlyB (707 amino acid residues) has been determined by fusion of ß-lactamase (Blight and Holland, 1990; Wang *et al.,* 1991) or of alkaline phosphatase- and ß-galactosidase to HlyB (Gentschev and Goebel, 1992a). The data obtained by these methods together with computer predictions of transmembrane domains from ABC transporters suggest that HlyB is inserted in the cytoplasmic membrane by eight hydrophobic, α-helical transmembrane domains (TMDs). These TMDs extend from the amino acid positions 38 to 432 of HlyB. The cytoplasmic loops between the TMDs are relatively large and carry an excess of positively charged amino acids, while the periplasmic loops are rather small. Interestingly, the first two TMDs can be deleted with no effect on secretion of hemolysin (Blight *et al.,* 1994b). Based on suppressor mutations, Zhang and co-workers proposed that TMD 3, 5 and 8 are close to sites on HlyB that interact with the C-terminal secretion signal of HlyA (Zhang *et al.,* 1993b). This interaction is likely to involve either binding of HlyA to HlyB and/or activation of the secretion process.

As shown by Koronakis *et al.* (1993) the purified ABC domain of HlyB displays ATPase activity and binding of ATP induces a conformational change in HlyB. Mutations in the Walker$_A$ and Walker$_B$ motifs (Walker *et al.,* 1982) of the HlyB-ABC domain block hemolysin secretion *in vivo* and inactivate the ATPase function *in vitro* (Koronakis *et al.,* 1995).

4.2 HlyD: a member of the membrane fusion protein (MFP) family

HlyD (478 amino acid residues) is one of the best characterized members of the membrane fusion protein (MFP) family (Dinh *et al.,* 1994). These proteins are of similar size (422 residues +/- 13%) and their predicted structural features are also quite similar. Many MFPs are anchored in the cytoplasmic membrane by a single TMD and possess a large periplasmic domain within the C-terminal 100 amino acid residues, which is highly conserved among these proteins (Schülein *et al.,* 1992, 1994; Dinh *et al.,* 1994). Using site-directed mutagenesis, Schülein and co-workers showed that some amino acid residues of the C-terminal region (Leu475, Glu477 and the last amino acid of HlyD, Arg 478) are essential for HlyD function, since mutations in either of these amino acids inhibit the secretion of hemolysin (Schülein *et al.,* 1994). In addition, a second region of HlyD was found between amino acids Leu127 and Leu170, that showed a high homology to TolC (Schülein *et al.,* 1994). Deletion of this region abolishes secretion of hemolysin. This sequence of HlyD also seems to interact with a component of the hemolysin secretion machinery, since a hybrid HlyD protein carrying the corresponding TolC sequence, although inactive in the transport of HlyA, is able to displace wild-type HlyD from the secretion apparatus (Schülein *et al.,* 1994).

Interestingly, using COILS algorithm (Lupas *et al.,* 1991) Pimenta *et al.* (1996) found that HlyD contains most likely coiled-coil structures. Coiled-coil domains are implicated in protein dynamics and may represent regions involved in subunit oligomerisation as shown for fibrous proteins (Young and Holland, 1999). The simian virus 5 protein belongs also to the MFP family (Zgurskaya and Nikaido, 1999). This viral protein has a proven membrane-fusion function, fusing adjacent host cells thus allowing the virus to spread.

A recent study grouped the members of the MFP family in the superfamily of periplasmic efflux proteins (PEP, Johnson and Church, 1999).

4.3   TolC: a member of outer membrane factor (OMF) protein family

TolC (471 animo acid residues) of *E. coli* is a member of OMF family (Paulsen *et al.,* 1997), that was recently arranged in the superfamily of outer membrane efflux proteins (OEPs) (Johnson and Church, 1999). Proteins of this family function in conjunction with an ABC transporter and a periplasmic efflux protein (PFP). Such three component complexes allow the transport of various substrates e.g. drugs, divalent metal cations, oligosaccharides and proteins (Paulsen *et al.,* 1997; Johnson and Church, 1999). Analysis of the phylogenetic tree of the OMF members suggests that these proteins can be grouped into at least three subfamilies, involved in protein secretion, drug efflux and cation efflux. TolC is one of the most multifunctional proteins known in gram-negative bacteria and appears to be rather conserved in these organisms (Spreng *et al.,* 1999). Interestingly, TolC is essential for hemolysin (Wandersman and Delepelaire, 1990) and colicin V (Gilson *et al.,* 1990) secretion, required for export of toxic compounds (Fralick, 1996), for import of colicin E1 (Otsuji *et al.,* 1982), for infection by phage U3 (Austin *et al.,* 1990) and for hypersensivity to hydrophobic antibiotics, dyes and detergents (Ma *et al.,* 1994). TolC is also involved in the secretion of heat-stable enterotoxin I (Yamanaka *et al.,* 1998) and the peptide antibiotics J 25 (Delgado *et al.,* 1999) and colicin 10 (Pilsl and Braun, 1995).

On artificial lipid bilayers Benz *et al.* (1993) first demonstrated the pore-forming property of TolC and proposed that TolC may function as gate for hemolysin through the outer membrane. The crystal structure of TolC, recently determined to 2.1-Å resolution (Koronakis *et al.,* 2000), shows that a TolC homotrimer forms a channel tunnel, 140 Å in length, that comprises a 100 Å long α-helical barrel (the tunnel domain) anchored in the outer membrane by 12 stranded β-barrel spans (the channel domain). This structure of TolC is a unique in bacterial outer membrane channel architecture (Andersen *et al.*, 2000; Lewis, 2000; Buchanan, 2001).

4.4   HlyA secretion signal (HlyAs)

The HlyA secretion apparatus recognizes a signal sequence, located at the C-terminus of HlyA, that comprises the last 46 (Koronakis *et al.*, 1989; Kenny *et al.*, 1992, 1994) to 60 amino acid residues of HlyA (Hess *et al.*, 1990, Gentschev *et al.*, 1990; Zhang *et al.*, 1993a; Hui *et al.*, 2000). This secretion signal (HlyAs) differs from the N-terminal signal sequences of proteins secreted via type II-IV pathways. Based on data obtained by site-directed mutagenesis, CD and NMR spectroscopy, several structural and sequence motifs in the HlyAs have been proposed as being essential for its signal function (Koronakis *et al.*, 1989; Hess *et al.*, 1990; Kenny *et al.*, 1992, 1994; Zhang *et al.*, 1993a, 1995; Chervaux and Holland, 1996; Hui *et al.*, 2000). However, the precise nature of HlyAs remains basically unknown. HlyAs seems to be necessary and sufficient for secretion of HlyA, since a peptide consisting of just the C-terminal 60 amino acids of HlyA is secreted in presence of HlyB/HlyD/TolC with the same efficiency as HlyA itself (Jarchau *et al.*, 1994). In addition a large number of hybrid proteins (more than 400) have been generated by gene fusion with the 3′end of *hlyA* that encodes HlyAs. For this purpose several different systems have been developed (e.g. plasmid vectors and a transposon encoding the HlyAs) that allow the generation of such hybrid proteins and their secretion (Hess *et al.*, 1990, 1996, 1997; Kenny *et al.*, 1992; Blight and Holland, 1994; Blight *et al.*, 1994a; Gentschev *et al*, 1994, 1996a, 1996b, 1997, 2000; Mollenkopf *et al.*, 1996b; Tzschaschel *et al.*, 1996a, b). From studies concerned with the expression and secretion of HlyAs fusion proteins (e. g. Blight and Holland, 1994; Gentschev *et al.*, 1994, 1996a, b; Mollenkopf *et al.*, 1996b) the following rather general rules can be deduced.
- Proteins which are normally transported via the Sec-transport pathway are also efficiently secreted by the HlyA secretion system provided the N-terminal signal sequence is removed.
- Proteins containing both an N-terminal signal (type II or type III) and the C-terminal HlyAs signal, are inefficiently transported across the membrane by any of the two protein transport systems unless one of the transport systems is inactivated, e.g. by a *secA* mutation (Gentschev *et al.*, 1990,

1996a) or a mutation in *hlyD* (Schülein *et al.,* 1994; Hess *et al.,* 1996; Gentschev *et al.,* unpublished data).
- In most cases the fusion of HlyAs to the C-terminus of a protein seems to slow down the rate of folding of cytoplasmic proteins thus rendering many cytoplasmic proteins with a C-terminal HlyAs sequence secretion-competent.
- The size of the fusion partner can be rather short. As stated above, it has been shown that the 60 amino acid signal sequence alone is secreted by the HlyA secretion system (Jarchau *et al.,* 1994). On the other end, ß-Gal-HlyAs fusion proteins (about 2000 amino acids) have been constructed which are larger in size than the natural substrate HlyA (1024 amino acids). These fusion proteins are still secreted (Kenny *et al.,* 1991, Gentschev *et al.,* 1996a).
- There are definitely sequences within a polypeptide chain which are inhibitory and sequences which are more favorable for HlyA-specific secretion (Gentschev *et al.* 1996a). This has been shown by the random insertion of a specifically constructed transposon Tn*hly*As (Gentschev *et al.* 1996a) into different parts of the *lacZ*- and the *phoA* gene.
-Some of HlyAs fusion proteins are still enzymatically active (Nakano *et al.,* 1992; Gentschev *et al.,* 1995; Hess *et al.,* 1995; Chervaux *et al.,* 1995; Kern and Ceglowski, 1995)
- The rate of secretion of a HlyAs fusion protein is rather similar in most gram-negative bacteria (e.g. *E. coli*, *Salmonella* spp., *Shigella* spp., *Yersinia* spp., Spreng *et al.,* 1999).

5. INTERACTIONS BETWEEN THE COMPONENTS OF THE HEMOLYSIN SECRETION SYSTEM

Earlier studies showed that HlyB alone seems to be able to recognize the HlyA signal (Oropeza-Wekerle *et al.,* 1990; Gentschev *et al.,* 1992a). This result is supported by the observation that mutations in the HlyA secretion signal are partially compensated by suppressor mutations in HlyB (Zhang *et al.,* 1993b; Sheps *et al.,* 1995). However, until now the binding site for HlyAs has not been identified.

Schlör *et al.* (1997) used a genetic approach to identify the possible sites of interaction between the components of the HlyA secretion apparatus. In this study, the authors exploited the fact that *in vivo* the sensitivity to vancomycin is enhanced in the presence of the hemolysin (*hlyCABD*) determinant of *E. coli* or its translocator portion *hlyBD* (Blight *et al.,* 1994b; Wandersman and Letoffe, 1993). The results are an important indication that

TolC interacts with the TolC homologous region of HlyD in order to form a pore in the OM and that the C-terminal sequence of HlyD is in contact with periplasmic loop(s) of HlyB. In additon, Schlör *et al.* (1997) proposed a hairpin conformation for HlyD, which allows these interactions. Interestingly, Johnson and Church (1999) demonstrated that such α-helical hairpin structures of paired coiled-coils exist in most PEPs. Moreover, it is also known that the HlyD region, which is predicted to form a coiled-coil (Pimenta *et al.*, 1996; Johnson and Church, 1999), is required for hemolysin export. Deletion of a portion (the TolC homologous domain of HlyD) of this region, or substitution with a similar sequence, abolishes either HlyA secretion *in vivo* (Schlülein *et al.*, 1994) or an HlyD interaction with TolC *in vitro* (Schlör *et al.*, 1997).

Using a combination of co-affinity protein purification and cross-linking, Thanabalu *et al.* (1998) demonstrated that HlyB and HlyD form (in the absence of HlyA) a stable inner membrane complex. This complex formation is independent of TolC. Binding of HlyA to the preformed HlyB/D complex induces contact to TolC via a HlyD trimer in order to form a trans-periplasmic export channel. Interestingly, HlyD can interact with HlyA even in the absence of HlyB. This is in contrast to results reported by Letoffe *et al.* (1996) who showed a sequential assembly of the metalloprotease (PrtDEF) exporter system which is only induced when the protease binds to the ABC transporter PrtD. These findings may be caused by functional differences between HlyB and PrtD (which belong to different ABC clusters) and/or the methods used.

One important question concerning the HlyA secretion arises: is there a chaperone involved in this process? Delepelaire and Wandersman (1998) showed that the secretion of the HasA protein is strongly dependent upon the SecB chaperone. However, this is possibly a specific case, because hemolysin and protease secretion are SecB independent (Gentschev *et al.*, 1990; Blight and Holland, 1994). Other unidentified chaperones may be involved, or the hemolysin secretion may be rapid with respect to the folding process, in which case HlyA would not need chaperones at all.

5.1   Model of HlyA secretion

HlyA, probably as an unfolded protein, interacts with the preformed HlyB/D complex in the cytoplasm (Fig. 1; Thanabalu *et al.*, 1998). After the binding of the HlyA secretion signal to the HlyB/HlyD complex, HlyD induces contact to TolC. Based on the new crystal structure data of TolC, Koronakis *et al.*, (2000) proposed a model for a possible TolC HlyD interaction. TolC trimer forms a cylinder, 140 Å long and about 35 Å in internal diameter. The HlyD trimer has a mass very similar to the TolC trimer and could form a cylinder with a similar diameter. The periplasmic end of the TolC tunnel is sealed by sets of coiled-coils that might untwist upon contact with HlyD to open the channel (Johnson and Church, 1999; Koronakis *et al.*, 2000). In this way hemolysin could pass into TolC from HlyD via a continuous tunnel made by the HlyD-TolC complex. Binding and hydrolysis of ATP via HlyB is directly coupled to the secretion of hemolysin (Koronakis *et al.*, 1995).

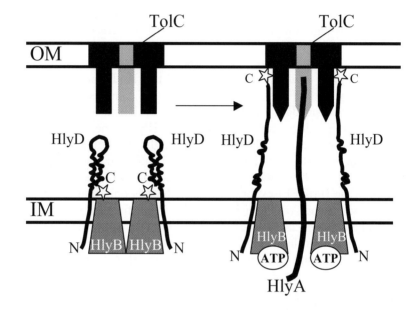

Fig. 1. Topological model of HlyA secretion across the two membranes of *Eschericha coli* (modified from Johnson and Church, 1999, Koronakis et al., 2000). In response to HlyA engagement, HlyD, which bends at the gap between the coiled-coil regions, contacts the trimeric TolC in order to form a continuous trans-periplasmic export channel. Only two HlyD molecules are shown for clarity (HlyD seems to be trimeric as shown by Thanabalu et al., 1998). The HlyD coiled-coil regions are indicated by a serpentine line, the asterisk marks the C-terminal domain of HlyD. The circles labeled ATP denote the ATP-binding sites of HlyB.

6.   APPLICATION OF HEMOLYSIN SECRETION SYSTEM

The *E. coli* hemolysin secretion apparatus recognizes a C-terminal signal sequence of about 50-60 amino acids in length, which is sufficient for secretion. Heterologous proteins covalently linked at their C-terminus to HlyAs are well recognized by this secretion apparatus (Blight and Holland, 1994; Blight *et al.*, 1994a; Gentschev *et al.*, 1996a, b; Tzschaschel *et al.*, 1996b). Most of these hybrid proteins (at present more than 400) are efficiently secreted by the hemolysin secretion pathway although they are entirely unrelated to HlyA.

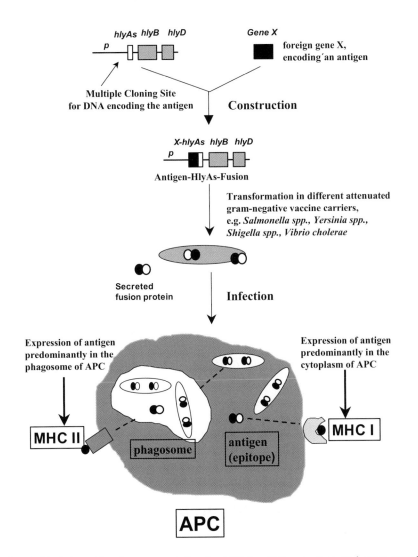

Fig. 2. Variation of antigen presentation in APC by different attenuated gram-negative vaccine carriers via the hemolysin secretion machinery.

Since HlyAs by itself represents a very weak used for production of polyclonal and for B-cells and T-cells, hemolysin-fused proteins were successfully used for production of polyclonal and monoclonal antibodies (Mollenkopf *et al.*, 1996a; Ludwig *et al.,* 1995), for detection of protective antigens of different pathogenic bacteria (Mollenkopf, 1996b; Gentschev *et al.*, 1997; Spreng *et al.*, 2000a) and for heterologous antigen presentation (Gentschev *et al.*, 1996b, 1997, 2000; Hess *et al.*, 1996, 1997, 1999; Tzschaschel *et al.*, 1996b; Orr *et al.*, 1999).

As recently shown, a vector encoding different protective protein antigens in frame with *hlyA*s can stably replicate in many Gram-negative bacteria, including different attenuated *Salmonella* serotypes, *Shigella spp.* and *Vibrio cholerae*, and the encoded HlyAs-fused antigen is secreted by these bacterial carriers (Spreng *et al.*, 1999, 2000c; Gentschev *et al.*, 2000). Since the latter bacteria colonize different compartments of mammalian cells, i.e. cell surface (*V. cholerae*), specialized phagosome of APCs (*Salmonella*) and cytosol (*Shigella*), the same antigen could be introduced into different parts of antigen-presenting cells and hence trigger a different immune response against the same antigen (Fig. 2). Interestingly, Spreng *et al.,* 1999, showed that TolC analogues, which can interact with HlyD are present in most Gram-negative bacteria, thus allowing the use of the Hly-System for antigen-secretion by a wide range of attenuated Gram-negative carrier bacteria.

On the basis of the hemolysin secretion system several successful live vaccines against *Listeria monocytogenes* (Hess *et al.*, 1996, 1997), *Theileria parva* (Gentschev *et al.* 1998), *Mycobacterium tuberculosis* (Hess *et al.*, 2000), *Clostridium difficile* (Ryan *et al.*, 1997), *Measles virus* (Spreng *et al.*, 2000b) and *Plasmodium falciparum* (Gomez-Duarte *et al.*, 2001) were developed using attenuated *Salmonella* spp. or *V. cholerae* carriers.

7. FUTURE RESEARCH

Several general questions remain for future research on hemolysin secretion.

- Are there chaperones responsible for the unfolding process of HlyA or other proteins excepting HlyB/HlyD/TolC involved in the structure of the translocator channel?
- How does the hemolysin signal recognize the components of the secretion apparatus?

- What is the chronological order of interaction between the components of the secretion apparatus?
- What is the stoichiometry of the components of the translocator channel for hemolysin and what is its architecture during the secretion process?
- What is the polarity of HlyA secretion ?
- Can TolC, as a component of different translocation channels, effect the simultaneous secretion of protein toxins and toxic drugs? Is there an order of substrate priority?

8.   ACKNOWLEDGEMENTS

We thank G. Dietrich and Z. Sokolovic for stimulating discussions. A.N. Rycroft and M. Dietrich for crtitical reading of the manuscript.

9.   REFERENCES

Akatsuka, H., Binet, R., Kawai, E., Wandersman, C. and Omori, K. (1997) J Bacteriol 179: 4754-4760.

Andersen, C., Hughes, C. and Koronakis, V. (2000) EMBO Reports 1: 313-318.

Austin, E.A., Graves, J.F., Hite, L.A., Parker C.T. and Schnaitman, C.A. (1990) J Bacteriol 172: 5312-5325.

Awram, P. and Smit, J. (1998) J Bacteriol 180: 3062-3069.

Bailey, M.J.A., Koronakis, V., Schmoll, T. and Hughes, C. (1992) Mol Microbiol 6: 1003-1012.

Bailey, M.J.A., Hughes, C. and Koronakis, V. (1997) Mol Microbiol 26: 845-851.

Baldwin, T.J., Knutton, S., Sellers L., Hernandez, H.A., Aitken, A. and Williams, P.H. (1992) Infect Immun 60: 2092-2095.

Benz, R., Maier, E. and Gentschev, I. (1993) Zentralbl Bakteriol 278: 187-196.

Billson, F.M., Harbour, C., Michalski, W.P., Tennent, J.M., Egerton, J.R. and Hodgson, J.L. (2000) Infect Immun 68: 3469-3474.

Binet, R., Létoffé, S., Ghigo, J.-M., Delepelaire, P. and Wandersman, C. (1997) Gene 192: 7-11.

Blight, M.A. and Holland, I.B. (1990) Mol Microbiol 4: 873-880.

Blight, M.A., Chervaux, C. and Holland, I.B. (1994a) Curr Opin Biotechnol 5: 468-474.

Blight, M.A. and Holland, I.B. (1994) Trends Biotechnol 12: 450-455.

Blight, M.A., Pimenta, A.L., Lazzaroni, J.C., Dando, C., Kotelevets, L., Seror, S.J. and Holland, I.B. (1994b) Mol Gen Genet 245: 431-440.

Buchanan, S.K. (2001) Trends Biochem Sci 26: 3-6.

Burrows, L.L. and Lo, R.Y. (1992) Infect Immun 60: 2166-2173.

Chang, Y.F., Ma, D.P., Shi, J. and Chengappa, M.M. (1993) Infect Immun 61: 2089-2095.

Chervaux, C., Sauvonnet, N., Le Clainche, A., Kenny, B., Hung, A.L., Broome-Smith, J.K. and Holland I.B. (1995) Mol Gen Genet 249: 237-245.

136

Chervaux, C. and Holland, I.B. (1996) J Bacteriol 178: 1232-1236.

Christie, P.J. and Vogel, J.P. (2000) Trends Microbiol. 8: 354-360.

Delpelaire, P., Wandersman, C. (1990) J Biol Chem 265: 9083-9089.

Delepelaire, P. and Wandersman, C. (1998) EMBO J 17: 936-944.

Delgado, M.A., Solbiati, J.O., Chiuchiolo, M.J., Farias, R.N. and Salomon, R.A. (1990) *J* Bacteriol 181: 1968-1970.

Dinh, T., Paulsen, I.T. and Saier, Jr. M.H. (1994) J Bacteriol 176: 3225-3231.

Dobrindt, U., Janke, B., Piechaczek, K., Nagy, G., Ziebuhr, W., Fischer, G., Schierhorn, A., Hecker, M., Blum-Oehler, G. and Hacker, J. (2000) Int J Med Microbiol 290: 307-311.

Duong, F., Soscia, C., Lazdunski, A. and Murgier, M. (1994) Mol Microbiol 11: 1117-1126.

Duong, F., Lazdunski, A. and Murgier M. (1996) Mol Microbiol 21: 459-470.

Duong, F., Bonnet, E., Geli, V., Lazdunski, A., Murgier, M. and Filloux, A. (2001) Gene 262: 147-153.

Economou, A., Hamilton, W.D., Johnston, A.W. and Downie, J.A. (1990) EMBO J 9: 349-354.

Fath, M.J., Skvirsky, R.C. and Kolter, R. (1991) J Bacteriol 173: 7549-7556.

Felmlee, T., Pellett, S. and Welch, R.A. (1985) J Bacteriol 163: 94-105.

Fralick, J.A. (1996) J Bacteriol 178: 5803-5805.

Frey, J., Bosse, J.T., Chang, Y. F., Cullen, J. M., Fenwick, B., Gerlach, G.F., Gygi, D., Haesebrouck, F., Inzana, T.J. and Jansen, R. (1993) J Gen Microbiol 139: 1723-1728.

Gentschev, I., Hess, J. and Goebel, W. (1990) Mol Gen Genet 222: 211-216.

Gentschev, I. and Goebel, W. (1992a) Mol Gen Genet 232: 40-48.

Gentschev, I., Sokolovic, Z., Köhler, S., Krohne, G.F., Hof, H., Wagner, J. and Goebel, W. (1992b) Infect. Immun. 60: 5091-5098.

Gentschev, I., Mollenkopf, H.-J., Sokolovic, Z., Ludwig, A., Tengel, C., Gross, R., Hess, J., Demuth, A. and Goebel, W. (1994) Behring Inst Mitt 95: 57-66.

Gentschev, I., Sokolovic, Z., Mollenkopf, H.-J., Hess J., Kaufmann, S.H.E., Kuhn, M., Krohne, G.F. and Goebel, W. (1995) Infect Immun 63: 4202-4205.

Gentschev, I., Maier, G., Kranig, A. and Goebel, W. (1996a) Mol Gen Genet 252: 266-274.

Gentschev, I., Mollenkopf, H., Sokolovic, Z., Hess, J., Kaufmann, S.H.E. and Goebel, W. (1996b) Gene 179: 133-140.

Gentschev, I., Dietrich, G., Mollenkopf, H.-J., Sokolovic, Z., Hess, J., Kaufmann, S.H.E. and Goebel, W. (1997) Behring Inst Mitt 98: 103-113.

Gentschev, I, Glaser, I., Goebel, W., McKeever, D.J. and Heussler, V. (1998) Infect Immun 66: 2060-2064.

Gentschev, I., Dietrich, G., Spreng, S., Kolb-Maurer, A., Daniels, J., Hess, J., Kaufmann, S.H. and Goebel, W. (2000) J Biotechnol 83: 19-26.

Ghigo, J.M. and Wandersman, C. (1992) Mol Gen Genet 236: 135-144.

Ghigo, J.M. and Wandersman, C. (1994) J Biol Chem 269: 8979-8985.

Gilson, L., Mahanty, H.K. and Kolter, R. (1990) EMBO J 9: 3875-3894.

Glaser, P., Sakamoto, H., Bellalou, J., Ullmann, A. and Danchin, A. (1998) EMBO J 7: 3997-4004.

Goebel, W. and Hedgpeth, J. (1982) J Bacteriol 151: 1290-1298.

Gomez-Duarte, O.G., Pasetti, M.F., Santiago, A., Sztein, M.B., Hoffman, S.L. and Levine M.M. (2001) Infect Immun 69: 1192-1198.

Gray, L., Baker, K., Kenny, B., Mackman, N., Haigh, R. and Holland I.B. (1989) J Cell Sci Suppl 11: 45-57.

Guzzo, J., Murgier, M., Filloux, A. and Lazdunski, A. (1990) J Bacteriol 172: 942-948.

Guzzo, J., Duong, F., Wandersman, C., Murgier, M. and Lazdunski, A. (1991) Mol Microbiol 5: 447-453.

Hacker, J., Hughes, C., Hof, H. and Goebel, W. (1993) Infect Immun 42: 57-63.

Hess, J., Wels, W., Vogel, M. and Goebel, W. (1986) FEMS Microbiol Lett 34: 1-11.

Hess, J., Gentschev, I., Goebel, W. and Jarchau, T. (1990) Mol Gen Genet 224: 201-208.

Hess, J, Gentschev, I, Szalay, G, Ladel, C, Bubert, A, Goebel, W. and Kaufmann, S.H.E. (1995) Infect Immun 63: 2047-2053

Hess, J., Gentschev, I., Miko, D., Welzel, M., Ladel, Ch., Goebel, W. and Kaufmann, S.H.E. (1996) Proc Natl Acad Sci USA 93:1458-1463.

Hess, J., Dietrich,G., Gentschev, I., Miko, D., Goebel, W. and Kaufmann, S.H.E. (1997) Infect Immun 65: 1286-1292.

Hess, J. and Kaufmann, S.H.E. (1999) FEMS Immunol Med Microbiol 23: 165-173.

Hess, J., Grode, L., Hellwig, J., Gentschev, I., Goebel, W., Ladel, C. and Kaufmann, S.H.E. (2000) FEMS Immunol Med Microbiol 27: 283-289.

Higgins, C.F. (1992) Annu Rev Cell Biol 8: 67-113.

Holland, I.B., Blight, M.A. and Kenny, B. (1990) J Bioenerg Biomem 22: 473-491.

Holland, I.B. and Blight, M.A. (1999) J Mol Biol 293: 381-399.

Hueck, C. (1998) Microbiol Mol Biol Rev 62: 379-433.

Hui, D., Morden, C., Zhang, F. and Ling, V. (2000) J Biol Chem 275: 2713-2720.

Issartel, J.P., Koronakis, V. and Hughes, C. (1991) Nature 351: 759-761.

Izadi-Pruneyre, N., Wolff, N., Redeker, V., Wandersman, C., Delepierre, M. and Lecroisey, A. (1992) Eur J Biochem 261: 562-568.

Jarchau, T., Chakraborty, T., Garcia, F. and Goebel, W. (1994) Mol Gen Genet 245: 53-60.

Johnson, J.M. and Church, G.M. (1999) J Mol Biol 287: 695-715.

Kawai, E., Akatsuka, H., Idei, A., Shibatani, T. and Omori, K. (1998) Mol Microbiol 27: 941-952.

Kenny, B., Haigh, R. and Holland, I.B. (1991) Mol Microbiol 5: 2557-2568.

Kenny, B., Taylor, S. and Holland, I.B. (1992) Mol Microbiol 6: 1477-1489.

Kenny, B., Chervaux, C. and Holland, I.B. (1994) Mol Microbiol 11: 99-109.

Kern, I. and Ceglowski, P. (1995) Gene 163: 53-57.

Knapp, S., Hacker, J., Jarchau, T. and Goebel, W. (1986) J Bacteriol 168: 22-30.

Koronakis, V., Cross, M., Senior, B., Koronakis, E. and Hughes, C. (1987) J Bacteriol 169:1509-1515.

Koronakis, V., Koronakis, E. and Hughes, C. (1989) EMBO J 8: 595-605.

Koronakis, V., Hughes, C. and Koronakis, E. (1993) Mol Microbiol 8:1163-1175.

Koronakis, E., Hughes, C., Milisav, I. and Koronakis, V. (1995) Mol Microbiol 16: 87-96.

Koronakis, V., Sharff, A., Koronakis E, Luisi, B. and Hughes, C. (2000) Nature 405: 914-919.

Kraig, E., Dailey, T. and Kolodrubetz, D. (1990) Infect Immun 58: 920-929.

Letoffe, S., Delepelaire, P, and Wandersman, C. (1990) EMBO J 9: 1375-1382.

Létoffé, S., Delepelaire, P. and Wandersman, C. (1991) J Bacteriol 173: 2160-2166.

Létoffé, S. and Wandersman, C. (1992) J Bacteriol 174: 4920-4927.

Létoffé, S., Ghigo, J.M. and Wandersman, C. (1994) Proc Natl Acad Sci USA 91: 9876–9880.

Létoffé, S., Delepelaire, P. and Wandersman, C. (1996) EMBO J 15: 5804-5811.

Létoffé, S., Redeker, V. and Wandersman, C. (1998) Mol Microbiol 28: 1223–1234.

Lewis, K. (2000) Curr Biol 10: 678-681.

Liao, C.H. and McCallus, D.E. (1998) Appl Environ Microbiol 64: 914-921.

Lo, R.Y.C. and Strahdee, C.A. (1997) Shewen, P.E. Infect Immun 55: 1987-1996.

Ludwig, A., Vogel, M. and Goebel, W. (1987) Mol Gen Genet 206: 238-245.

Ludwig, A., Tengel, C., Bauer, S., Bubert, A., Benz, R., Mollenkopf, H.J. and Goebel, W. (1995) Mol Gen Genet 249: 474-486.

Ludwig, A. and Goebel, W. (1999) In: *The Comprehensive Sourcebook of Bacterial Protein Toxins*, Joseph, E. Alouf and John, H. Freer, eds. Academic Press.

Lupas, A., Van Dyke, M. and Stock J. (1991) Science 252:1162-1164.

138

Ma, D., Cook, D.N., Hearst, J.E. and Nikaido, H. (1994) Trends Microbiol 2: 489-493.

Masure, H.R., Au, D.C., Gross, M.K., Donovan, M.G. and Storm, D.R. (1990) Biochemistry 29: 140-145.

Mollenkopf, H.-J., Gentschev, I., Bubert, A. and Goebel, W. (1996a) Applied Microbiol Biotechn 45: 629-637.

Mollenkopf, H.-J., Gentschev, I. and Goebel, W. (1996b) Biotechniques 21: 854-860.

Nakano, H., Kawakami, Y. and Nishimura, H. (1992) Appl Microbiol Biotechnol 37: 765-771.

Oropeza-Wekerle, R.L., Speth, W., Imhof, B., Gentschev, I. and Goebel, W. (1990) J Bacteriol 172: 3711-3717.

Orr, N., Galen, J.E. and Levine, M.M. (1999) Infect Immun 67: 4290-4294.

Otsuji, N., Soejima, T., Maki, S. and Shinagawa, H. (1982) Mol Gen Genet 187: 30-36.

Palacios, J.P., Zaror, I., Martínez, P., Uribe, F., Opazo, P., Socías, T., Gidekel, M. and Venegas, A. (2001) J Bacteriol 183: 1346-1358.

Paulsen, I.T., Park, J.H., Choi, P.S. and Saier, M.H.Jr. (1997) FEMS Microbiol Lett 156: 1-8.

Pilsl, H. and Braun, V. (1995) Mol Microbiol 16: 57-67.

Pimenta, A., Blight, M, Clarke, D. and Holland, I.B. (1996) Mol Microbiol 19: 643-645.

Prada, J. and Beutin, L. (1991) FEMS Microbiol Lett 63:111-114.

Pugsley, A.P. (1993) Microbiol Rev 57: 50-108.

Pugsley, A.P., Francetic, O., Possot, O.M., Sauvonnet, N. and Hardie, K.R. (1997) Gene 192: 13-19.

Ryan, E.T., Butterton, J.R.,. Smith, R.N., Carroll, P.A., Crean, T.I. and Calderwood, S.B. (1997) Infect Immun 65: 2941-2949.

Saurin, W., Hofnung, M. and Dassa, E. (1999) J Mol Evol 48: 22-41.

Schlör, S., Schmidt, A., Maier, E., Benz, R., Goebel, W. and Gentschev, I. (1997) Mol Gen Genet 256: 306-320.

Schmidt, H., Beutin, L. and Karch, H. (1995) Infect Immun 63: 1055-1061.

Schmidt, H., Kernbach, C. and Karch, H. (1996) Microbiology 142: 907-914.

Schülein, R., Gentschev, I., Mollenkopf, H.-J. and Goebel, W. (1992) Mol Gen Genet 234: 155-163.

Schülein, R., Gentschev, I., Schlör, S., Gross, R. and Goebel, W. (1994) Mol Gen Genet 245: 203-211.

Sebo, P. and Ladant, C. (1993) Mol Microbiol 9: 999-1009.

Sheps, J.A., Cheung, I. and Ling, V. (1995) J Biol Chem 270: 14829-14834.

Spreng, S., Dietrich, G., Goebel, W. and Gentschev, I. (1999) Mol Microbiol 31: 1596-1598.

Spreng, S., Gentschev, I., Goebel, W., Mollenkopf, H., Eck, M., Muller-Hermelink, H.K. and Schmausser, B. (2000a) FEMS Microbiol Lett 186: 251-256.

Spreng, S., Gentschev, I., Goebel, W., Weidinger, G., ter Meulen, V. and Niewiesk, S. (2000b) Microbes Infect 2: 1687-1689.

Spreng, S., Dietrich, G., Niewiesk, S., ter Meulen, V., Gentschev, I. and Goebel, W. (2000c) FEMS Immunol Med Microbiol 27: 299-304.

Thanabalu, T., Koronakis, E., Hughes, C. and Koronakis, V. (1998) EMBO J 17: 6487-6496.

Thanassi, D.G. and Hultgren, S.J. (2000) Curr Opin Cell Biol 12: 420-430.

Thompson, S.A., Wang, L.L. and Sparling, P.F. (1993) Mol Microbiol 9: 85-96.

Tzschaschel, B.D., Guzman, C.A., Timmis, K.N. and de Lorenzo, V. (1996a) Nat Biotechnol 14: 765-769.

Tzschaschel, B.D., Klee, S.R., de Lorenzo, V., Timmis, K.N. and Guzman, C.A. (1996b) Microb Pathog 21: 277-288.

van Belkum, M.J., Worobo, R.W. and Stiles, M.E. (1997) Mol Microbiol 23: 1293-1301.

Vogel, M., Hess, J., Then, I., Juarez, A. and Goebel, W. (1988) Mol Gen Genet 212: 76-84.

Wagner, W., Vogel, M. and Goebel, W. (1983) J Bacteriol 154: 200-2010.

Wandersman, C., Delepelaire, P., Létoffé, S., Schwartz, M. (1987) J Bacteriol 169: 5046-5053.

Wandersman, C. and Delepelaire, P. (1990) Proc Natl Acad Sci USA 87: 4776-4780.

Wandersman, C. and Letoffe, S. (1993) Mol Microbiol 7:141-150.

Wandersman, C. (1996) In: *Escherichia coli* and *Salmonella typhimurium*. Cellular and molecular biology. Frederick C. Neidhardt, ed. ASM Press. Washington, D.C.

Wang, R., Seror, S.J., Blight, M.A., Pratt, J.M., Broome-Smith, J.K and Holland, I.B. (1991) J Mol Biol 217: 441-454.

Wassif, C., Cheek, D. and Belas, R. (1995) J Bacteriol 177: 5790-5798.

Weiss, A.A. and Hewlett E.L. (1986) Ann Rev Microbiol 40: 661-681.

Welch, R.A. (1987) Infect Immun 55: 2183-2190.

Welch, R.A. and Pellett, S. (1988) J Bacteriol 170: 1622-1630.

Welch, A.R. (1995) In: Virulence mechanisms of bacterial pathogens. Richard A. Roth, ed. ASM Press, Washington, D.C.

Wolff, N., Delepelaire, P., Ghigo, J.M. and Delepierre, M. (1997) Eur J Biochem 243: 400–407.

Wolff, N., Ghigo, J.M., Delepelaire, P., Wandersman, C. and Delepierre, M. (1994) Biochemistry 33: 6792–6801.

Yamanaka, H., Nomura, T., Fujii, Y. and Okamoto, K. (1998) Microb Pathog 25:111-120.

Yin, Y., Zhang, F., Ling, V. and Arrowsmith, C.H. (1995) FEBS Lett 366: 1-5.

Young, J. and Holland, I.B. (1995) Biochim Biophys Acta 1461: 177-200.

Zgurskaya, H.I. and Nikaido, H. (1999) J Mol Biol 285: 409-420.

Zgurskaya, H..L. and Nikaido, H. (2000) Mol Microbiol 37: 219-225.

Zhang, F., Greig, D.I. and Ling, V. (1993a) Proc Natl Acad Sci USA 90: 4211-4215.

Zhang, F., Sheps, J.A. and Ling, V. (1993b) J Biol Chem 268: 19889-19895.

Zhang, F., Yin, Y., Arrowsmith, C.H. and Ling, V. (1995) Biochemistry 34:4193-4201.

Chapter 8

TYPE II PROTEIN SECRETION

Alain Filloux and Manon Gérard-Vincent

Laboratoire d'Ingénierie des Systèmes Macromoléculaires
CNRS-IBSM-UPR9027
31 Chemin Joseph Aiguier
13402 Marseille Cedex 20, France

1.   INTRODUCTION

Bacterial membranes are essential barriers, preserving the integrity of the organism. However, the cell envelope must be sufficiently permeable to allow the traffic of molecules into and also out of the cells. This is essential for the acquisition of nutrients, the uptake of DNA, the liberation of enzymes or toxins, and the assembly of organelles such as pili and flagella on the cell surface. The cell envelope of gram-negative bacteria consists of two membranes delimiting the periplasm. Nutrient molecules are small enough to diffuse through outer membrane porins and are subsequently actively transported across the inner membrane into the cytoplasm. Extracellular enzymes and toxins are much larger molecules that use dedicated secretory pathways.

The general secretory pathway (GSP) is a protein translocation pathway, in which exoproteins cross successively, in two steps, the inner and outer membrane. The first step, involves translocation across the inner (cytoplasmic) membrane. The transported protein is synthesized as a precursor containing an N-terminal cleavable signal peptide. This precursor is targeted and transported through the inner membrane *via* a proteinaceous complex, the Sec translocon (see chapter 2). The signal peptide is then cleaved by leader peptidase and the mature protein is released into the periplasm. This first series of events is called the general secretory pathway (GSP). The exoprotein then requires another machinery, an extension of the GSP, to assist its translocation across the outer membrane. This second event is called the terminal branch of the GSP. Several branches have been identified.

B. Oudega (ed.), Protein Secretion Pathways in Bacteria, 141–165.
© 2003 *Kluwer Academic Publishers. Printed in the Netherlands.*

142

   This chapter will focus on a particular transport process, the type II secretion mechanism or main terminal branch (MTB) of the GSP, which involves 12-14 different proteins that constitutes the so-called **"SECRETON"**. The components of this pathway were first discovered in *Klebsiella oxytoca* (d'Enfert *et al.*, 1987). Their subsequent identification in *Pseudomonas aeruginosa* demonstrated that they are conserved among most gram-negative bacteria (Table 1) (Filloux *et al.*, 1990). Gsp is currently the general term used to describe those proteins of the secreton that will be described in further detail in the following part of this chapter.

Table 1: Characteristics of Gsp components[a] in the type II secretory pathway and their presence in various gram-negative bacteria[b].

Characteristic	P.ae Xcp	P.al c Xcp	P.p u Xcp	K.o x Pul	E.c h Out	E.c a Out	A.hy Exe	V.ch Eps	X.ca Xps	E.c o Gsp
ATP binding motif/Interacts with B Inner membrane							A	A		A
[c]TonB homologue?/Interacts with A Inner membrane/Energy transducer				B	B	B	B	B		B
Inner membrane/Bitopic	P	P	P	C	C	C	C	C		C
Secretin channel Outer membrane/Homomultimers	Q	Q	Q	D	D	D	D	D	D	D
ATP binding motif/Interacts with L (Y) Inner membrane-associated	R	R	R	E	E	E	E	E	E	E
Inner membrane/Polytopic 3 transmembrane segments	S	S	S	F	F	F	F	F	F	F
Pseudopilin	T	T	T	G	G	G	G	G	G	G
Pseudopilin	U	U	U	H	H	H	H	H	H	H

Pseudopilin	V	V	V	I	I	I	I	I	I	I
Pseudopilin	W	W	W	J	J	J	J	J	J	J
Atypical pseudopilin/Lacks E+5	X	X		K	K	K	K	K	K	K
Inner membrane/Bitopic Docks E (R) to the inner membrane Interacts with M (Z)	Y	Y	Y	L	L	L	L	L	L	L
Inner membrane/Bitopic/Interacts with L (Y)	Z	Z	Z	M	M	M	M	M	M	M
Inner membrane/Bitopic				N	N		N	N	N	N
Prepilin peptidase/Inner membrane 8 transmembrane segments	A (PilD)	A	A	O	O	O	O (Tap D)	O (Vcp D)	O (PilD)	O
Chaperone/Outer membrane/Lipoprotein Secretin pilot				S	S	S				S

[a]Each line represents a family of homologous proteins, with their characteristics summarized in the left column. Proteins are identified by the fourth letter of their designation, the first three letters being indicated in the top line.
[b]Components from *P. aeruginosa* (*P. ae*), *P. alcaligenes* (*P. alc*), *P. putida* (*P. pu*), *K. oxytoca* (*K. ox*), *E. carotovora* (*E. ca*), *E. chrysanthemi* (*E. ch*), *A. hydrophila* (*A. hy*), *V. cholerae* (*V. ch*), *X. campestris* (*X. ca*) and *E. coli* (*E. co*) are listed.
[c]Homology with TonB has been detected only for ExeB, not for the other B proteins.

2.   THE GSP PROTEINS

The genetic determinant of all known type II systems is a set of genes mostly clustered and organized into large operons including *gspC$_P$-O$_A$* and *gspS*. Mutations in most of these genes abort the secretion process resulting in the accumulation of exoproteins within the periplasm. Thus, *gsp* genes are essential for outer membrane translocation. Surprisingly, the deduced amino acid sequences and initial characterization of the proteins indicated that only two of the fourteen are effectively located in the outer membrane. All but one of the other proteins are anchored in the inner membrane, with the remaining protein peripherally associated at the cytoplasmic side of the inner membrane.

2.1 Pseudopilins and their dedicated peptidase

2.1.1 GspO$_A$, a specialized peptidase

P. aeruginosa and *Aeromonas hydrophila gsp* gene clusters have an unusual feature in that *gspO$_A$*, usually the last gene in the operon, is not clustered with the other *gsp* genes. It is found at another chromosomal location, clustered with genes required for the assembly of type IV pili, called *pilD* and *tapD*, respectively (Nunn *et al.*, 1990; Pepe *et al.*, 1996). Type IV pili were originally defined as long cell surface appendages located at the poles and involved in a particular type of motion called twitching motility (Henrichsen, 1983). PilD and TapD are the prepilin peptidases required for the processing of the N-terminal leader peptide of the pilin subunit before its assembly into pilus (Nunn and Lory, 1991; Pepe, *et al.*, 1996) (Fig. 1). The leader peptide of these subunits consists of a stretch of 6-7 residues with an overall positive charge, preceding a hydrophobic domain (Strom and Lory, 1993) (Fig. 1). The processing site is located immediately before the hydrophobic region, after a highly conserved glycine residue (G-1) (Fig. 1). Mature pilin subunits are then helically packed *via* interactions between their hydrophobic N-terminal domains and are assembled into pilus. A conserved glutamate residue, at position +5 within the hydrophobic domain (E+5) (Fig. 1), may be a key element in the registration mechanism associated with the assembly of the pilin subunits (Parge *et al.*, 1995). The new N-terminal residue of pilin, a conserved phenylalanine (F+1) (Fig. 1) or a hydrophobic residue such as the methionine of the TcpA pilin of *Vibrio cholerae*, is methylated during processing. This second post-translational modification is also catalyzed by the prepilin peptidase, which is thus a bifunctional enzyme (Strom and Lory, 1993). The G-1 residue is essential for cleavage of the pilin subunits and assembly into a pilus structure (Strom and Lory, 1991). There is no strict requirement for phenylalanine at position +1 or for glutamate at position +5 for prepilin processing. However, although an E+5V substitution did not affect leader peptide cleavage, it did abolish methylation and piliation. In case of an F+1S substitution, pilin was assembled into pili despite the complete absence of N-methylation. This observation raises questions as to the function of methylation during pilus biogenesis.

Mutations in the prepilin peptidase *pilD* and *tapD* genes affect not only type IV piliation but also abolish the protein secretion process. These proteins, despite the particular location of their genes within the genome are thus real Gsp components, namely GspO$_A$. The GspO$_A$ proteins are polytopic inner membrane proteins with eight transmembrane segments, as demonstrated for *Erwinia chrysanthemi* OutO (Reeves *et al.*, 1994). The first cytoplasmic loop of these proteins is relatively large and contains a tetracysteine consensus motif, C-X-X-C....X$_{21}$....C-X-X-C. The enzymatic activity of the peptidase, including its methyl-transferase activity, involves these conserved cysteine residues (Strom and Lory, 1993). This implies that the enzyme interacts with its substrate from the cytoplasmic side of the membrane (Fig. 1).

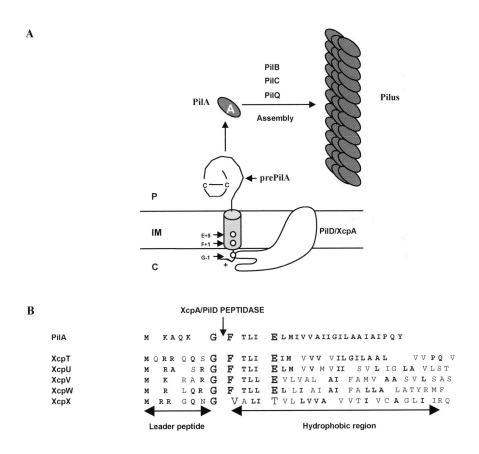

Fig. 1. **A**. The successive series of events during assembly of the pilin subunit, PilA, of *P. aeruginosa* into pilus. The prepilin peptidase, XcpA/PilD, cleaves the positively charged (+)

146

N-terminal leader peptide of prePilA between a glycine (G-1) and a phenylalanine (F+1) residue. The mature subunit (PilA) is then assembled into a pilus with the help of several accessory proteins including PilB, -C and -Q. This representation of the bacterial inner membrane (IM) shows the topology of prePilA and the location of XcpA/PilD, but is not intended to depict the location of the mature PilA subunit and the pilus. The white circles mark the approximate positions of the conserved G-1, F+1 and E+5 residues within the PilA protein. The cys-bond is also shown (C-C). P: periplasm; C: cytoplasm.
B. Sequence alignment of the N-terminal domains of the PilA pilin subunit and pseudopilins of the Xcp-type II secretory pathway from *P. aeruginosa*. XcpT,-U, -V, -W and –X are also more generally named GspG, -H, -I, -J and –K, respectively. The residues in the pseudopilins that are identical to those in the PilA sequence are shown in bold. Here R and K, and also I and L, are treated as "identical" residues. The conserved G-1, F+1 and E+5 residues are shown in uppercase. The position of the leader sequence cleavage site is indicated by an arrow. The leader peptide and the hydrophobic region are shown.

2.1.2 GspG$_T$, H$_U$, I$_V$, J$_W$ and K$_X$, the so-called pseudopilins

Five of the Gsp proteins, GspG$_T$-K$_X$, have a N-terminus similar to type IV pilins (Fig. 1), except for GspK$_X$ that lacks the E+5 residue (see next section). This sequence similarity has led to name these proteins **PSEUDOPILINS**, despite the C-terminal domain of pseudopilins and pilins being rather different. One relevant difference is that the two conserved cysteines at the extreme C-terminal end of the pilin subunit, which form a disulphide bridge, are not present in Gsp pseudopilins. The cys-bond formed in pilins is important for bacterial adhesion involving type IV pili (Hahn, 1997).

The GspO$_A$-dependent cleavage of N-terminal pseudopilin leader peptides has been demonstrated in several organisms (Bally *et al.*, 1992; Nunn and Lory, 1992; Pugsley and Dupuy, 1992; Howard *et al.*, 1993). As with the pilins, the new N-terminal residue is methylated (Nunn and Lory, 1993). Topology studies of the pseudopilin indicate that they are bitopic inner membrane proteins with a single transmembrane segment (Reeves *et al.*, 1994). The *in vivo* location of these proteins is more ambiguous, since upon overproduction they are partly found in the outer membrane fraction, particularly the mature form (Bally *et al.*, 1992). This suggests that there may be a redistribution of the pseudopilin after its processing. This may involve the relocation of the molecule to the outer membrane, or the assembly of pseudopilins into a macromolecular complex, thereby changing fractionation behaviour. The similarity between pilins and pseudopilins suggests that pseudopilins assemble into a pilus-like structure tentatively called **"PSEUDOPILUS"** (Fig. 2) (Hobbs and Mattick, 1993). Pseudopilins, like pilins, have been reported to form homomultimers stabilized by chemical cross-linking (Pugsley, 1996). *In vivo* cross-linking experiments with *P. aeruginosa* cells expressing the *gsp* gene cluster from the chromosome have detected mainly pseudopilin dimers (Lu *et al.*, 1997). These dimers are $GspG_T$ homodimers and heterodimers containing the other pseudopilins, $GspH_U$, I_V or J_W. It has been suggested that the dimeric forms are intermediates between the membrane pools of monomers and the fully assembled pseudopilus. In addition, the composition of the cross-linked species depends on the size of the individual pools, which may reflect an active process involving the assembly and disassembly of individual subunits. The relative amounts of *P. aeruginosa* $GspG_T$, $GspH_U$, $GspI_V$ and $GspJ_W$ are approximately 16:1:1:4, respectively (Nunn and Lory, 1993).

Depolymerization of the assembled type IV pilus is thought to be involved in twitching motility and during the infection of *P. aeruginosa* by pili-specific phages (Whitchurch and Mattick, 1994). Evidence for $GspG_T$ forming a pseudopilus have now been obtained (see below). Therefore, an analogous dynamic process may drive the secretion of proteins from the periplasm, the assembling pseudopilus pushing exoproteins out of the cell (Fig. 2).

2.1.3   Relationship between pilus and pseudopilus

Recent work has shown that pseudopilins are indeed able to form a pilus-like structure (Sauvonnet *et al.*, 2000). This has been found by over-expressing the *K. oxytoca* Pul secreton in the heterologous *E. coli* K12 host. A thicker structure as compared to type IV pili is visible at the cell surface. Immunogold-labeling showed that this structure is composed of the major pseudopilin $GspG_T$. It has to be noted that the pseudopilus is not observed when *pul* genes are chromosomally encoded. Similar observations have been made in *P. aeruginosa* (unpublished results). Consequently, one may think that the long cell surface pseudopilus is artificial, but it reveals that pseudopilins have the ability to pack into helical complex similar to pilus. It is possible to imagine that the complex involved in type II secretion spans the periplasm without extruding out of the cell (Fig. 2).

The four pseudopilins, $GspG_T$-J_W, have the G-1, F+1 and E+5 residues conserved (Fig. 1) (although some differences are permitted at position F+1, which can be replaced by hydrophobic residues). Interestingly, in recent years, five genes (*fimT, fimU, pilV, pilW* and *pilE*), the products of which have characteristics in common with Gsp pseudopilins, have shown to be involved in fimbrial biogenesis in *P. aeruginosa* (Alm and Mattick, 1997), even though they are not found in the pilus structure itself. It is tempting to suggest that these pilin subunits form a cell envelope complex similar to the pseudopilus involved in type II secretion, whereas PilA is the

148

subunit which forms the pilus structure present at the cell surface (Fig. 2). Furthermore it can be proposed that exoproteins use the pseudopilus in type II secretion, like PilA uses it for cell surface exposure.

GspK$_X$ is considered to be atypical, because it lacks the E+5 residue and has a higher relative molecular mass (>30 kDa) than classical pilins and pseudopilins (15-20 kDa). Such atypical subunits have also been found in the type IV piliation system, as PilX in *P. aeruginosa*. It is not clear whether E+5, which is required for the methylation of the pilin subunit, is required in the case of pseudopilins, because the E+5V substitution in *K. oxytoca* GspG$_T$ did not affect N-methylation (Pugsley, 1993). The significance of this difference between the two systems is not understood. The E+5 residue has also been suggested to play a registration role in assembly of the pilus, by participating in the formation of a salt bridge between the negatively charged glutamate residue of subunit N and the positively charged methylated N-terminus of subunit N^{-1} (Parge *et al.*, 1995). The absence of E+5 in the newly incorporated pilin or pseudopilin subunit may prevent incorporation of the subsequent subunit, preventing elongation of the pilus or pseudopilus and initiating disassembly of the structure. This speculation provides clues about the particular role of GspK$_X$ in protein secretion (Bleves *et al.*, 1998).

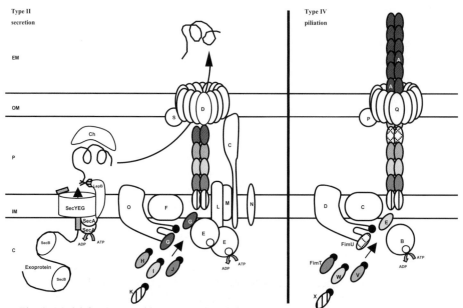

Fig. 2. Model for Gsp machinery assembly (left), and comparison with a model for type IV pilus assembly in *P. aeruginosa* (right). On the left, the Gsp-dependent exoprotein, shown as a black line with an N-terminal signal peptide (dotted rectangle) is exported across the inner

membrane (IM), from the cytoplasm (C) to the periplasm (P), *via* the Sec machinery. The exoprotein is kept unfolded in the cytoplasm, by binding to SecB, and folds within the periplasm after peptide signal cleavage by the leader peptidase, LepB (scissors). Folding may involve the function of a chaperone protein (Ch). The exoprotein is subsequently recognised by the Gsp machinery and translocated across the outer membrane (OM) *via* the secretin, GspD, and into the extracellular medium (EM). The secretin GspD is shown as a homomultimeric ring forming a channel with a large central opening. Most Gsp proteins are shown according to their membrane topology, and for those for which an interaction has been proposed, and supported by experimental evidence, the corresponding parts are shown in contact. The GspG-K pseudopilins are processed (removal of the leader peptide, shown as a small black circle) by the leader peptidase, GspO. The pseudopilus is arbitrarily represented as a succession of different pseudopilins dimers. The over-expression of GspG may lead to the appearance of the pseudopilus at the cell surface. The GspA and GspB proteins have not been represented. For more details about the Gsp machinery, see the text. The homologous components of the type IV piliation system in *P. aeruginosa* (Pil) are shown on the right for comparison. The various pilins (PilE, PilV-X and FimT-U) involved in the type IV pilus assembly but that are not found in the extracellular structure are arbitrarily represented in a similar manner as compared to the Gsp pseudopilus. The PilA subunit is represented within the pilus structure. Similarity in functioning between PilP and GspS has been shown for *N. gonorrhoeae*.

2.2   The outer membrane complex

2.2.1   GspD$_Q$, the secretin

GspD$_Q$ proteins are associated with the outer membrane and more generally called **SECRETINS** (Bitter *et al.*, 1998; Hardie *et al.*, 1996a; Hu *et al.*, 1995). Sequence comparisons have shown that in all cases, the C-terminal domain is highly conserved whereas the N-terminal domain is variable (Genin and Boucher, 1994). The conserved C-terminal region covers 200-300 residues and contains a very highly conserved block of about 60 amino acid residues. This conserved block contains invariant glycine and proline residues that have been shown to be functionally important (Russel, 1994a). The GspD$_Q$ proteins are multimers consisting of 12-15 subunits (Kazmierczak *et al.*, 1994; Chen *et al.*, 1996; Hardie *et al.*, 1996a; Bitter *et al.*, 1998). They are extremely stable and in most cases are heat- and detergent-resistant. In case of the *P. aeruginosa* GspD$_Q$, thirteen putative transmembrane β-strands were found in the C-terminal domain of the protein (Bitter *et al.*, 1998). This observation suggests that the C-terminal domain of GspD$_Q$ is required for the insertion of the protein into the outer membrane, whereas the N-terminal domain extends into the periplasm to facilitate interactions with other proteins. This is confirmed by the fact that the C-terminus of secretins is a protease-resistant domain (Brok *et al.*, 1999; Nouwen *et al.*, 2000).

The multimerization of $GspD_Q$'s makes it possible to envisage the formation of a pore-like structure in the outer membrane. Biochemical and electron microscopy studies have shown that the *P. aeruginosa* $GspD_Q$ multimer can indeed adopt a ring-shaped structure with a large central cavity of about 9.5 nm in diameter (Bitter *et al.*, 1998). The C-terminal domain of the secretin alone has the property to form these oligomeric rings. Reconstitution in planar lipid bilayers showed that secretins have a pore forming activity with higher single-channel conductance when compared to classical outer membrane porins (Brok *et al.*, 1999; Nouwen *et al.*, 1999; Marciano *et al.*, 1999). The large size of the channel is consistent with type II secretion-dependent exoproteins being brought across the outer membrane in a folded conformation. For example, the folded elastase from *P. aeruginosa* is 6.0 nm across in its wider dimension (Thayer *et al.*, 1991). The presence of such a large pore in the outer membrane could lead to cell death, so the opening of this pore is probably controlled. One possibility is that the N-terminus of secretins fold back into the cavity formed by the C-terminus (Nouwen *et al.*, 2000) and/or that interaction with other proteins controls the channel gating. Other gated outer membrane channels are known, such as the iron-siderophore complex, vitamin B12 and colicin uptake receptors. The mechanism involves a conformational change in the receptor and an energy-transducing process requiring proton-motive-force (pmf) (Moeck and Coulton, 1998). The mechanism of energy transduction to the outer membrane is unclear but it involves a bitopic cytoplasmic membrane protein called TonB. TonB has a large periplasmic domain, and its energized form interacts with the outer membrane receptor. On substrate binding, TonB induces a conformational change of the receptor, leading to the entry of the substrate into the periplasm. It has been shown that the translocation of proteins across the outer membrane using the type II secretory apparatus is dependent on pmf (Wong and Buckley, 1989; Letellier *et al.*, 1997; Possot *et al.*, 1997). An analogous system to the TonB mechanism may be involved in protein secretion. In *P. aeruginosa*, *tonB* mutant strains are not secretion-defective (unpublished observation). However, other proteins, particularly bitopic cytoplasmic membrane proteins of the type II secretion machinery may have a similar function as TonB. $GspC_P$, N or B proteins are good candidates for this function.

The $GspD_Q$ or secretin family of proteins is involved not only in type II secretion but also in other unrelated membrane translocation systems (Filloux *et al.*, 1998; Hobbs and Mattick, 1993). Its members are involved in type III secretion, type IV pilus assembly, DNA uptake, assembly of S-layers and filamentous phage assembly and extrusion.

2.2.2 GspS, the secretin-specific chaperone-like protein

In some cases, the insertion of secretin into the outer membrane depends on another protein, GspS (Hardie *et al.*, 1996a). GspS is a small peripheral outer membrane lipoprotein present in the type II secretory systems of *K. oxytoca* and *Erwinia* species (Table 1). It is essential for the outer membrane insertion of $GspD_Q$ and also protects $GspD_Q$ against proteolytic degradation (Hardie *et al.*, 1996b; Shevchik and Condemine, 1998). In other bacteria, one cannot exclude the possibility that the *gspS* gene does not belong to the *gsp* gene cluster and may be present at another chromosomal location as in *E. coli* K12 (Francetic *et al.*, 2000). The protective effect of GspS against proteolytic degradation has been shown using hybrid proteins containing the $GspD_Q$ C-terminus, which is thought to be the GspS binding site. The addition of this domain to the pIV secretin or to unrelated proteins such as maltose-binding protein or pectate lyase PelD rendered these proteins dependent on GspS for stability (Daefler *et al.*, 1997; Shevchik and Condemine, 1998). GspS-devoid type II systems may compensate by efficiently partitioning $GspD_Q$ into the outer membrane. In contrast, if kinetic partitioning is too slow, the binding of GspS to $GspD_Q$ may prevent the degradation of the protein prior to membrane insertion and may accelerate insertion as well. In unrelated transport processes involving secretins, a GspS-like function has been assigned to PilP, in type IV piliation in *N. gonorrhoeae* (Drake *et al.*, 1997), and InvH, in type III secretion in *Salmonella typhimurium* (Daefler and Russel, 1998). These two proteins are also small outer membrane lipoproteins.

2.3 $GspE_R$, the putative ATP-binding protein

Protein secretion is an active process that requires energy sources. $GspE_R$ proteins have sequences similar to the Walker A motif and, to a lesser extent, to the Walker B motif of ATPases (Walker *et al.*, 1982). Despite the presence of these motifs, it was not possible to demonstrate the binding of ATP to these proteins. However, mutation of the conserved glycine residue within the Walker A box of $GspE_R$ from *P. aeruginosa*, *K. oxytoca*, *E. chrysanthemi* or *V. cholerae* causes the bacteria to become secretion-defective (Turner *et al.*, 1993; Possot and Pugsley, 1994; Py *et al.*, 1999; Sandkvist *et al.*, 1995). ATP-binding activity could not be demonstrated but autokinase activity has been detected in the case of *V. cholerae* $GspE_R$ (Sandkvist *et al.*, 1995). Mutations in the less well conserved Walker B box have little or no effect on the secretion process (20-30 % reduction in pullulanase secretion for mutations in the *K. oxytoca*

152

GspE$_R$) (Possot and Pugsley, 1994). The Walker B box is thought to be the determinant for nucleoside recognition, and the weak requirement for this motif may account for the failure to demonstrate ATP-binding activity in GspE$_R$ proteins. Alternatively, binding may be stimulated when GspE$_R$ interacts with other components of the type II secretion machinery or other unknown molecules. ATP binding and hydrolysis are, for example, stimulated when SecA associates with the preprotein (Lill *et al.*, 1989).

Within the GspE$_R$ family, there is a highly conserved central region between the Walker A and B boxes consisting of two short aspartate-rich motifs called aspartate boxes (Possot and Pugsley, 1994). They are required for the function of GspE$_R$ in the secretion process, as shown by the substantial decrease, 80-90 %, in pullulanase secretion if the aspartate residues are replaced by asparagine residues in *K. oxytoca* GspE$_R$ (Possot and Pugsley, 1994). The aspartate residues may be involved in the formation and stabilization of the nucleotide-binding fold by interacting with Mg^{2+}. Another motif, a tetracysteine motif identical to that described for the GspO$_A$ peptidases, is present in members of the GspE$_R$ family (Possot and Pugsley, 1997). It appears to be essential for function, because replacement of any of the cysteine residues by a serine within the *K. oxytoca* GspE$_R$ leads to a large decrease in pullulanase secretion (80 %). This similarity between GspO$_A$ and GspE$_R$ proteins may be purely coincidental or may reflect, as suggested by Possot and Pugsley, a common function such as co-ordination of a Zn^{2+} ion. An additional feature of proteins involved in transport processes is their association with the membrane (Higgins, 1992). The deduced amino-acid sequence of GspE$_R$ proteins shows that they are mainly hydrophilic and do not possess any hydrophobic domains that could anchor the protein in the membrane. This observation is consistent with the observation that GspE$_R$ proteins are present in the cytoplasm if produced in *E. coli* (Possot *et al.*, 1992; Sandkvist *et al.*, 1995; Thomas *et al.*, 1997; Ball *et al.*, 1999). However, these proteins are associated with the cytoplasmic membrane in their original host (Possot and Pugsley, 1994; Sandkvist *et al.*, 1995; Ball *et al.*, 1999). This suggests that GspE$_R$ probably interacts with other Gsp components to form a functional machinery. It appears that GspE$_R$'s are bound to the inner face of the inner membrane thanks to an interaction with GspL$_Y$. This is demonstrated by the co-expression of the genes encoding GspE$_R$ and GspL$_Y$ from *V. cholerae* and *P. aeruginosa* in *E. coli*, and by the low level of membrane association of GspE$_R$ in a corresponding *gspL$_Y$* mutant strain (Sandkvist *et al.*, 1995; Ball *et al.*, 1999). The GspE$_R$-GspL$_Y$ interaction drives a conformational change of GspL$_Y$, revealing the dynamic of secreton assembly and functioning (Py *et al.*, 1999). GspL$_Y$ is a bitopic inner membrane protein with an N$_{in}$-C$_{out}$ topology (Bleves *et al.*, 1996), and a large cytoplasmic domain, which interacts with

GspE$_R$ proteins (Ball *et al.*, 1999; Py *et al.*, 1999). Finally, traffic ATPases function as dimers in a wide range of transport systems (Higgins, 1992). GspE$_R$ proteins also appear to associate as homodimers, as shown using a domain of the lambda phage cI repressor as a reporter for dimerization or the yeast two-hybrid system (Turner *et al.*, 1997; Py *et al.*, 1999). It is not yet possible to exclude the possibility that GspE$_R$ assembles into higher-order multimers. The GspE$_R$ homologue involved in *H. pylori* type IV secretion was shown to form homohexamers (Krause *et al.*, 2000).

The exact role of GspE$_R$ is unknown, but the various observations discussed above suggest that it acts as a traffic ATPase for transport across the inner membrane. The GspE$_R$ homologue PilB is required for the assembly of type IV pili in *P. aeruginosa* (Nunn *et al.*, 1990). This suggests that GspE$_R$ is an ATP-binding protein involved in the translocation of pseudopilins through the inner membrane before their assembly into a "pseudopilus" structure. Unlike *P. aeruginosa* PilD (GspO$_A$), PilB is exclusively involved in type IV piliation and is not required for type II secretion (Lu *et al.*, 1997). In addition, the effects of thermosensitive (*ts*) mutations within the *P. aeruginosa* GspG$_T$ pseudopilin are suppressed by a secondary mutation within GspE$_R$ (Kagami *et al.*, 1998). This strongly supports the idea that pseudopilins are substrates for GspE$_R$. Moreover, the *ts* mutations affects the periplasmic domain of the pseudopilin, suggesting that the interaction takes place before translocation, or that GspE$_R$ pushes the pseudopilins through the membrane in a manner similar to the insertion-deinsertion cycles of SecA (Economou and Wickner, 1994). One interesting result is that it is not possible to rescue pullulanase secretion in *K. oxytoca* cells that have synthesized pullulanase at the same time as expressing all of the *gsp* secretion genes except *gspE$_R$*, by subsequent expression of *gspE$_R$* alone (Possot *et al.*, 1992). This could indicate that *gspE$_R$* must be co-expressed with other *gsp* genes in order to participate in the formation of a functional secretion apparatus. We could distinguish two modules in the type II secretion machinery: (i) assembly of the pseudopilus, which requires GspE$_R$ as an "energiser", and (ii) transport of the exoproteins via the secreton through the outer membrane, which may require a different energy source.

2.4 The GspE$_R$, F$_S$, L$_Y$, M$_Z$ inner membrane platform

Comparison between type II protein secretion and type IV piliation has already suggested that homologous components, such as PilD/GspO$_A$, PilB/GspE$_R$, PilA/GspG$_T$-J$_W$, PilX/GspK$_X$ from *P. aeruginosa* are involved in these processes. Another Pil protein, PilC, is essential in type IV piliation.

It has a Gsp homologue, GspF$_S$. GspF$_S$ is a polytopic integral inner membrane protein with a small periplasmic loop and two larger cytoplasmic domains connected by three transmembrane regions (Thomas *et al.*, 1997). PilB, C, D and A are clustered in most type IV piliated bacteria, suggesting that PilB and C are simultaneously involved in the translocation/assembly of PilA after processing by PilD (Strom and Lory, 1993). GspF$_S$ may have a similar function to PilC with respect to pseudopilins, even though GspL$_Y$ is required for the association of GspE$_R$, the PilB homologue, with the membrane. However, recent works proved that the GspF$_S$ N-terminus (172 aa) interacts both with GspE$_R$ and GspL$_Y$ from *E. chrysanthemi* (Py *et al.*, 2001). GspF$_S$ may thus strengthen the association of GspE$_R$ with the membrane, and/or be involved in pore formation in the inner membrane, allowing pseudopilin translocation.

No homologues of GspL$_Y$, GspM$_Z$ and GspC$_P$ specifically involved in type IV piliation have yet been identified. These Gsp components are bitopic inner membrane proteins with an N$_{in}$-C$_{out}$ topology (Bleves *et al.*, 1996). Homologs of the *gspC$_P$*, *gspL$_Y$* and *gspM$_Z$* genes have been identified by analyzing the DNA sequence of *E. coli* K12, close to a *gspO$_A$* gene homologue, *pppa*, which encodes prepilin peptidase (Francetic *et al.*, 1998). These genes have unknown functions in *E. coli*.

The precise function of GspM$_Z$ is unknown, but it was shown to be crucial for the stability of Gsp$_L$ (Michel *et al.*, 1998). The stabilization process is reciprocal because the abundance of GspM$_Z$ in the cell depends on Gsp$_L$, indicating that these two Gsp components interact with each other. Many other studies have since confirmed this observation (Sandkvist *et al.*, 1999; Possot *et al.*, 2000). It is unknown whether and how the cascade of protein-protein interactions is kinetically organized during the assembly of the machinery. The GspM$_Z$ protein may determine the membrane location of the secretion site, recruit the GspL$_Y$ component which in turn brings GspE$_R$ into association with the membrane.

2.5 GspC and GspN the connecting components?

The function of the GspC$_P$ protein is unknown. Its principal feature is the presence of a PDZ domain in the C-terminal region of the protein (Pallen and Ponting, 1997). PDZ domains are named after the three eukaryotic proteins (post synaptic density protein, disc large and zo-1 proteins), in which they were first discovered. They mediate a variety of protein-protein interactions by binding to short sequences (X-T/S-X-V-COO$^-$), usually at the C termini of target polypeptides, but may also bind homotypically (Ponting *et al.*, 1997). It is widely thought that PDZ domain-containing proteins mediate the organization of multimolecular complexes at sites of membrane specialization. Some GspC$_P$'s have no PDZ domain. This is true of the GspC$_P$'s of *P. aeruginosa* and *P. alcaligenes*. However, in both cases, the PDZ region is replaced by a coiled-coil structure, a motif also known to be involved in protein-protein interactions. Hence, GspC$_P$ proteins may form homomultimers or interact with other proteins by one of two mechanisms, one involving a coiled-coil and the other, PDZ domains.

The topology of GscpC$_P$ is similar to that of TonB (Bleves *et al.*, 1996). The *P. aeruginosa* GspC$_P$ protein has been proposed to interact with the secretin, and in case of *K. oxytoca* it has been shown to partly fractionate into the outer membrane (Bleves *et al.*, 1999; Possot *et al.*, 1999). In addition, in *P. aeruginosa* and *P. alcaligenes*, the genes encoding GspC$_P$ and GspD$_Q$ are organized in a separate operon, suggesting their co-ordinated action. These observations lead to propose a role for GspC$_P$ in energy transduction and channel gating of the secretin. This hypothesis may be supported by the fact that GspC$_P$ probably connect the two sub-secreton complexes, the inner membrane platform GspE$_R$-F$_S$-L$_Y$-M$_Z$ and the outer membrane complex GspD$_Q$-S (Possot *et al.*, 2000).

The GspN protein is, like GspC$_P$, a bitopic inner membrane protein. It is absent in many known type II systems, and has been identified only in *K. oxytoca*, *E. carotovora*, *Xanthomonas campestris* and *P. putida*. In case of the *X. campestris* GspN, an interaction with the secretin has been suggested (Lee *et al.*, 2000). In addition it was recently proposed that *X. campestris* GspN participates to the formation of the GspL$_Y$-GspM$_Z$ complex (Lee *et al.*, 2001). However, in the case of *K. oxytoca* GspN is not required for type II secretion (Possot *et al.*, 2000).

2.6   GspA and GspB, "energizer" or "regulator"

We have seen that there may be small differences in the composition of type II secretory machineries. Two additional Gsp proteins, bringing the number of Gsp components up to sixteen, have been described in *A. hydrophila*. These two proteins, GspA and GspB, are essential to the protein secretion process (Schoenhofen *et al.*, 1998). The GspB protein has a sequence and structure similar to that of TonB, whereas GspA is a membrane protein with a consensus ATP binding site. GspA and GspB form a complex within the inner membrane and it has been suggested that such a complex transduces energy from ATP hydrolysis, to the protein secretion process. If this is the case, the complex plays a crucial role, and it is not clear why it is not found in most type II systems. In *A. hydrophila*, secretion from the periplasm requires pmf and ATP (Letellier *et al.*, 1997), whereas in

K. oxytoca, it requires only pmf (Possot *et al.*, 1997). This may partly explain the role of GspA and B in *A. hydrophila*.

GspB homologues are found in *Erwinia* species (Out) and *K. oxytoca* (Pul). PulB is not required for secretion (Possot *et al.*, 2000), but OutB is required and may interact with the OutD secretin (Condemine and Shevchik, 2000).

GspA and GspB homologues have also been identified in the *E. coli gsp* gene cluster (Francetic and Pugsley, 1996). Recent work showed that the *E. coli* secreton is functional and allows secretion of a chitinase under particular growth conditions (Francetic *et al.*, 2000). If the GspC$_P$-O$_A$ secreton is alone sufficient to achieve chitinase secretion, expression of the GspA-B homologues increases the level of secretion (Francetic *et al.*, 2000). This raises the question whether GspA-B have a regulatory role, are involved in energy transduction, or both.

3. EXOPROTEIN RECOGNITION

Secretion by the type II system is a two-step process. The first step, translocation across the inner membrane is controlled by the Sec machinery, which recognizes signal peptide-bearing exoproteins. Once in the periplasm, the mature polypeptide becomes the substrate for the Gsp machinery, which should discriminate between exoproteins and periplasmic or outer membrane proteins. This process may be based on the recognition of a secretion motif by some of the Gsp proteins.

3.1   Species-specific recognition of exoproteins

The Gsp secretion apparatus is widespread in Gram-negative bacteria and a wide variety of enzymes and toxins use this pathway. However, Gsp-dependent exoproteins such as *K. oxytoca* pullulanase, are not recognized by the Gsp machinery of *P. aeruginosa*, demonstrating that the process is specific (de Groot *et al.*, 1991). These two bacteria are not closely related and heterologous secretion has been described in closely related organisms. For example, the *Burkholderia glumae* lipase, and the *A. hydrophila* aerolysin are secreted by *P. aeruginosa* and *Vibrio*, respectively (Frenken *et al.*, 1993a; Wong *et al.*, 1990). In addition, the *P. alcaligenes* lipase (Gerritse *et al.*, 1998) is secreted by *P. aeruginosa*, whereas *P. aeruginosa* elastase is secreted by *P. alcaligenes* (de Groot *et al.*, 2001). In contrast, even though the cellulases, Cel5 (ex EGZ) and CelV, from *E.*

chrysanthemi and *E. carotovora*, respectively, are very similar (40% identity), these two proteins are recognized exclusively by their own secretion machinery (Py *et al.*, 1991). Thus, the principle on which the exoprotein recognition process is based is unclear but it may rely on the presence of a secretion motif within the secreted protein. *P. aeruginosa* secretes enzymes as diverse as lipase, elastase, alkaline phosphatase, phospholipases and the ADP-ribosylating exotoxin A, all of which use the same secretion machinery. These proteins should, therefore, have a common secretion motif. However, sequence analyses have not identified a linear motif of residues and led researchers in the field to suggest that there is a non-linear motif, a conformational signal constructed by folding.

3.2   Translocation of folded proteins

Two major observations suggest that Gsp-dependent exoproteins are translocated across the outer membrane in a folded conformation. First, the studies with *E. chrysanthemi* cellulases and pectate lyases and *K. oxytoca* pullulanase have demonstrated that disulphide bridges are formed in exoproteins during secretion (Bortoli-German *et al.*, 1994; Pugsley, 1992). The cysteine bond is formed in the periplasm by disulphide bond isomerases (Dsb), particularly by DsbA (Shevchik *et al.*, 1995, Urban *et al.*, 2001). Second, studies with *V. cholerae* and *A. hydrophila* have shown that the B pentamer of the cholera toxin is formed in the periplasm before its secretion (Hirst and Holmgren, 1987), and that proaerolysin is secreted as a dimer (Hardie *et al.*, 1995). Thus, the secretion motif may be formed and presented to the secretory apparatus on the folding of the exoprotein. It is clear that it is not the disulphide bond itself that is recognized by the machinery. CelV, secreted in a Gsp-dependent manner by *E. carotovora* contains no cysteines (Cooper and Salmond, 1993), and an engineered cysteine-free pullulanase is efficiently secreted (Sauvonnet and Pugsley, 1998). Secretion of the cysteine-free pullulanase is still DsbA-dependent, raising questions about the role of DsbA. Two secreton components, $GspK_X$ and GspS possess disulphide bonds. The formation of these bonds have been shown to be crucial for $GspK_X$ activity and GspS stability in *K. oxytoca* (Pugsley *et al.*, 2001). However no GspS is found in *P. aeruginosa* and $GspK_X$ contains only one single cysteine. In this case, DsbA is still require for stability of secreted lipase (Liebeton *et al.*, 2001). The results of studies on pullulanase conflict with the work of Bortoli-German and collaborators (Bortoli-German *et al.*, 1994), who showed that site-directed mutagenesis of either of the cysteines in Cel5 cellulase prevented secretion. The cysteine bond may thus act as a clip, stabilizing and fixing the protein in its folded conformation.

The requirement for the disulphide bond therefore depends on the stability of the structure adopted by the reduced exoprotein.

The folding of exoprotein within the periplasm seems, in some cases, to involve not only general catalysts such as DsbA, but also specifically dedicated chaperones. This is the case for the propeptide of *P. aeruginosa* elastase (Braun *et al.*, 1996), and for the Lif proteins, which are required for folding and activation of *Pseudomonas* lipases (Frenken *et al.*, 1993b, El Khattabi *et al.*, 1999). Recent studies proposed that PlcR is required for full hemolytic activity of *P. aeruginosa* phospholipase C, suggesting that it act as well as an helper for secretion. (Cota-Gomez *et al.*, 1997).

3.3   The secretion motif

Gsp-dependent exoproteins acquire a highly ordered structure after reaching the periplasm and before outer membrane translocation. In the absence of a linear motif, the secretion signal for the various exoproteins is therefore probably a patch signal involving distal regions brought into close proximity by the folding of the protein. If this is true, the motif should be very sensitive to any changes in the structure of the exoproteins. This is consistent with the observation made by Py and collaborators (Py *et al.*, 1993), who showed that slight modifications in the *E. chrysanthemi* Cel5 protein result in non-secreted derivatives. Moreover, increased changes in the molecular dimension of exotoxin A coincide with a decrease of its secretion efficiency (Voulhoux *et al.*, 2000). In *A. hydrophila*, a single substitution of the W227 residue yields a protein that is not secreted (Wong and Buckley, 1991). In contrast linker insertion experiments within *K. oxytoca* pullulanase, resulted in the construction of 23 derivatives, all of which were efficiently secreted (Sauvonnet *et al.*, 1995). Some of these derivatives were inactive, indicating changes in the enzyme structure, thereby challenging the existence of the conformational motif. However, it is possible that the motif is formed at an intermediate state of folding rather than in the fully active native protein. The fold to a specific secretion-competent conformation prior to secretion may thus be different from the folding state of the secreted species. An interesting model, in which the secretion of Cel5 involves a transient intra-molecular interaction between the cellulose-binding domain (CBD) and a region close to the active site, has been proposed (Chapon *et al.*, 2000). Once secreted, the protein may then vary its fold to allow the CBD to interact with the cellulose substrate.

Two regions in the *K. oxytoca* pullulanase, A and B, which are well separated from each other in the primary protein sequence, together direct the secretion of a PulA-Bla hybrid (Sauvonnet and Pugsley, 1996). These two domains are therefore probably part of a structural motif, which is still formed and recognized within such a hybrid. Studies with the *P. aeruginosa* exotoxin A identified a single stretch of 60 amino acids (60-120) which were sufficient to direct Bla secretion (Lu and Lory, 1996). This region appears to be rich in anti-parallel β-sheets but has no apparent sequence similarity with any region of the other type II secretion-dependent exoproteins of *P. aeruginosa* or the A and B motifs of pullulanase. Moreover, another study with *P. aeruginosa* exotoxin A has shown that a truncated protein retaining the first 30 and last 305 residues, but not residues 60-120, is still efficiently secreted (McVay and Hamood, 1995). The nature of the secretion signal is therefore unclear.

One alternative to a unique structural motif is that successive interactions lead to the secretion of exoproteins. These interactions may involve different secretion signals that are not essential individually but are required simultaneously for optimal secretion. Finally, the secretion motif may not be recognized by a common component of the secretion machinery but by a specific intermediate, such as the previous described dedicated chaperone, which will in turn be recognized by a Gsp component of the secretion machinery.

3.4 Recognition by GspD$_Q$

One of the Gsp component is involved in the cascade of recognition events required to direct the exoprotein is the GspD$_Q$ secretin. The pectate lyase PelB of *E. chrysanthemi* binds to the N-terminus of GspD$_Q$ (Shevchik *et al.*, 1997). The N-terminus is the most variable region of the protein, consistent with the notion of specificity for recognition of the exoprotein. Even though contradictory results have been obtained with *K. oxytoca* GspD$_Q$ (Guilvout *et al.*, 1999), this observation is consistent with the fact that of all the Gsp components of *E. chrysanthemi* required for pectate lyase secretion, only two cannot be replaced by Gsp homologues from *E. carotovora*, namely GspC$_P$ and GspD$_Q$ (Lindeberg *et al.*, 1996). This suggests that these two proteins are intimately involved in the species-specific recognition of the exoprotein by the Gsp system. Another possibility, however, is that these proteins are not incorporated into the heterologous system, because they fail to make specific interactions with the other Gsp components that are required for the assembly of the secretion machinery. This second hypothesis is supported by the work of De Groot and collaborators who showed that *P. aeruginosa* and *P. alcaligenes* GspC$_P$ or GspD$_Q$ proteins cannot be exchanged even though the two species secrete

160

each other exoproteins. Because $GspC_P$ and $GspD_Q$ can be exchanged simultaneously one may suspect a specific interaction between these two components.

4. CONCLUSIONS AND FUTURE PERSPECTIVES

Based on the observations described in this chapter, it is possible to formulate ideas about several of the main features of the type II machinery (Fig. 2). Secretion occurs in two steps, as shown by Poquet and collaborators (Poquet *et al.*, 1993), and the Gsp components of the secretory apparatus are involved in the translocation of the transient periplasmic form across the outer membrane. However it is important to say that the periplasmic form might correspond to an extremely short period, both steps of membrane translocation being tightly connected. The elastase fused to *E. coli* colicin A is efficiently secreted by *P. aeruginosa* and there is no time for colicin A to insert in the membrane from the periplasmic side and exerts its toxic effect (Voulhoux *et al.*, 2001). The assembly of the secretion machinery, and the recognition of the secreted proteins can be summarized in several points:

(1) One can imagine the existence of a trans-periplasmic structure, formed of helically packed pseudopilins that was named pseudopilus. Assembly would require a peptidase, an ATP-binding "energizer" protein and possibly several inner membrane proteins to achieve the translocation of pseudopilins across the inner membrane. The assembly of the "pseudopilus" may be a key element, helping to push the exoprotein through the secretin for its final release. Type IV pili are retractile organelles, and by comparison, pseudopilus retraction might correlate cycles of pushing off. Alternatively, the pseudopilus may act like a cork, blocking the secretin channel, with secretion occuring only if the structure is retracted. The pseudopilus may also serve as a guide for routing exoproteins to the secretin channel.

(2) The final channel, allowing exoproteins to reach the extracellular environment, is made of a so-called secretin ($GspD_Q$), which forms multimers (12-15 subunits) and may form a large hole (95 nm) within the outer membrane. In some cases, $GspD_Q$ outer membrane insertion requires the GspS outer membrane lipoprotein.

(3) The controlled opening of the secretin channel may require energy, from pmf or ATP. Transduction of this energy to the outer membrane may be achieved *via* large periplasmic domains of cytoplasmic membrane proteins, interacting with the secretin. $GspC_P$ and GspA/GspB

complex are good candidates for this function. GspN has also been shown to be interacting with GspD$_Q$.

(4) Despite the conservation of Gsp systems, most Gsp components are not exchangeable, suggesting that there is specificity within the machinery itself. The probability of producing a functional hybrid machinery depends on the phylogenetic distance between the organisms involved. As described above, only GspC$_P$ and GspD$_Q$ are not exchangeable. More generally, the GspO$_A$ peptidase may be functionally moved from one species to another (Dupuy *et al.*, 1992; de Groot *et al.*, 1994), probably because it has an enzymatic rather than a structural function. This idea is supported by the observation that GspE$_R$ hybrids may be functional if the variable N-terminus is in a homologous context, whereas the ATP binding site-containing C terminus may be exchanged between species (Sandkvist *et al.*, 1995; de Groot *et al.*, 1996). Apart from that, the most permissive Gsp component is GspG$_T$, which can be exchanged, more or less efficiently, in several cases (Francetic and Pugsley, 1996; Lindeberg *et al.*, 1996; Pugsley, 1996; de Groot *et al.*, 1999).

(5) The secreted proteins must have features (secretion motifs) that mediate specific recognition by Gsp components, such as the secretin GspD$_Q$. The exoproteins are folded in the periplasm and the secretion motif is probably conformational rather than sequential. As they are folded, the exoproteins need the large hole formed by the secretin for their translocation across the outer membrane. This translocation of folded exoproteins contrasts with the "dogmatic" unfolded competent state for translocation across the bacterial inner membrane *via* the Sec machinery. However, a novel Sec-independent translocation pathway has been shown to translocate metalloenzymes into the periplasm in a stably folded conformation (Santini *et al.*, 1998). Interestingly, recent work indicates that the Tat system might in some cases be used as the first step of the general secretory pathway instead of the Sec system (unpublished results).

Finally, it is clear that the components of the type II secretion machinery have now been well established. However, we do not yet understand the precise organization of the machinery and the mechanisms controlling its secretion function. The function of the "pseudopilus" is probably one of the key feature of the type II secretion apparatus. Looking at the multiple interactions between the Gsp components, and looking also at the fact that most of them are included in the complex as homomultimers, the secreton appears to be an extremely large structure of the bacterial cell envelope. The proper incorporation of the complex might involve the interaction with additional constituents of the envelope like it was suggested with lipopolysaccharides (Michel *et al.*, 2000).

Gsp-like components are also involved in the transport and assembly of various macromolecules through the membranes of Gram-negative bacteria. Therefore, specific mechanisms may be required to adapt regions of the cell envelope, including the peptidoglycan layer, to such transport processes. An important question, which should be addressed in the near future, concerns the location and the distribution of the secretion sites within the bacterial cell envelope. It is possible that these sites are not spread out evenly over the envelope but are rather confined to particular areas, such as the pole of the cell. This distribution would be reminiscent of that for type IV pili, and would limit leakage zones to a part of the bacterial cell in which the periplasm is compartmented (Foley *et al.*, 1989), reducing the effect of the transient opening of large holes in the outer membrane on the integrity of the cell. In this region the peptidoglycan layer may be looser, making the assembly of the machinery and secretion of the exoproteins easier.

5. REFERENCES

Alm, R. A. and Mattick, J. S. (1997) Gene 192: 89-98.
Ball, G., Chapon-Hervé, V., Bleves, S., Michel, G. and Bally, M. (1999) J Bacteriol 181: 382-388.
Bally, M., Filloux, A., Akrim, M., Ball, G., Lazdunski, A. and Tommassen, J. (1992) Mol Microbiol 6: 1121-1131.
Bitter, W., Koster, M., Latijnhouwers, M., de Cock, H. and Tommassen, J. (1998) Mol Microbiol 27: 209-219.
Bleves, S., Lazdunski, A. and Filloux, A. (1996) J Bacteriol 178: 4297-300.
Bleves, S., Voulhoux, R., Michel, G., Lazdunski, A., Tommassen, J. and Filloux, A. (1998) Mol Microbiol 27: 31-40.
Bleves, S., Gerard-Vincent, M., Lazdunski, A. and Filloux, A. (1999) J Bacteriol 181: 4012-4019.
Bortoli-German, I., Brun, E., Py, B., Chippaux, M. and Barras, F. (1994) Mol Microbiol 11: 545-553.
Braun, P., Tommassen, J. and Filloux, A. (1996) Mol Microbiol 19: 297-306.
Brok, R., van Gelder, P., Winterhalter, M., Ziese, U., Koster, A.J., de Cock, H., Koster, M., Tommassen, J. and Bitter, W. (1999) J Mol Microbiol 294: 1169-1179.
Browne, B. L., McClendon, V. and Bedwell, D. M. (1996) J Bacteriol 178: 1712-1719.
Chapon, V., Simpson, H.D., Morelli, X., Brun, E. and Barras, F. (2000) J Mol Biol 303: 117-123.
Chen, L. Y., Chen, D. Y., Miaw, J. and Hu, N. T. (1996) J Biol Chem 271: 2703-2708.
Condemine, G. and Shevchik V.E. (2000) Microbiol 146: 639-647.
Cooper, V. J. and Salmond, G. P. (1993) Mol Gen Genet 241: 341-350.
Cota-Gomez, A., Vasil, A. I., Kadurugamuwa, J., Beveridge, T. J., Schweizer, H. P. and Vasil, M. L. (1997) Infect Immun 65: 2904-2913.
d'Enfert, C., Ryter, A. and Pugsley, A. P. (1987) EMBO J 6: 3531-3538.

Daefler, S., Guilvout, I., Hardie, K. R., Pugsley, A. P. and Russel, M. (1997) Mol Microbiol 24: 465-475.

Daefler, S. and Russel, M. (1998) Mol Microbiol 28: 1367-1380.

De Groot, A., Filloux, A. and Tommassen, J. (1991) Mol Gen Genet 229: 278-284.

De Groot, A., Heijnen, I., de Cock, H., Filloux, A. and Tommassen, J. (1994) J Bacteriol 176: 642-650.

De Groot, A., Krijger, J. J., Filloux, A. and Tommassen, J. (1996) Mol Gen Genet 250: 491-504.

De Groot, A., Gerritse, G., Tommassen, J., Lazdunski, A. and Filloux, A. (1999) Gene 226: 35-40.

De Groot, A., Koster, M, Gerard-Vincent, M., Gerritse, G., Lazdunski, A., Tommassen, J. and Filloux, A. (2001) J Bacteriol 183: 959-967.

Drake, S. L., Sandstedt, S. A. and Koomey, M. (1997) Mol Microbiol 23: 657-668.

Dupuy, B., Taha, M. K., Possot, O., Marchal, C. and Pugsley, A. P. (1992) Mol Microbiol 6: 1887-1894.

Economou, A. and Wickner, W. (1994) Cell. 78: 835-843.

El Khattabi, M., Ockhuijsen, C., Bitter, W., Jaeger, K.E and Tommassen, J. (1999) Mol Gen Genet 261: 770-776.

Feng, J. N., Russel, M. and Model, P. (1997) Proc Nat Acad Sci USA 94: 4068-4073.

Filloux, A., Bally, M., Ball, G., Akrim, M., Tommassen, J. and Lazdunski, A. (1990) EMBO J 9: 4323-4329.

Filloux, A., Michel, G. and Bally, M. (1998) FEMS Microbiol Rev 22: 177-198.

Foley, M., Brass, J. M., Birmingham, J., Cook, W. R., Garland, P. B., Higgins, C. F. and Rothfield, L. I. (1989) Mol Microbiol 3: 1329-1336.

Francetic, O. and Pugsley, A. P. (1996) J Bacteriol 178: 3544-3549.

Francetic, O., Lory, S. and Pugsley, A. P. (1998) Mol Microbiol 27: 763-775.

Francetic, O., Belin, D., Badaut, C. and Pugsley, A.P. (2000) EMBO J 19: 6697-6703.

Frenken, L. G., Bos, J. W., Visser, C., Muller, W., Tommassen, J. and Verrips, C. T. (1993b) Mol Microbiol 9: 579-589.

Frenken, L. G., de Groot, A., Tommassen, J. and Verrips, C. T. (1993a) Mol Microbiol 9: 591-599.

Genin, S. and Boucher, C. A. (1994) Mol Gen Genet 243: 112-118.

Gerritse, G., Ure, R., Bizoullier, F. and Quax, W. J. (1998) J Biotechnol 64: 23-38.

Guilvout, I., Hardie, K. R., Sauvonnet, N. and Pugsley, A.P. (1999) J Bacteriol 181: 7212-7220.

Hahn, H. P. (1997) Gene 192: 99-108.

Hardie, K. R., Schulze, A., Parker, M. W. and Buckley, J. T. (1995) Mol Microbiol 17: 1035-1044.

Hardie, K. R., Lory, S. and Pugsley, A. P. (1996a) EMBO J 15: 978-988.

Hardie, K. R., Seydel, A., Guilvout, I. and Pugsley, A. P. (1996b) Mol Microbiol 22: 967-976.

Henrichsen, J. (1983) Ann Rev Cell Biol 8: 67-113.

Hirst, T. R. and Holmgren, J. (1987) Proc Nat Acad Sci USA 84: 7418-7422.

Hobbs, M. and Mattick, J. S. (1993) Mol Microbiol 10: 233-243.

Howard, S. P., Critch, J. and Bedi, A. (1993) J Bacteriol 175: 6695-6703.

Hu, N. T., Hung, M. N., Liao, C. T. and Lin, M. H. (1995) Microbiol 141: 1395-1406.

Kagami, Y., Ratliff, M., Surber, M., Martinez, A. and Nunn, D. N. (1998) Mol Microbiol 27: 221-333.

Kazmierczak, B. I., Mielke, D. L., Russel, M. and Model, P. (1994) J Mol Biol 238: 187-198.

Krause, S., Barcena, M., Pansegrau, W., Lurz, R., Carazo, J. M. and Lanka, E. (2000) Proc Nat Acad Sci USA 97: 3067-3072.

164

Lee, H. M., Wang, K. C., Liu, Y. L., Yew, H. Y., Chen, L. Y., Leu, W. M., Chen, D. C. and Hu, N. T. (2000) J Bacteriol 182: 1549-1557.

Lee, H.M., Tyan, S. W., Leu, W. M., Chen, L. Y., Chen, D. C. and Hu, N. T. (2001) J Bacteriol 183: 528-535.

Letellier, L., Howard, S. P. and Buckley, J. T. (1997) J Biol Chem 272: 11109-11113.

Liebeton, K., Zacharias, A. and Jaeger, K.E. (2001) J Bacteriol 183: 597-603.

Lill, R., Cunningham, K., Brundage, L. A., Ito, K., Oliver, D. and Wickner, W. (1989) EMBO J 8: 961-966.

Lindeberg, M., Salmond, G. P. and Collmer, A. (1996) Mol Microbiol 20: 175-190.

Lu, H. M. and Lory, S. (1996) EMBO J 15: 429-436.

Lu, H. M., Motley, S. T. and Lory, S. (1997) Mol Microbiol 25: 247-259.

Marciano, D. K., Russel, M. and Simon, S.M. (1999) Science 284: 1516-1519.

McVay, C. S. and Hamood, A. N. (1995) Mol Gen Genet 249: 515-525.

Michel, G., Bleves, S., Ball, G., Lazdunski, A. and Filloux, A. (1998) Microbiol 144: 3379-3386.

Michel, G., Ball, G., Goldberg, J. B. and Lazdunski, A. (2000) J Bacteriol 182: 696-703.

Moeck, G. S. and Coulton, J. W. (1998) Mol Microbiol 28: 675-681.

Nouwen, N., Ranson, N., Saibil, H., Wolpensinger, B., Engel, A., Ghazi, A. and Pugsley, A.P. (1999) Proc Nat Acad Sci USA 96: 8173-8177.

Nouwen, N., Stahlberg, H., Pugsley, A. P. and Engel, A. (2000) EMBO J 19: 2229-2236.

Nunn, D., Bergman, S. and Lory, S. (1990) J Bacteriol 172: 2911-2919.

Nunn, D. N. and Lory, S. (1991) Proc Nat Acad Sci USA 88: 3281-3285.

Nunn, D. N. and Lory, S. (1992) Proc Nat Acad Sci USA 89: 47-51.

Nunn, D. N. and Lory, S. (1993) J Bacteriol 175: 4375-4382.

Pallen, M. J. and Ponting, C. P. (1997) Mol Microbiol 26: 411-413.

Parge, H. E., Forest, K. T., Hickey, M. J., Christensen, D. A., Getzoff, E. D. and Tainer, J. A. (1995) Nature 378: 32-38.

Pepe, C. M., Eklund, M. W. and Strom, M. S. (1996) Mol Microbiol 19: 857-869.

Ponting, C. P., Phillips, C., Davies, K. E. and Blake, D. J. (1997) Bioessays 19: 469-479.

Poquet, I., Faucher, D. and Pugsley, A. P. (1993) EMBO J 12: 271-278.

Possot, O., d'Enfert, C., Reyss, I. and Pugsley, A. P. (1992) Mol Microbiol 6: 95-105.

Possot, O. and Pugsley, A. P. (1994) Mol Microbiol 12: 287-299.

Possot, O. M., Letellier, L. and Pugsley, A. P. (1997) Mol Microbiol 24: 457-464.

Possot, O. M. and Pugsley, A. P. (1997) Gene 192: 45-50.

Possot, O. M., Gerard-Vincent, M. and Pugsley, A. P. (1999) J Bacteriol 181: 4004-4011.

Possot, O. M., Vignon, G., Bomchil, N., Ebel, F. and Pugsley, A. P. (2000) J Bacteriol 182: 2142-2152.

Pugsley, A. P. (1992) Proc Nat Acad Sci USA 89: 12058-12062.

Pugsley, A. P. and Dupuy, B. (1992) Mol Microbiol 6: 751-760.

Pugsley, A. P. (1993) Mol Microbiol 9: 295-308.

Pugsley, A. P. (1996) Mol Microbiol 20: 1235-1245.

Pugsley, A. P., Bayan, N. and Sauvonnet, N. (2001) J Bacteriol 183: 1312-1319.

Py, B., Salmond, G. P. C., Chippaux, M. and Barras, F. (1991) FEMS Microbiol Lett 79: 315-322.

Py, B., Chippaux, M. and Barras, F. (1993) Mol Microbiol 7: 785-793.

Py, B., Loiseau, L. and Barras, F. (1999) J Mol Biol 289: 659-670.

Py, B., Loiseau, L. and Barras, F. (2001) EMBO Reports 2: 244-248.

Reeves, P. J., Douglas, P. and Salmond, G. P. (1994) Mol Microbiol 12: 445-457.

Russel, M. (1994a) Mol Microbiol 14: 357-369.

Sandkvist, M., Bagdasarian, M., Howard, S. P. and DiRita, V. J. (1995) EMBO J 14: 1664-1673.

Sandkvist, M., Hough, L. P., Bagdasarian, M. M. and Bagdasarian, M. (1999) J Bacteriol 181: 3129-3135.

Santini, C. L., Ize, B., Chanal, A., Muller, M., Giordano, G. and Wu, L. F. (1998) EMBO J 17: 101-112.

Sauvonnet, N., Poquet, I. and Pugsley, A. P. (1995) J Bacteriol 177: 5238-5246.

Sauvonnet, N. and Pugsley, A. P. (1996) Mol Microbiol 22: 1-7.

Sauvonnet, N. and Pugsley, A. P. (1998) Mol Microbiol 27: 661-667.

Sauvonnet, N., Vignon, G., Pugsley, A. P. and Gounon, P. (2000) EMBO J 19: 2221-2228.

Schoenhofen, I. C., Stratilo, C. and Howard, S. P. (1998) Mol Microbiol 29: 1237-47.

Shevchik, V. E., Bortoli-German, I., Robert-Baudouy, J., Robinet, S., Barras, F. and Condemine, G. (1995) Mol Microbiol. 16, 745-753.

Shevchik, V. E., Robert-Baudouy, J. and Condemine, G. (1997) EMBO J 16: 3007-3016.

Shevchik, V. E. and Condemine, G. (1998) Microbiol 144: 3219-3228.

Strom, M. S. and Lory, S. (1991) J Biol Chem 266: 1656-1664.

Strom, M. S. and Lory, S. (1993) Ann Rev Microbiol 47: 565-596.

Thayer, M. M., Flaherty, K. M. and McKay, D. B. (1991) J Biol Chem 266: 2864-2871.

Thomas, J. D., Reeves, P. J. and Salmond, G. P. (1997) Microbiol 143: 713-720.

Turner, L. R., Lara, J. C., Nunn, D. N. and Lory, S. (1993) J Bacteriol 175: 4962-4969.

Turner, L. R., Olson, J. W. and Lory, S. (1997) Mol Microbiol 26: 877-887.

Urban, A., Leipelt, M., Eggert, T. and Jaeger, K. E. (2001) J Bacteriol 183: 587-596.

Von Heijne, G. (1985) J Mol Biol 184: 99-105.

Voulhoux, R., Taupiac, M.P., Czjzek, M., Beaumelle, B. and Filloux, A. (2000) J Bacteriol 182: 4051-4058.

Voulhoux, R., Lazdunski, A. and Filloux, A. (2001) EMBO Reports 2: 49-54.

Walker, J. E., Saraste, M., Runswick, M. J. and Gay, N. J. (1982) EMBO J 1: 945-951.

Whitchurch, C. B. and Mattick, J. S. (1994) Mol Microbiol 13: 1079-1091.

Wong, K. R. and Buckley, J. T. (1989) Science 246: 654-656.

Wong, K. R., McLean, D. M. and Buckley, J. T. (1990) J Bacteriol 172: 372-376.

Wong, K. R. and Buckley, J. T. (1991) J Biol Chem 266: 14451-14456.

Chapter 9

THE TYPE III SECRETION PATHWAY IN PATHOGENIC BACTERIA

Claude Parsot

Unité de Pathogénie Microbienne Moléculaire
INSERM U389, Institut Pasteur
25 rue du Docteur Roux
75724 Paris Cedex 15, France

1. INTRODUCTION

The findings that *Yersinia* spp. release Yop proteins into the culture medium even though these proteins do not contain a classical N-terminal signal sequence (Heesemann *et al.*, 1986; Michiels *et al.*, 1990) and that their secretion requires a large set of "accessory" proteins (Michiels *et al.*, 1991), together with the observation that proteins required for the virulence of other bacteria exhibit striking sequence similarities with proteins required for Yop secretion (Andrews and Maurelli, 1992; Gough *et al.*, 1992), led to the characterization of what is now designated the Type III secretion (TTS) pathway (van Gijsegem *et al.*, 1993). This secretion pathway is present in numerous Gram-negative bacteria that are pathogenic for humans, animals, or plants. The pathway is required for pathogenicity of these bacteria. Some characteristic features of this secretion pathway include (1), the nature of the components and the structure of the secretion apparatus; (2), the regulation of the activity of the secretion apparatus by external signals; (3), the storage of the "secreted" proteins in the bacterial cytoplasm prior to their secretion; (4), the regulation of gene transcription by the activity of the secretion apparatus; and (5), the insertion of some proteins into the membrane of eukaryotic target cells and the injection of other proteins from the bacterial cytoplasm into the host cell cytoplasm, which is the real purpose of this secretion pathway. It should be emphasized that not all the aforementioned

B. Oudega (ed.), Protein Secretion Pathways in Bacteria, 167–190.

features have been demonstrated for, and might not even be common to all bacteria that have been proposed to use the TTS pathway.

An exhaustive and very comprehensible review covering all topics of the TTS pathway has been published by Hueck (1998). This represents a gold mine of references of original work and a synthesis of sometimes controversial results and interpretations from different laboratories. Numerous reviews have been published recently on the involvement of the TTS pathway in the pathogenicity of various bacteria (Collmer et al., 2000; Cornelis et al., 1998; Dale and Welburn, 2001; Darwin and Miller, 1999; Galan and Collmer, 1999; Plano et al., 2001; Rahme et al., 1995; Rockey et al., 2000), to cite a few.

2.   OVERVIEW

Given the variety of pathogens that use the TTS pathway for virulence and the variety of outcomes of the interactions between these pathogens and host cells, all TTS systems do not function exactly the same way. Nevertheless, data obtained from studies on various bacteria suggest a general model for the TTS pathway. There is no need to emphasize that the model presented below is a caricature and that it is proposed only as a general framework for the topics that are described in the following chapters. Approximately twenty genes encode the components of the secretion apparatus and specific transcriptional regulator(s). These genes are clustered on the chromosome (or on a plasmid) and exhibit a GC content that is different from the rest of the chromosome Expression of the specific regulator(s) is controlled by global regulators and is induced under conditions that might be relevant to conditions encountered by the pathogen in its host. Under these conditions, the specific regulator is produced and it activates transcription of genes encoding components of the secretion apparatus. The secretion apparatus is assembled in the bacterial envelope, albeit in an inactive, or poorly active, form. Adjacent to the operons encoding the secretion apparatus is a cluster of genes that exhibit the same GC content as that of secretion genes. These genes are transcribed under the same conditions as secretion genes and encode specific cytoplasmic chaperones and proteins that are secreted by the secretion apparatus. Since the secretion apparatus is not active, the "secreted" proteins are stored in the cytoplasmic compartment, some of them associated with chaperones. The secretion apparatus can be activated by external signals, such as contact of bacteria with eukaryotic cells, and targeting of proteins to the secretion apparatus allows their (supposedly) ordered passage through the conduit of the apparatus. Proteins are delivered to the immediate vicinity of the cell membrane, where some of them associate to form a channel - the translocator - within that membrane. Other secreted proteins - the effectors - follow and are transported in a one step process from the bacterial cytoplasm to the cell cytoplasm through the secretion apparatus and the translocator. Injected effectors interfere with cellular signaling cascades to subvert the cell defences and to elicit responses that are favorable for the pathogen. In addition, the increased activity of the secretion machinery leads to the activation of transcription of a set of genes, not necessarily linked to the secretion genes, that encode other effectors. These effectors are also injected into the eukaryotic cell, possibly to counteract the effects induced by the first wave of effectors or to modify the environment of the bacterium.

3. THE TTS APPARATUS

3.1 Distribution

The signature of the TTS pathway is in the sequence conservation exhibited by some components of the secretion apparatus, rather than in the behavior of bacteria that use this pathway to interact with their hosts or the activities of secreted proteins. There are numerous structural and functional similarities between components of the TTS apparatus and the flagellar export apparatus and the two systems appear to be variations on the same theme (Macnab, 1999; Nguyen *et al.*, 2000). Components of the TTS apparatus have been detected in numerous Gram-negative pathogenic bacteria, including pathogenic *Yersinia* spp.; *Shigella* spp., as well as the closely related enteroinvasive *Escherichia coli* (EIEC), various enteropathogenic *E. coli* spp. (EPEC, RDEC, STEC), pathogenic serotypes of *Salmonella enterica*, pathogenic *Bordetella* spp., *Pseudomonas aeruginosa*, *Chlamydia* spp., the insect endosymbiont *Sodalis glossinidus* (Dale *et al.*, 2001), various plant pathogens including *Pseudomonas syringae*, *Ralstonia solanacearum*, *Xanthomonas campestris*, and *Erwinia amylovora*, and the plant symbiont *Rhizobium*. These bacteria have different lifestyles within their hosts, ranging from extracellular (*Yersinia* and *Bordetella* spp., EPEC, plant pathogens), to facultative intracellular (*Salmonella* and *Shigella* spp.), obligate intracellular (*Chlamydia* spp.), and even endosymbiotic (*S. glossinidus*). Except for the related flagellar apparatus, the TTS pathway has not been detected in Gram-positive bacteria, which suggests that this pathway evolved in Gram-negative bacteria after they diverged from Gram-positive bacteria (Nguyen *et al.*, 2000).

Genes encoding components of the TTS apparatus are, in most cases, clustered on a pathogenicity island, *i. e.* a hot spot of integration of foreign elements (*Salmonella* spp. and EPEC) or on a plasmid (*Yersinia*, *Shigella*, *Ralstonia*, *Rhizobium*, and EIEC spp.). In contrast, genes for components of the TTS apparatus of *Chlamydia* spp. are dispersed over the chromosome. Both *Salmonella* and *Yersinia* spp. contain genes for two TTS machineries (Shea *et al.*, 1996; Haller *et al.*, 2000). The GC content of genes encoding components of the TTS apparatus is often lower than that of the rest of the chromosome, especially in the case of *Shigella* spp. (34% vs 52%) and EPEC (38% vs 52%). These observations, and the fact that the phylogeny derived from components of the TTS apparatus is not consistent with that derived from the 16 S RNA (Nguyen *et al.*, 2000) have led to the proposal that these secretion genes were acquired by horizontal transfer.

3.2 Components of the TTS apparatus

Each TTS apparatus is encoded by approximately 20 genes that are clustered in large operons. As discussed by Hueck (1998), the encoded proteins can be grouped into several classes, depending on the degree of conservation, *i. e.* proteins (1) present in all systems, (2) proteins shared only by closely related systems, and (3) proteins unique to each system. Extensive comparisons between components of the TTS machineries and the flagellar assembly machineries have been recently reviewed (Nguyen *et al.*, 2000).

According to the nomenclature in use for *Yersinia* spp., proteins that are conserved in all systems are the following: (i), YscC is an outer membrane protein that belongs to the family of secretins - the purified YscC protein forms a ring structure with an external diameter of 200 Å and a central pore of 50 Å (Koster *et al.*, 1997); (ii), YscJ is a lipoprotein that contains a hydrophobic C-terminal segment and might be anchored in both the outer and the inner membrane; (iii), LcrD is an inner membrane protein that contains 6 or 8 trans-membrane segments and a large C-terminal domain probably located within the cytoplasm; (iv), YscR, YscS, YscT, and YscU are proteins that are predicted to be localized in the inner membrane; (v), and YscN is a cytoplasmic protein that exhibits extensive sequence similarity with the β subunit of the F1 component of the bacterial F0F1 ATPase and that is likely to be involved in providing energy for secretion (Hueck, 1998; Cornelis and van Gijsegem, 2000). Another component that might be shared by all systems, although there are no obvious sequence similarities, is the protein corresponding to PrgH in *S. typhimurium* and MxiG in *S. flexneri*, since PrgH and MxiG are components of the needle complex (see below). They contain an internal hydrophobic segment that spans the inner membrane and separates the cytoplasmic N-terminal and the periplasmic C-terminal domains. With the exception of the secretin YscC, homologues of these conserved proteins are also present in the flagellar apparatus, where they are components of MS and C rings of the basal body. The function and localization of other proteins, which are unique to each TTS system or shared by only the most closely related ones, are, in most cases, not known. Some are components of the external part of the secretion apparatus or regulate the activity of secretion or the length of the external part of the apparatus.

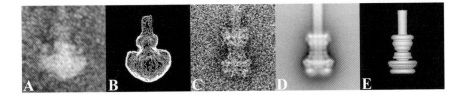

Fig. 1: Structural analysis of the *S. flexneri* TTS apparatus and needle complex by electron microscopy. **A**: Negative staining of one TTS apparatus on osmotically shocked bacteria. **B**: Deduced projection density map of an averaging of images as shown in A. **C**: Negative staining of one purified needle complex. **D**: Average image of a set of purified needle complexes. **E**: Surface representation of the volume of the needle complex, assuming cylindrical symmetry. This figure was designed by Eric Larquet and Pierre Gounon (Station Centrale de Microscopie Electronique, Institut Pasteur, Paris) and used materials presented in Blocker *et al.* (1999, 2000).

3.3   Structure of the TTS apparatus

The TTS apparatus of *Salmonella* and *Shigella* spp. has been visualized by electron microscopic analysis of osmotically-shocked bacteria (Kubori *et al.*, 2000; Tamano *et al.*, 2000; Blocker *et al.*, 2001). In *Shigella* spp., there are approximately 50 machineries that are evenly distributed on the surface of the bacterium. The structure consists of three parts; (1), an external "needle", 10 nm wide and 50 nm long, which protrudes from the surface of the bacterium; (2), a central neck, 21 nm wide and 10 nm long, which might span the bacterial envelope; and (3), a proximal bulb, 44 nm wide and 27 nm long, which is probably located within the cytoplasm (Blocker *et al.*, 1999). A part of the secretion apparatus, designated the needle complex, has been purified from the wild-type strain and various secretion-deficient mutants of *Salmonella* and *Shigella* spp. (Kimbrough and Miller, 2000; Kubori *et al.*, 2000; Tamano *et al.*, 2000; Blocker *et al.*, 2001; Sukhan *et al.*, 2001). The complex contains the needle attached to a structure composed of two rings, which is itself attached to a base composed of two rings. The needle complex contains a hollow conduit through which secreted proteins probably transit. SDS-PAGE analysis of the needle complex indicates that it is composed of four major proteins, which are conserved in *S. typhimurium* and *S. flexneri*: InvG/MxiD and PrgK/MxiJ, which belong to the YscC and YscJ families mentioned above, respectively; PrgH/MxiG, which might be the functional equivalent of YscD; and PrgI/MxiH, which is a 8 kDa protein related to YscF. In addition, the needle complex from *S. flexneri* contains MxiI, which has a size similar to and exhibits sequence similarities with MxiH. The equivalent of MxiI in *S. typhimurium* is PrgJ and that in *Yersinia* spp. might be YscJ. PrgI and MxiH are probably the main components of the needle, as mutants in which the genes encoding these proteins have been inactivated produce secretion machineries that do not have a needle. Since the purification procedure involves the use of detergents, the purified needle complex did not contain some components of the secretion apparatus located in the inner membrane and represents only the central part of the TTS apparatus.

Systematic electron microscopic analyses and biochemical characterization of the needle complex produced by numerous mutants gave indications on the protein interactions that occur within the needle complex and on the discrete steps of its assembly (Kimbrough and Miller, 2000; Sukhan *et al.*, 2001). Formation of a structure containing InvG, PrgH, and PrgK is not, or only moderately, affected by inactivation of genes encoding other components of the apparatus. In contrast, PrgI is absent in all the mutants that are defective for secretion, except the *invJ* mutant. Thus, the core of the needle complex consists of InvG, PrgH, and PrgK and can assemble in the absence of other components, whereas incorporation of PrgI, which is a substrate of the TTS apparatus, requires all other components. The *S. typhimurium invJ* mutant produces abnormally long needles, which might account for the secretion defect of this mutant. This suggests that the length of the needle is controlled by InvJ, which is also secreted by the TTS apparatus (Sukhan *et al.*, 2001). A similar phenotype has been observed in a *S. flexneri* strain overproducing MxiH (Tamano *et al.*, 2000).

Whereas the needle that extends from the surface of *Salmonella* and *Shigella* spp., is rather short (50 nm), much longer structures (up to 1 μm) have been detected at the surface of EPEC (Ebel *et al.*, 1998; Knutton *et al.*, 1998) and plant pathogens, including *P. syringae*, *R. solanacearum*, and *E. amylovora* (Roine *et al.*, 1997; van Gijsegem *et al.*, 2000; Jin *et al.*, 2001). Formation of these structures, which have been designated the EspA filaments (EPEC) and the Hrp pili (plant pathogens), requires a functional TTS apparatus. The EspA filament contains EspA and inactivation of *espA* abolishes induction of the attaching and effacing lesions on epithelial cells, although it does not affect secretion of other Esp proteins. In contrast, inactivation of *hrpA* in *E. amylovora*, which encodes one component of the Hrp pilus, abolished secretion of Hrp-secreted proteins. Labelling of Hrp pili by antibodies raised against the effector proteins HrpN and DspA/E confirmed that these structures are involved in the TTS pathway (Jin *et al.*, 2001). These long filaments might represent extensions of the TTS apparatus that allow these bacteria to interact with cells from a greater distance.

4.   REGULATION OF THE TTS APPARATUS

4.1   Expression of secretion genes

Transcription of secretion genes is tightly regulated and under the control of both global and specific regulators. A common environmental signal is the temperature of growth, *e. g.*, components of the TTS apparatus of *Yersinia*, *Shigella*, and EPEC spp. are expressed at 37°C, but not at temperatures below 30°C (Adler *et al.*, 1989; Lambert de Rouvroit *et al.*, 1992; Rosenshine *et al.*, 1996). On the other hand, expression of secretion genes in plant pathogens is often induced at lower temperatures. Osmolarity and pH of the growth medium also modulate expression of secretion genes.

Global regulators include (i), histone-like proteins, such as H-NS and IHF in EPEC and *Shigella* spp. (Hromockyj *et al.*, 1992; Porter and Dorman, 1997; Friedberg *et al.*, 1999; Bustamante *et al.*, 2001), YmoA in *Yersinia* spp. (Cornelis *et al.*, 1991), and Fis in *Salmonella* spp. (Wilson *et al.*, 2001); and (ii), two-components systems, such as OmpR/EnvZ and CpxR/CpxA in *Shigella* spp. (Bernardini *et al.*, 1990; Nakayama and Watanabe, 1995; Nakayama and Watanabe, 1998), OmpR/EnvZ, PhoP/PhoQ, and CsrA/CsrB in *Salmonella* spp. (Pegues *et al.*, 1995; Arricau *et al.*, 1998; Lucas *et al.*, 2000), or BvgA/BvgS in *Bordetella* spp. (Yuk *et al.*, 1998). Genes encoding these regulators are not linked to secretion genes and these proteins are thought to integrate diverse environmental parameters to signal to the pathogen that it has reached its host.

In addition to these global regulators, expression of secretion genes requires specific factors that are encoded within or in the immediate vicinity of operons for the TTS apparatus. These are; (i), transcriptional activators of the AraC family, such as VirF in *Yersinia* spp. (Cornelis *et al.*, 1989), ExsA in *P. aeruginosa* (Hovey and Frank, 1995), InvF, HilC, and HilD in *Salmonella* spp. (Kaniga *et al.*, 1994; Eichelberg *et al.*, 1999; Schechter *et al.*, 1999; Lucas and Lee, 2001), VirF in *Shigella* spp. (Sakai *et al.*, 1988; Kato *et al.*, 1989), PerA/BfpW in EPEC (Tobe *et al.*, 1996), HrpB in *R. solanacearum* (Genin *et al.*, 1992), and HrpXv in *X. campestris* (Wengelnik and Bonas, 1996); (ii), response regulators of two components systems, such as HilA and SsrB in *Salmonella* spp. (Bajaj *et al.*, 1995; Valdivia and Falkow, 1997), HrpR and HrpS in *P. syringae* and *E. amylovora* (Grimm and Panopoulos, 1989), and HrpX and HrpY in *E. amylovora* (Wei *et al.*, 2000); (iii), alternative sigma factors, such as HrpL in *P. syringae* and *E. amylovora* (Xiao *et al.*, 1994; Wei and Beer, 1995; Wei *et al.*, 2000); and (iv), H-NS related proteins, such as Ler in EPEC (Bustamante *et al.*, 2001). Expression of these specific regulators is under the control of the aforementioned global regulators and the environmental signals, if any, which might modulate their activity, are not known. It is conceivable that the main control is on expression of the specific regulators and that, once they are produced, they activate transcription of their target genes without further control.

4.2   Regulation of the activity of secretion

The secretion apparatus is not, or only weakly, active when bacteria are grown in laboratory media. In the case of *Shigella* spp., only a small proportion (approximately 5%) of IpaA-D proteins is secreted by bacteria during growth in laboratory media (Allaoui *et al.*, 1993) and secretion of pre-synthesized Ipa proteins is induced upon contact of bacteria with epithelial cells (Menard *et al.*, 1994a). By using a luciferase reporter system controlled by the activity of secretion, Pettersson *et al.* (1996) showed that the TTS apparatus of *Yersinia* spp. is activated in bacteria that are in contact with target cells but not in bacteria that adhere to the plastic dish. However, because not all bacteria do adhere to cells and the bacterial population is heterogeneous, quantitative studies of secretion have been carried out mostly with bacteria growing in laboratory media and, in this case, the activity of the TTS apparatus is induced by various artificial means. Inducing conditions include the removal of calcium from the growth medium of *Yersinia* spp. (Michiels *et al.*, 1990), the addition of the dye Congo red to the growth medium of *Shigella* spp. (Parsot *et al.*, 1995), and even the exposure to bovine serum albumine at an acidic pH for *X. campestris* (Rossier *et al.*, 1999). These inducing conditions have little biological relevance and the mechanisms by which these "inducers" activate the TTS apparatus have not been characterized.

Genetic analyses led to the characterization of mutants in which the secretion apparatus was no longer regulated. For example, inactivation of *yopN* or *lcrG* in *Yersinia* spp. led to secretion of Yop proteins in the presence of calcium (Forsberg *et al.*, 1991; Skryzpek and Straley, 1993), and inactivation of *ipaB* or *ipaD* in *Shigella* spp. led to secretion of Ipa proteins in the absence of Congo red (Menard *et al.*, 1994a). The YopN and LcrG proteins and the IpaB and IpaD proteins are themselves secreted by the type III pathway, and it has been proposed that these proteins are either components of or required for the formation of a plug for the secretion apparatus. YopN is exposed on the surface of bacteria prior to contact with target cells and might act as a sensor regulating the opening of the secretion apparatus (Forsberg *et al.*, 1991). Assuming that TTS machineries are evenly distributed on the bacterial surface, interaction of a bacterium with a cell could lead to activation of either only the machineries that contact the cell or all the machineries that are present in this bacterium. Results obtained with *Yersinia* spp. indicate that proteins are transferred mostly to the target cell and not secreted to the external milieu, which suggests that secretion is directional, *i. e.*, that the TTS machineries that are not in contact with the cell are not activated. (Rosqvist *et al.*, 1994; Persson *et al.*, 1995).

4.3   Regulation of gene expression by the activity of secretion

In *S. typhimurium*, expression of genes involved in late stages of assembly of the flagellum is controlled by the activity of the hook-basal body complex that exhibits structural and functional similarities with the TTS systems used by pathogenic bacteria. Transcription of the flagellin gene is dependent on a dedicated sigma factor, σ^{28}, the activity of which is regulated by an anti-sigma factor, FlgM. FlgM binds to σ^{28} and prevents its association with the core RNA polymerase. Upon completion of the hook-basal body complex, FlgM is secreted and σ^{28} can associate with the core RNA polymerase to transcribe the late genes (Chilcott and Hughes, 2000; Hughes *et al.*, 1993).

In the case of *Yersinia* spp., contact of bacteria with target cells and incubation in a medium depleted of calcium not only lead to secretion (or translocation) of Yop proteins but also to an increased transcription of *yop* genes (Forsberg and Wolf-Watz, 1988). Moreover, in contrast to the wild-type strain, secretion deficient mutants do not express Yop proteins in the absence of calcium (Allaoui *et al.*, 1995). The control of transcription of *yop* genes by the activity of secretion has been designated the "feed-back control". Repression of *yop* transcription in wild-type *Y. tuberculosis* growing in the presence of calcium or in secretion deficient mutants requires the LcrQ protein that can be secreted by the TTS apparatus (Rimpilainen *et al.*, 1992). Under conditions of non-secretion, the *lcrQ* gene is expressed and LcrQ accumulates in the cytoplasm, whereas, under conditions of secretion, LcrQ is secreted, which leads to a decreased concentration of the protein in the cytoplasm (Pettersson *et al.*, 1996). In *Y. enterocolitica*, the *lcrQ* gene is designated *yscM1* and the pYV plasmid also encodes YscM2 that exhibits extensive sequence similarities with YscM1. Inactivation of both the *yscM1* and *yscM2* is required to deregulate transcription of *yop* genes in a secretion deficient mutant (Stainier *et al.*, 1997). Although LcrQ is clearly involved in the feed-back control, its effect on transcription might be indirect since this protein does not exhibit features that are characteristic of a DNA binding protein. In the case of *S. flexneri*, transcription of a subset of genes encoding secreted proteins is also regulated by the activity of the TTS apparatus. In addition to IpaA, B, C, and D, approximately 15 proteins are secreted by the TTS apparatus, including VirA and members of the IpaH family (Demers *et al.*, 1998). Whereas the *ipaA-D* genes are transcribed whether or not secretion is active, the *virA* and *ipaH* genes are transcribed only in conditions of active secretion, *i. e.* following contact with epithelial cells, exposure to Congo red, or inactivation of *ipaB* or *ipaD* (Demers *et al.*, 1998). Regulatory proteins involved in this control have yet to be identified.

In each of the systems described above, transcription of genes encoding some (not all) secreted proteins appears to be controlled by the activity of secretion. The mechanisms of this control are probably different in each case, since no proteins homologous to σ^{28} or FlgM are encoded by the virulence plasmids of *Yersinia* and *Shigella* spp. and, likewise, no proteins homologous to LcrQ are encoded by the *S. flexneri* plasmid. Although the mechanisms might be different, the differential expression of secreted proteins might be a general feature of the TTS pathway.

5.   THE CYTOPLASMIC CHAPERONES

The fact that the activity of secretion is regulated, imposes particular constraints, among which is the storage of proteins in the cytoplasm prior to their secretion. In some cases, this storage involves specific chaperones that associate with secreted proteins until they are secreted. Genes encoding the chaperones are located in the vicinity of genes encoding the proteins to which the chaperones bind. In *Yersinia* spp., the first chaperone to be identified was SycE (YerA) that associates with YopE (Wattiau and Cornelis, 1993; Frithz-Lindsten *et al.*, 1995). Chaperones for YopH, YopT, and YopN have been designated SycH, SycT, and SycN (Wattiau *et al.*, 1994; Day and Plano, 1998; Iriarte and Cornelis, 1998; Iriarte and Cornelis, 1999). These chaperones bind only one protein, or two proteins that exhibit sequence similarities (Cambronne *et al.*, 2000). In contrast, the SycD chaperone binds two unrelated proteins, YopB and YopD (Neyt and Cornelis, 1999b). Dependent on the function of the protein with which they associate, chaperones might be classified into two families: the SycE family includes chaperones that bind to effector proteins and the SycD family includes chaperones that bind to translocators. Although these chaperones do not share sequence similarities, they exhibit common features such as a size of approximately 15 kDa, a pI of approximately 5.0 , and a putative amphiphilic α-helix in their C-terminal part (Wattiau *et al.*, 1996).

The binding sites of SycE and SycH are located between residues 20 to 70 of YopE and YopH, respectively (Frithz-Lindsten *et al.*, 1995; Sory *et al.*, 1995; Schesser *et al.*, 1996). Inactivation of the gene for the chaperone impairs secretion of the protein(s) with which the chaperone is normally associated, and, in most cases, affects its stability (Wattiau *et al.*, 1994). In contrast to the wild-type protein, a recombinant YopE protein lacking the binding site for SycE is secreted by a *sycE* mutant and, likewise, a YopH protein lacking the SycH binding site is secreted by a *sycH* mutant (Woestyn *et al.*, 1996). Accordingly, chaperones do not act as pilots that are absolutely required to present their substrates to the secretion apparatus. Indeed, the first 20 residues of YopE are sufficient to allow secretion of recombinant proteins, whether or not SycE is present. However, in the absence of this N-terminal secretion signal, a substantial amount of YopE can still be secreted and, in this case, secretion is dependent on SycE (Cheng *et al.*, 1997; Cheng and Schneewind, 1999). Therefore, depending on the context, the chaperone may or may not be involved in secretion of its target protein. Chaperones of the SycE family that have been identified in other bacteria include: (i), SicP, the chaperone for StpP, and SigE, the chaperone for SigD (SopB), in *Salmonella* spp. (Fu and Galan, 1998; Darwin *et al.*, 2001); (ii), IpgE, the chaperone of IpgD in *Shigella* spp. (Niebuhr *et al.*, 2000); and (iii), CesT, the chaperone for intiminin in EPEC (Abe *et al.*, 1999; Elliott *et al.*, 1999).

In contrast to chaperones of the SycE family that associate with only one protein and bind a discrete region of this protein, chaperones of the SycD family associate with two unrelated proteins, or a complex of several proteins, and might bind to several regions of these proteins (Francis *et al.*, 2000). Proteins of the SycD family share significant sequence similarities and have been detected in *Yersinia* spp. (SycD/LcrH), *Shigella* spp. (IpgC), *Salmonella* spp. (SicA in SPI1 and SscB in SPI2), *P. aeruginosa* (PrcH), EPEC (CesD), and *Chlamydia* spp. (SycD), but not in plant pathogens. Despite these sequence similarities and the fact that these chaperones are required for stability of proteins with which they associate, results obtained on different systems are surprisingly different. In *Yersinia* spp., SycD is associated with a complex that contain both YopB and YopD (Neyt and Cornelis, 1999b); in *Shigella* spp., IpgC is associated independently with IpaB and IpaC (Menard *et al.*, 1994b); in *Salmonella* spp., SicA is associated with SipB and SipC and, in addition, is required for the activity of the transcriptional activator InvF (Darwin and Miller, 1999; Tucker and Galan, 2000); and, in EPEC, CesD is associated only with EspD, although it is also required for stability of EspB (Wainwright and Kaper, 1998).

Although chaperones have essential roles, their function has yet to be elucidated. The initial proposals that chaperones are pilot proteins that address secreted proteins to the TTS apparatus (Wattiau *et al.*, 1994) or anti-scaffolding proteins that prevent the premature association of proteins within the cytoplasm prior to their secretion (Menard *et al.*, 1994b) have not received much support, and have not been really contradicted by recent studies. Actually, chaperones might have different roles at different times, *i. e.* during storage and secretion. Indeed, delivery of pre-synthesized YopE within target cells is strictly dependent on SycE (YerA), whereas delivery of the YopE molecules that are synthesized following contact of bacteria with cells does not require SycE (Lloyd *et al.*, 2001). In the former case, YopE is stored prior to its translocation whereas, in the latter case, YopE is addressed to the active secretion apparatus as soon as it is synthesized. Finally, it should be emphasized that not all proteins that are secreted by the TTS pathway have a dedicated chaperone. In plant pathogens, the only example of a chaperone is DspB that is required for secretion of DspA by *E. amylovora* (Gaudriault *et al.*, 1997). The virulence plasmid of *Shigella* spp. encodes 25 secreted proteins and only 2 specific chaperones (Buchrieser *et al.*, 2000; Niebuhr *et al.*, 2000). Therefore, secreted proteins might also be classified into two groups depending on whether or not they associate with a chaperone. Schematically, components of the translocator seem to have a chaperone of the SycD family and only a subset of effectors do have a chaperone of the SycE family. One of the functions of chaperones might be to regulate the order in which proteins are addressed to the secretion apparatus (Boyd *et al.*, 2000).

6.   THE SECRETION SIGNAL(S)

Numerous proteins have been identified as being secreted by TTS machineries from various bacteria, and heterologous secretion has been observed in several cases; YopE from *Yersinia* spp. can be secreted by *S.*

typhimurium and by *X. campestris* (Rosqvist *et al.*, 1995a; Rossier *et al.*, 1999), some proteins of *C. pneumoniae* can be secreted by *S. flexneri* and *Y. enterocolitica* (Fields and Hackstadt, 2000; Subtil *et al.*, 2001), IpaB from *S. flexneri* can be secreted by *Y. pseudotuberculosis* and SipB from *S. typhimurium* can be secreted by *S. flexneri* (Hermant *et al.*, 1995; Rosqvist *et al.*, 1995a), and AvrB and AvrPto from *P. syringae* can be secreted by *Yersinia* spp. (Anderson *et al.*, 1999). This indicates that the secretion signal carried by these proteins is recognized by distantly related systems. However, the nature of this secretion signal is still controversial and independent signals might even be present on the same protein.

Initial studies indicated that the first 15 to 17 residues of YopH, YopE, YopN, or YopQ are sufficient to allow hybrid proteins to be secreted by *Yersinia* spp. (Sory *et al.*, 1995; Anderson and Schneewind, 1997, 1999). Likewise, the signal necessary for secretion of AvrB and AvrPto by the *E. chrysanthemi* TTS apparatus expressed in *E. coli* is contained within the first 15 residues of the proteins (Anderson *et al.*, 1999). The N-terminal secretion signal of YopE and YopH does not encompass the binding site for the SycE and SycH chaperones and the efficiency of secretion of hybrid proteins carrying this signal is not dependent on the presence of the chaperones (Sory *et al.*, 1995). However, removal of this N-terminal signal from YopE did not completely abolish secretion of the recombinant protein, which led to the identification of a second secretion signal corresponding to the SycE binding site on YopE (Cheng *et al.*, 1997). Secretion mediated by this second signal is dependent on SycE (YerA).

No conserved motif has been identified between the N-terminal regions of proteins that are secreted by TTS machineries. Furthermore, mutations that change the reading frame of the DNA region encoding the N-terminal part of YopE, YopN, or YopQ hybrid proteins do not affect secretion of encoded proteins (Anderson and Schneewind, 1997, 1999). This was taken as evidence that the secretion signal is present in the messenger RNA rather than in the N-terminal part of the secreted protein and that secretion of the protein is coupled to translation of the mRNA. However, this hypothesis has recently been challenged by Lloyd *et al.* (2001) who mutagenized the first 10 codons of YopE to change the nucleotide sequence of the 5' region of the mRNA without changing the N-terminal sequence of the protein and showed that this does not affect secretion of the encoded protein. Accordingly, neither the sequence of the 5' part of the messenger RNA nor the sequence of the N-terminal part of the protein seem to be essential features of the secretion signal! Furthermore, the N-terminal sequence of YopE can be replaced by a synthetic sequence consisting of seven alternating serine and isoleucine residues without impairing secretion of the protein (Lloyd *et al.*, 2001). Thus, the N-terminal secretion signal can be constituted of an unstructured amphipathic segment. These results raise a lot of questions: to what extent does this segment have to be amphipathic and are there any positions that should or should not contain specific residues? How can such a signal confer specificity? Are differences between the N-terminal sequences of secreted proteins involved in the differential sorting of these proteins to the secretion apparatus? In other words, although we know "where" the secretion signal is, we still do not know "what" it is.

An implication that the secretion signal might be located within the 5' end of the mRNA was that secretion should be co-translational, *i. e.* that translation of the mRNA for secreted proteins should occur in the immediate vicinity of the secretion apparatus and proteins should not be stored prior to secretion. This is not consistent with the observation that pre-synthesized Yop and Ipa proteins are secreted when bacteria are exposed to secretion inducing conditions. However, results on secretion of FlgM by the flagellar system suggest that, in some cases, secretion is co-translational (Aldridge and Hughes, 2001; Karlinsey *et al.*, 2000). Whether or not this example is unique to the flagellar system or reflects a general mechanism has to await further investigation.

7. THE SECRETED PROTEINS

Proteins that transit through the secretion apparatus can be considered as secreted by (or targets of) the TTS pathway. However, these proteins belong to very different functional categories. Some are components of the external part of the TTS apparatus or are involved in its regulation and these are addressed to the secretion apparatus during its construction or prior to its completion. The other secreted proteins include components of the translocator that insert within the cell membrane to form a channel and effectors that are delivered to the cell cytoplasm through that channel. Accordingly, translocators should transit prior to effectors. However, little is known on the hierarchy by which proteins are addressed to the secretion apparatus.

7.1 The translocator

YopE purified from the culture medium does not exhibit cytotoxic activity towards cultured cells, whereas YopE is cytotoxic when micro-injected into cells (Rosqvist *et al.*, 1991). These observations led to the concept that YopE should be injected by extracellular bacteria into the target cell. Injection, or translocation, of Yop proteins by *Yersinia* spp. has been demonstrated by pioneering studies that involved the use of either immunofluorescence analyses (Rosqvist *et al.*, 1994; Persson *et al.*, 1995) or hybrid proteins consisting of Yop proteins fused to the adenylate cyclase (Cya) of *B. pertussis* (Sory and Cornelis, 1994). Translocation of hybrid Yop-Cya proteins was monitored by assaying the intracellular cAMP resulting from activation of Cya by the intracellular calmodulin. Genetic analyses pointed to genes whose products are necessary for translocation to occur (Rosqvist *et al.*, 1995b; Boland *et al.*, 1996). These included components of the TTS apparatus, the inactivation of which abolished secretion of Yop proteins *in vitro*, and proteins that are encoded by the *lcrG-lcrV-sycD-yopB-yopD* operon. LcrG, LcrV, YopB, and YopD are secreted by the TTS pathway. Analysis of the function of these proteins in translocation has been complicated by the fact that these proteins, together with SycD, are also involved in the control of the activity of secretion or in the negative feed-back control on transcription of *yop* genes.

YopB and YopD have hydrophobic regions that could potentially allow these proteins to interact with or insert into lipid membranes. Similarly, IpaB and IpaC of *Shigella* spp., SipB and SipC in *Salmonella* spp., and EspB and EspD of EPEC have central hydrophobic regions and, as indicated above, these proteins are associated with chaperones of the SycD family. This suggests that these proteins might be functionally equivalent, even though they share little sequence similarity. Two series of experiments support the hypothesis that YopB and YopD, or the equivalent proteins in other bacteria, form a pore in the membrane of target cells. Firstly, the wild-type strains of *Yersinia*, *Shigella*, and EPEC spp. possess a contact-dependent haemolytic activity on erythrocytes and haemolysis can be prevented by addition of sugars of a defined radius, which suggests that it results from insertion of a pore into the membrane of erythocytes (Hakansson *et al.*, 1996b; Blocker *et al.*, 1999; Neyt and Cornelis, 1999a). Secretion deficient and *yopB* and *yopD* mutants are devoid of haemolytic activity. Secondly, association of YopB and YopD with artificial liposomes was demonstrated following incubation of bacteria in the presence of lipids in secretion-inducing conditions. Moreover, the liposome-associated YopB and YopD proteins were able to induce the formation of channels within a planar bilayer (Tardy *et al.*, 1999). These channels have a 105-pS conductance and no ionic selectivity. The number of YopB and YopD subunits that constitute such a channel is not known. It is likely that these channels correspond to the pores responsible for haemolysis and the proposed translocator.

As an extension of the secretion apparatus, the translocator is proposed to insert into the cytoplasmic membrane of host cells, or into the membrane of the vacuole in which bacteria reside. This would form a continuous conduit through which effectors transit from the cytoplasm of the bacterium to that of the cell. In theory, once this structure is installed, proteins engaged at one end of the secretion apparatus should end up at the other end. If this happens for components of the translocator, these proteins would also be translocated and could have additional functions as effectors. Indeed, YopD, a component of the translocator, is translocated within epithelial cells (Francis and Wolf-Watz, 1998). IpaB and IpaC of *Shigella* spp. are associated with the membrane of erythrocytes (Blocker *et al.*, 1999) and are the putative components of the translocator. Each of these proteins has additional functions; purified IpaB is able to bind and activate the cysteine-protease caspase I (ICE), which induces apoptosis in infected macrophages (Chen *et al.*, 1996), and purified IpaC is able to activate rearrangements of the cell cytoskeleton in semi-permeabilized cells (Tran Van Nhieu *et al.*, 1999). In *Salmonella* spp., SipB and SipC are also involved in apoptosis and actin rearrangements (Hayward and Koronakis, 1999; Hersh *et al.*, 1999). It is not known whether the cellular effects induced by these components of the translocator are due to some of their domains that are exposed to the cytoplasm of the cell once the translocator is inserted in the cell membrane or to molecules that are themselves injected within the cell.

7.2   The repertoire of effectors

Effectors are proteins that are injected into host cells and a number of these has been identified using various means, including the use immunochemistry and recombinant proteins fused to Cya, the analysis of mutants, and sequence comparisons (see Cornelis, 2000 #1373, for a review). For pathogens of humans and animals, these include YopE, YopH, YopM, YopJ/P, YopO/YpkA, and YopT of *Yersinia* spp., AvrA, SipA, SopB/SigD, SopD, SopE, SopE2, SptP, SspH1, and SspH2 of *Salmonella* spp., IpaA and IpgD of *Shigella* spp., ExoS, ExoT, ExoU, ExoY of *P. aeruginosa*, and Tir/EspE from EPEC/EHEC. For pathogens of plants, the numerous Avr proteins that, depending on the race of the host, are involved either in virulence of bacteria or in triggering a resistance response from the plant are also likely effectors, even though their injection has not yet been visualized.

In contrast to genes for components of the TTS apparatus and translocators that are clustered together within a pathogenicity island, genes for the known or suspected effectors are often dispersed over the chromosome on small pathogenicity islets and one is even carried by a phage (Hardt *et al.*, 1998; Mirold *et al.*, 2001). There is no common motif for the secretion signal of secreted proteins, all effectors do not have a dedicated chaperone, and, except in some cases, effectors used by various pathogens do not exhibit sequence similarities. Accordingly, even when the entire sequence of the genome of the bacterium has been elucidated, sequence analysis alone is not sufficient to identify genes encoding potential effectors. Proteins secreted by bacteria growing in broth are likely effectors but might represent only a subset of these, since some genes encoding secreted proteins might be expressed only in conditions of active secretion (which might not be obtained in laboratory media) or only in the host. Finally, effectors are often encoded by multigene families and inactivation of the gene encoding one member of the family might not lead to a recognizable phenotype. Therefore, in most cases, the full repertoire of effectors that are used by a bacterium during infection is not known. In both *Yersinia* and *Shigella* spp., the TTS machineries are encoded by a virulence plasmid and all the proteins that are secreted by these machineries are also encoded by the plasmids (Cornelis *et al.*, 1998; Buchrieser *et al.*, 2000). However, genes for a second TTS apparatus have been identified on the chromosome of *Yersinia* spp. (Haller *et al.*, 2000), and DNA hybridizing with virulence plasmid-borne genes that encode secreted proteins are present on the chromosome of *Shigella* spp. (Buysse *et al.*, 1995).

7.3   The activities of effectors

Various activities have been characterized for a number of effectors and are described in recent reviews (Bogdanove and Martin, 2000; Collmer *et al.*, 2000; Cornelis, 2000; Galan and Zhou, 2000; Tran Van Nhieu *et al.*, 2000; Vallance and Finlay, 2000). Briefly, YopH of *Yersinia* spp. and the C-terminal domain of StpP of *Salmonella* spp. are phosphotyrosine phosphatases (Persson *et al.*, 1997; Fu and Galan, 1999); SopB of *Salmonella* spp. is an inositol phosphate phosphatase (Norris *et al.*, 1998); YopO/YpkA of *Yersinia* spp. is a serine-threonine kinase (Hakansson *et al.*, 1996a); ExoS and ExoT of *P. aeruginosa* and SpvB of *Salmonella* spp. are ADP-ribosyltransferases (Finck-Barbancon *et al.*, 1997; Tezcan-Merdol *et al.*, 2001); and ExoY of *P. aeruginosa* is an adenylate cyclase (Yahr *et al.*, 1998). In addition, some effectors act by interacting with cellular proteins and modulating their activity. The N-terminal domain of StpP of *Salmonella* spp. corresponds to a GTPase activating protein (GAP) for Cdc42 and Rac1 (Fu and Galan, 1999) and the crystal structure of the StpP-Rac1 complex has been recently determined (Stebbins and Galan, 2000). A similar GAP activity, although with a different substrate specificity, has been detected for

YopE and ExoS, both of which share sequence similarities with StpP (Goehring *et al.*, 1999; Von Pawel-Rammingen *et al.*, 2000). SopE and SopE2 of *Salmonella* spp. are exchange factors that catalyze GDP/GTP exchange on Cdc42 and Rac1 (Rudolph *et al.*, 1999; Stender *et al.*, 2000). SipA of *Salmonella* and IpaA of *Shigella* spp. interact directly with actin and vinculin, respectively (Tran van Nhieu *et al.*, 1997; Bourdet-Sicard *et al.*, 1999; Zhou *et al.*, 1999). In plant pathogens, there is strong evidence that some Avr proteins, including AvrBs3, are transported to the cell nucleus (van den Ackerveken *et al.*, 1996), and this might also be the case for effectors of animal pathogens, such as YopM of *Yersinia* spp. and related proteins from *Salmonella* and *Shigella* spp. (Skrzypek *et al.*, 1998).

Whereas effectors injected by *Salmonella* and *Shigella* spp. lead to bacterial internalization in epithelial cells, those injected by *Yersinia* spp. prevent phagocytosis of bacteria by macrophages. Remarkably, *Salmonella* spp. has effectors that up- and down-regulate the activity of Cdc42 and Rac1, two molecular switches that are involved in organization of the actin cytoskeleton (Hall, 1998). Injection of SopE in epithelial cells triggers actin rearrangements that lead to internalization of bacteria and injection of StpP inhibits these rearrangements and allows the cell cytoskeleton to recover (Fu and Galan, 1999). The temporally co-ordinated and antagonistic actions of SopE and StpP strongly suggest that their delivery to the host cell is temporally co-ordinated.

The effector Tir of EPEC represents a fascinating example of a bacterial protein that is inserted within the cell membrane to serve as a receptor for another protein, intimin, which is located in the outer membrane of the bacterium (Kenny *et al.*, 1997). Tir has two predicted trans-membrane segments and the N- and the C-terminal regions of Tir are exposed on the cytoplasmic side of the cell membrane whereas the domain located between the trans-membrane regions is exposed on the external side and interacts with intimin (de Grado *et al.*, 1999; Hartland *et al.*, 1999). It is not known whether Tir is inserted within the membrane of the cell following its translocation within the cytoplasm of the cell or by lateral transfer from the translocator. Insertion of effectors within the cell membrane might not be restricted to Tir and could occur for the Inc proteins of *Chlamydia* spp. that are associated with the membrane of the inclusion in which bacteria multiply. Indeed, these Inc proteins also have an internal hydrophobic segment and have been proposed to be secreted by the TTS pathway (Subtil *et al.*, 2001).

8.   CONCLUSIONS AND PERSPECTIVE

The elucidation of the TTS pathway has been carried out at a tremendous pace, going from the analysis of Yop secretion in the early 90s to the visualization of the TTS apparatus in *Salmonella* and *Shigella* spp., from the demonstration of the injection of Yop proteins within the cell to the detection of the YopB-YopD channel in lipids, and, more recently, to the elucidation of the structure of StpP associated with Rac1. Results, methods, and concepts issued from the study of a particular pathogen have been rapidly transposed to the study of other pathogens and, notwithstanding the differences that exist between the various systems, this approach has proven to be very successful.

We have now a general view on how the TTS pathway functions, but there are numerous points that remain to be elucidated, including, to cite only a few, what are the protein interactions within the TTS apparatus and what is the mechanism by which proteins transit through the apparatus? How is the activity of the TTS apparatus regulated and what are the signals that activate secretion upon contact? How is the activity of secretion perceived and transmitted to the transcription apparatus to activate expression of effectors? What exactly are the roles of chaperones? What are the secretion signals carried by secreted proteins and how are these signals recognized by the TTS apparatus? What is the hierarchy in addressing proteins to the secretion apparatus and how is it achieved? What are the functions of effectors and why are so many effectors produced by a bacterium?

The TTS pathway injects bacterial proteins into cells upon contact and has been referred to as a device for close combat (Cornelis, 2000). This is clearly different from other secretion pathways that release toxins in the external milieu. These toxins are long range weapons acting at distance from the site of infection. In contrast, effectors of the TTS pathway, which are sometimes presented as toxins, act only on cells to which bacteria adhere. In addition, these effectors are rarely lethal for the cell. Therefore, one may wonder whether the TTS pathway could be a device for "communication with" rather than "combat against" eukaryotic cells. The number of effectors produced by bacteria using the TTS pathway suggests that this dialogue is not restricted to a deadly kiss, and the observation that the TTS pathway is present in symbionts opens new perspectives concerning its use outside the (quite restricted) field of pathogenesis. In most cases, there is evidence that TTS systems had been acquired by lateral transfer, which suggests that many more bacteria have a TTS pathways to interact with eukaryotic organisms. It is estimated that less than 1% of bacterial species have been identified. There is no doubt that research in the next few years will reveal

186

new fascinating aspects of this communication/combat device between
bacteria and cells.

9. REFERENCES

Abe, A., de Grado, M., Pfuetzner, R.A., Sanchez-Sanmartin, C., Devinney, R., Puente, J.L.,
  Strynadka, N.C. and Finlay, B.B. (1999) Mol Microbiol 33: 1162-1175.
Adler, B., Sasakawa, C., Tobe, T., Makino, S., Komatsu, K. and Yoshikawa, M. (1989) Mol
  Microbiol 3: 627-635.
Aldridge, P. and Hughes, K.T. (2001) Trends Microbiol 9: 209-214.
Allaoui, A., Sansonetti, P.J. and Parsot, C. (1993) Mol Microbiol 7: 59-68.
Allaoui, A., Schulte, R. and Cornelis, G.R. (1995) Mol Microbiol 18: 343-355.
Anderson, D.M., Fouts, D.E., Collmer, A. and Schneewind, O. (1999) Proc Natl Acad Sci
  USA 96: 12839-12843.
Anderson, D.M. and Schneewind, O. (1997) Science 278: 1140-1143.
Anderson, D.M. and Schneewind, O. (1999) Mol Microbiol 31: 1139-1148.
Andrews, G.P. and Maurelli, A.T. (1992) Infect Immun 60: 3287-3295.
Arricau, N., Hermant, D., Waxin, H., Ecobichon, C., Duffey, P.S. and Popoff, M.Y. (1998)
  Mol Microbiol 29: 835-850.
Bajaj, V., Hwang, C. and Lee, C.A. (1995) Mol Microbiol 18: 715-727.
Bernardini, M.L., Fontaine, A. and Sansonetti, P.J. (1990) J Bacteriol 172: 6274-6281.
Blocker, A., Gounon, P., Larquet, E., Niebuhr, K., Cabiaux, V., Parsot, C. and Sansonetti, P.
  (1999) J Cell Biol 147: 683-693.
Blocker, A., Jouihri, N., Larquet, E., Gounon, P., Ebel, F., Parsot, C., Sansonetti, P. and
  Allaoui, A. (2001) Mol Microbiol 39: 652-663.
Bogdanove, A.J. and Martin, G.B. (2000) Proc Natl Acad Sci USA 97: 8836-8840.
Boland, A., Sory, M.P., Iriarte, M., Kerbourch, C., Wattiau, P. and Cornelis, G.R. (1996)
  EMBO J 15: 5191-5201.
Bourdet-Sicard, R., Rudiger, M., Jockusch, B.M., Gounon, P., Sansonetti, P.J. and Nhieu,
  G.T. (1999) EMBO J 18: 5853-5862.
Boyd, A.P., Lambermont, I. and Cornelis, G.R. (2000) J Bacteriol 182: 4811-4821.
Buchrieser, C., Glaser, P., Rusniok, C., Nedjari, H., D'Hauteville, H., Kunst, F., Sansonetti, P.
  and Parsot, C. (2000) Mol Microbiol 38: 760-771.
Bustamante, V., Santana, F., Calva, E. and Puente, J. (2001) Mol Microbiol 39: 664-678.
Buysse, J.M., Hartman, A.B., Strockbine, N. and Venkatesan, M. (1995) Microb Pathog 19:
  335-349.
Cambronne, E.D., Cheng, L.W. and Schneewind, O. (2000) Mol Microbiol 37: 263-273.
Chen, Y., Smith, M.R., Thirumalai, K. and Zychlinsky, A. (1996) EMBO J 15: 3853-3860.
Cheng, L.W., Anderson, D.M. and Schneewind, O. (1997) Mol Microbiol 24: 757-765.
Cheng, L.W. and Schneewind, O. (1999) J Biol Chem 274: 22102-22108.
Chilcott, G.S. and Hughes, K.T. (2000) Microbiol Mol Biol Rev 64: 694-708.
Collmer, A., Badel, J.L., Charkowski, A.O., Deng, W.L., Fouts, D.E., Ramos, A.R., Rehm,
  A.H., Anderson, D.M., Schneewind, O., van Dijk, K. and Alfano, J.R. (2000) Proc Natl
  Acad Sci USA 97: 8770-8777.
Cornelis, G., Sluiters, C., de Rouvroit, C.L. and Michiels, T. (1989) J Bacteriol 171: 254-262.
Cornelis, G.R. (2000) Proc Natl Acad Sci USA 97: 8778-8783.
Cornelis, G.R., Boland, A., Boyd, A.P., Geuijen, C., Iriarte, M., Neyt, C., Sory, M.P. and
  Stainier, I. (1998) Microbiol Mol Biol Rev 62: 1315-1352.

Cornelis, G.R., Sluiters, C., Delor, I., Geib, D., Kaniga, K., Lambert de Rouvroit, C., Sory, M.P., Vanooteghem, J.C. and Michiels, T. (1991) Mol Microbiol 5: 1023-1034.

Dale, C. and Welburn, S.C. (2001) Int J Parasitol 31: 627-630.

Dale, C., Young, S.A., Haydon, D.T. and Welburn, S.C. (2001) Proc Natl Acad Sci USA 98: 1883-1888.

Darwin, K.H. and Miller, V.L. (1999) Clin Microbiol Rev 12: 405-428.

Darwin, K.H., Robinson, L.S. and Miller, V.L. (2001) J Bacteriol 183: 1452-1454.

Day, J.B. and Plano, G.V. (1998) Mol Microbiol 30: 777-788.

De Grado, M., Abe, A., Gauthier, A., Steele-Mortimer, O., DeVinney, R. and Finlay, B.B. (1999) Cell Microbiol 1: 7-17.

Demers, B., Sansonetti, P.J. and Parsot, C. (1998) EMBO J 17: 2894-2903.

Ebel, F., Podzadel, T., Rohde, M., Kresse, A.U., Kramer, S., Deibel, C., Guzman, C.A. and Chakraborty, T. (1998) Mol Microbiol 30: 147-161.

Eichelberg, K., Hardt, W.D. and Galan, J.E. (1999) Mol Microbiol 33: 139-152.

Elliott, S.J., Hutcheson, S.W., Dubois, M.S., Mellies, J.L., Wainwright, L.A., Batchelor, M., Frankel, G., Knutton, S. and Kaper, J.B. (1999) Mol Microbiol 33: 1176-1189.

Fields, K.A. and Hackstadt, T. (2000) Mol Microbiol 38 1048-1060.

Finck-Barbancon, V., Goranson, J., Zhu, L., Sawa, T., Wiener-Kronish, J.P., Fleiszig, S.M., Wu, C., Mende-Mueller, L. and Frank, D.W. (1997) Mol Microbiol 25: 547-557.

Forsberg, A., Viitanen, A.M., Skurnik, M. and Wolf-Watz, H. (1991) Mol Microbiol 5: 977-986.

Forsberg, A. and Wolf-Watz, H. (1988) Mol Microbiol 2: 121-133.

Francis, M.S., Aili, M., Wiklund, M.L. and Wolf-Watz, H. (2000) Mol Microbiol 38: 85-102.

Francis, M.S. and Wolf-Watz, H. (1998) Mol Microbiol 29: 799-813.

Friedberg, D., Umanski, T., Fang, Y. and Rosenshine, I. (1999) Mol Microbiol 34: 941-952.

Frithz-Lindsten, E., Rosqvist, R., Johansson, L. and Forsberg, A. (1995) Mol Microbiol 16: 635-647.

Fu, Y. and Galan, J.E. (1998) J Bacteriol 180: 3393-3399.

Fu, Y. and Galan, J.E. (1999) Nature 401: 293-297.

Galan, J.E. and Collmer, A. (1999) Science 284: 1322-1328.

Galan, J.E. and Zhou, D. (2000) Proc Natl Acad Sci USA 97: 8754-8761.

Gaudriault, S., Malandrin, L., Paulin, J.P. and Barny, M.A. (1997) Mol Microbiol 26: 1057-1069.

Genin, S., Gough, C.L., Zischek, C. and Boucher, C.A. (1992) Mol Microbiol 6: 3065-3076.

Goehring, U.M., Schmidt, G., Pederson, K.J., Aktories, K. and Barbieri, J.T. (1999) J Biol Chem 274: 36369-36372.

Gough, C.L., Genin, S., Zischek, C. and Boucher, C.A. (1992) Mol Plant Microbe Interact 5: 384-389.

Grimm, C. and Panopoulos, N.J. (1989) J Bacteriol 171: 5031-5038.

Hakansson, S., Galyov, E.E., Rosqvist, R. and Wolf-Watz, H. (1996a) Mol Microbiol 20: 593-603.

Hakansson, S., Schesser, K., Persson, C., Galyov, E.E., Rosqvist, R., Homble, F. and Wolf-Watz, H. (1996b) EMBO J 15: 5812-5823.

Hall, A. (1998) Science 279: 509-514.

Haller, J.C., Carlson, S., Pederson, K.J. and Pierson, D.E. (2000) Mol Microbiol 36: 1436-1446.

Hardt, W.D., Urlaub, H. and Galan, J.E. (1998) Proc Natl Acad Sci USA 95: 2574-2579.

Hartland, E.L., Batchelor, M., Delahay, R.M., Hale, C., Matthews, S., Dougan, G., Knutton, S., Connerton, I. and Frankel, G. (1999) Mol Microbiol 32: 151-158.

Hayward, R.D. and Koronakis, V. (1999) EMBO J 18: 4926-4934.

Heesemann, J., Gross, U., Schmidt, N. and Laufs, R. (1986) Infect Immun 54: 561-567.

188

Hermant, D., Menard, R., Arricau, N., Parsot, C. and Popoff, M.Y. (1995) Mol Microbiol 17: 781-789.

Hersh, D., Monack, D.M., Smith, M.R., Ghori, N., Falkow, S. and Zychlinsky, A. (1999) Proc Natl Acad Sci USA 96: 2396-2401.

Hovey, A.K. and Frank, D.W. (1995) J Bacteriol 177: 4427-4436.

Hromockyj, A.E., Tucker, S.C. and Maurelli, A.T. (1992) Mol Microbiol 6: 2113-2124.

Hueck, C.J. (1998) Microbiol Mol Biol Rev 62: 379-433.

Hughes, K.T., Gillen, K.L., Semon, M.J. and Karlinsey, J.E. (1993) Science 262: 1277-1280.

Iriarte, M. and Cornelis, G.R. (1998) Mol Microbiol 29: 915-929.

Iriarte, M. and Cornelis, G.R. (1999) J Bacteriol 181: 675-680.

Jin, Q., Hu, W., Brown, I., McGhee, G., Hart, P., Jones, A.L. and He, S.Y. (2001) Mol Microbiol 40: 1129-1139.

Kaniga, K., Bossio, J.C. and Galan, J.E. (1994) Mol Microbiol 13: 555-568.

Karlinsey, J.E., Lonner, J., Brown, K.L. and Hughes, K.T. (2000) Cell 102: 487-497.

Kato, J., Ito, K., Nakamura, A. and Watanabe, H. (1989) Infect Immun 57: 1391-1398.

Kenny, B., DeVinney, R., Stein, M., Reinscheid, D.J., Frey, E.A. and Finlay, B.B. (1997) Cell 91: 511-520.

Kimbrough, T.G. and Miller, S.I. (2000) Proc Natl Acad Sci USA 97: 11008-11013.

Knutton, S., Rosenshine, I., Pallen, M.J., Nisan, I., Neves, B.C., Bain, C., Wolff, C., Dougan, G. and Frankel, G. (1998) EMBO J 17: 2166-2176.

Koster, M., Bitter, W., de Cock, H., Allaoui, A., Cornelis, G.R. and Tommassen, J. (1997) Mol Microbiol 26: 789-797.

Kubori, T., Sukhan, A., Aizawa, S.I. and Galan, J.E. (2000) Proc Natl Acad Sci USA 97: 10225-10230.

Lambert de Rouvroit, C., Sluiters, C. and Cornelis, G.R. (1992) Mol Microbiol 6: 395-409.

Lloyd, S.A., Norman, M., Rosqvist, R. and Wolf-Watz, H. (2001) Mol Microbiol 39: 520-531.

Lucas, R.L. and Lee, C.A. (2001) J Bacteriol 183: 2733-2745.

Lucas, R.L., Lostroh, C.P., DiRusso, C.C., Spector, M.P., Wanner, B.L. and Lee, C.A. (2000) J Bacteriol 182: 1872-1882.

Macnab, R.M. (1999) J Bacteriol 181: 7149-7153.

Menard, R., Sansonetti, P. and Parsot, C. (1994a) EMBO J 13: 5293-5302.

Menard, R., Sansonetti, P., Parsot, C. and Vasselon, T. (1994b) Cell 79: 515-525.

Michiels, T., Vanooteghem, J.C., Lambert de Rouvroit, C., China, B., Gustin, A., Boudry, P. and Cornelis, G.R. (1991) J Bacteriol 173: 4994-5009.

Michiels, T., Wattiau, P., Brasseur, R., Ruysschaert, J.M. and Cornelis, G. (1990) Infect Immun 58: 2840-2849.

Mirold, S., Ehrbar, K., Weissmuller, A., Prager, R., Tschape, H., Russmann, H. and Hardt, W.D. (2001) J Bacteriol 183: 2348-2358.

Nakayama, S. and Watanabe, H. (1995) J Bacteriol 177: 5062-5069.

Nakayama, S. and Watanabe, H. (1998) J Bacteriol 180: 3522-3528.

Neyt, C. and Cornelis, G.R. (1999a) Mol Microbiol 33: 971-981.

Neyt, C. and Cornelis, G.R. (1999b) Mol Microbiol 31: 143-156.

Nguyen, L., Paulsen, I.T., Tchieu, J., Hueck, C.J. and Saier, M.H., Jr. (2000) J Mol Microbiol Biotechnol 2: 125-144.

Niebuhr, K., Jouihri, N., Allaoui, A., Gounon, P., Sansonetti, P.J. and Parsot, C. (2000) Mol Microbiol 38: 8-19.

Norris, F.A., Wilson, M.P., Wallis, T.S., Galyov, E.E. and Majerus, P.W. (1998) Proc Natl Acad Sci USA 95: 14057-14059.

Parsot, C., Menard, R., Gounon, P. and Sansonetti, P.J. (1995) Mol Microbiol 16: 291-300.

Pegues, D.A., Hantman, M.J., Behlau, I. and Miller, S.I. (1995) Mol Microbiol 17: 169-181.

Persson, C., Carballeira, N., Wolf-Watz, H. and Fallman, M. (1997) EMBO J 16: 2307-2318.

Persson, C., Nordfelth, R., Holmstrom, A., Hakansson, S., Rosqvist, R. and Wolf-Watz, H. (1995) Mol Microbiol 18: 135-150.

Pettersson, J., Nordfelth, R., Dubinina, E., Bergman, T., Gustafsson, M., Magnusson, K.E. and Wolf-Watz, H. (1996) Science 273: 1231-1233.

Plano, G.V., Day, J.B. and Ferracci, F. (2001) Mol Microbiol 40: 284-293.

Porter, M.E. and Dorman, C.J. (1997) J Bacteriol 179: 6537-6550.

Rahme, L.G., Stevens, E.J., Wolfort, S.F., Shao, J., Tompkins, R.G. and Ausubel, F.M. (1995) Science 268: 1899-1902.

Rimpilainen, M., Forsberg, A. and Wolf-Watz, H. (1992) J Bacteriol 174: 3355-3363.

Rockey, D.D., Lenart, J. and Stephens, R.S. (2000) Infect Immun 68: 5473-5479.

Roine, E., Wei, W., Yuan, J., Nurmiaho-Lassila, E.L., Kalkkinen, N., Romantschuk, M. and He, S.Y. (1997) Proc Natl Acad Sci USA 94: 3459-3464.

Rosenshine, I., Ruschkowski, S. and Finlay, B.B. (1996) Infect Immun 64: 966-973.

Rosqvist, R., Forsberg, A. and Wolf-Watz, H. (1991) Biochem Soc Trans 19: 1131-1132.

Rosqvist, R., Hakansson, S., Forsberg, A. and Wolf-Watz, H. (1995a) EMBO J 14: 4187-4195.

Rosqvist, R., Magnusson, K.E. and Wolf-Watz, H. (1994) EMBO J 13: 964-972.

Rosqvist, R., Persson, C., Hakansson, S., Nordfeldt, R. and Wolf-Watz, H. (1995b) Contrib Microbiol Immunol 13: 230-234.

Rossier, O., Wengelnik, K., Hahn, K. and Bonas, U. (1999) Proc Natl Acad Sci USA 96: 9368-9373.

Rudolph, M.G., Weise, C., Mirold, S., Hillenbrand, B., Bader, B., Wittinghofer, A. and Hardt, W.D. (1999) J Biol Chem 274: 30501-30509.

Sakai, T., Sasakawa, C. and Yoshikawa, M. (1988) Mol Microbiol 2 589-597.

Schechter, L.M., Damrauer, S.M. and Lee, C.A. (1999) Mol Microbiol 32 629-642.

Schesser, K., Frithz-Lindsten, E. and Wolf-Watz, H. (1996) J Bacteriol 178 7227-7233.

Shea, J.E., Hensel, M., Gleeson, C. and Holden, D.W. (1996) Proc Natl Acad Sci USA 93: 2593-2597.

Skryzpek, E. and Straley, S.C. (1993) J Bacteriol 175: 3520-3528.

Skrzypek, E., Cowan, C. and Straley, S.C. (1998) Mol Microbiol 30: 1051-1065.

Sory, M.P., Boland, A., Lambermont, I. and Cornelis, G.R. (1995) Proc Natl Acad Sci USA 92: 11998-12002.

Sory, M.P. and Cornelis, G.R. (1994) Mol Microbiol 14: 583-594.

Stainier, I., Iriarte, M. and Cornelis, G.R. (1997) Mol Microbiol 26: 833-843.

Stebbins, C.E. and Galan, J.E. (2000) Mol Cell 6: 1449-1460.

Stender, S., Friebel, A., Linder, S., Rohde, M., Mirold, S. and Hardt, W.D. (2000) Mol Microbiol, 36: 1206-1221.

Subtil, A., Parsot, C. and Dautry-Varsat, A. (2001) Mol Microbiol 39: 792-800.

Sukhan, A., Kubori, T., Wilson, J. and Galan, J.E. (2001) J Bacteriol 183: 1159-1167.

Tamano, K., Aizawa, S., Katayama, E., Nonaka, T., Imajoh-Ohmi, S., Kuwae, A., Nagai, S. and Sasakawa, C. (2000) EMBO J 19: 3876-3887.

Tardy, F., Homble, F., Neyt, C., Wattiez, R., Cornelis, G.R., Ruysschaert, J.M. and Cabiaux, V. (1999) EMBO J 18: 6793-6799.

Tezcan-Merdol, D., Nyman, T., Lindberg, U., Haag, F., Koch-Nolte, F. and Rhen, M. (2001) Mol Microbiol 39: 606-619.

Tobe, T., Schoolnik, G.K., Sohel, I., Bustamante, V.H. and Puente, J.L. (1996) Mol Microbiol 21: 963-975.

Tran Van Nhieu, G., Ben-Ze'ev, A. and Sansonetti, P.J. (1997) EMBO J 16: 2717-2729.

Tran Van Nhieu, G., Bourdet-Sicard, R., Dumenil, G., Blocker, A. and Sansonetti, P.J. (2000) Cell Microbiol 2: 187-193.

Tran Van Nhieu, G., Caron, E., Hall, A. and Sansonetti, P.J. (1999) EMBO J 18: 3249-3262.

Tucker, S.C. and Galan, J.E. (2000) J Bacteriol 182: 2262-2268.

190

Valdivia, R.H. and Falkow, S. (1997) Science 277: 2007-2011.

Vallance, B.A. and Finlay, B.B. (2000). Proc Natl Acad Sci USA 97: 8799-8806.

Van den Ackerveken, G., Marois, E. and Bonas, U. (1996) Cell 87: 1307-1316.

Van Gijsegem, F., Genin, S. and Boucher, C. (1993) Trends Microbiol 1: 175-180.

Van Gijsegem, F., Vasse, J., Camus, J.C., Marenda, M. and Boucher, C. (2000) Mol Microbiol 36: 249-260.

Von Pawel-Rammingen, U., Telepnev, M.V., Schmidt, G., Aktories, K., Wolf-Watz, H. and Rosqvist, R. (2000) Mol Microbiol 36: 737-748.

Wainwright, L.A. and Kaper, J.B. (1998) Mol Microbiol 27: 1247-1260.

Wattiau, P., Bernier, B., Deslee, P., Michiels, T. and Cornelis, G.R. (1994) Proc Natl Acad Sci USA 91: 10493-10497.

Wattiau, P. and Cornelis, G.R. (1993) Mol Microbiol 8: 123-131.

Wattiau, P., Woestyn, S. and Cornelis, G.R. (1996) Mol Microbiol 20: 255-262.

Wei, Z., Kim, J.F. and Beer, S.V. (2000) Mol Plant Microbe Interact 13: 1251-1262.

Wei, Z.M. and Beer, S.V. (1995) J Bacteriol 177: 6201-6210.

Wengelnik, K. and Bonas, U. (1996) J Bacteriol 178: 3462-3469.

Wilson, R.L., Libby, S.J., Freet, A.M., Boddicker, J.D., Fahlen, T.F. and Jones, B.D. (2001) Mol Microbiol 39: 79-88.

Woestyn, S., Sory, M.P., Boland, A., Lequenne, O. and Cornelis, G.R. (1996) Mol Microbiol 20: 1261-1271.

Xiao, Y., Heu, S., Yi, J., Lu, Y. and Hutcheson, S.W. (1994) J Bacteriol 176: 1025-1036.

Yahr, T.L., Vallis, A.J., Hancock, M.K., Barbieri, J.T. and Frank, D.W. (1998) Proc Natl Acad Sci USA 95: 13899-13904.

Yuk, M.H., Harvill, E.T. and Miller, J.F. (1998) Mol Microbiol 28: 945-959.

Zhou, D., Mooseker, M.S. and Galan, J.E. (1999) Science 283: 2092-2095.

Chapter 10

AUTOTRANSPORTERS

Ben R. Otto

Department of Molecular Microbiology, IMBW/BioCentrum Amsterdam
Faculty of Biology, Vrije Universiteit Amsterdam
De Boelelaan 1087
1081 HV Amsterdam, NL

1.   INTRODUCTION

Gram-negative bacteria export many different proteins to their cell surface. The proteins remain anchored to the cell surface, are assembled into cell-surface organelles, or are secreted into the extracellular environment. Protein secretion is necessary for a variety of reasons, including nutrient acquisition, for the expression of proteinaceous virulence factors and for organelle biogenesis. Secretion in Gram-negative bacteria is complicated by the fact that proteins have to cross the periplasm as well as the outer membrane (OM), in addition to the inner membrane (IM). In this chapter secretion is defined as any process in which a protein crosses the OM barrier. The autotransporter secretion pathway sometimes referred to as the type V secretion pathway (Henderson et al., 2000). is a highly efficient route. The pathway was first described for the immunoglobulin A1 protease (IgA1 protease) in Neisseria gonorrhoeae (Pohlner et al., 1987) and at this moment the pathway has been identified for an increasing number of proteins and in an increasing number of pathogens (Henderson and Nataro, 2001) (Table 1).

The autotransporter secretion pathway is distinct from the other known pathways in that the protein moiety mediating export through the OM is contained within the secreted protein itself, hence the nickname "autotransporters" for proteins secreted this way (Jose et al., 1995).

This chapter will present an overview of the autotransporter secretion pathway, highlighting the requirements for secretion and processing. Readers are referred to the other chapters of the book to find

B. Oudega (ed.), Protein Secretion Pathways in Bacteria, 191–205.

detailed information about the other secretion pathways in Gram-negative bacteria.

Table 1. Autotransporter proteins of microbial pathogens (Henderson and Nataro, 2001)

Pathogen	Proteins	Function passenger domains
Bordetella spp.	Pertactin, BrkA, TcfA, Vag8	Adhesins or serum resistance
Dichelobacter nodosus	BprV, BprB, AprV2, BprX	Elastases
Escherichia coli	EspP, Pet, Sat, Tsh, Hbp, Pic, AIDA-I, TibA, Ag43	Proteolytic toxins, proteases or adhesins
Haemophilus influenzae	IgA1, Hap, Hia, Hsf	Proteases or adhesins
Helicobacter spp.	Hsr, VacA, BabA	Toxins or adhesins
Moraxella catarrhalis	UspA1, UspA2, UspA2h	Adhesins or serum resistance
Neisseria spp.	IgA1	Protease
Pasteurella haemolytica	Ssa1	Protease
Pseudomonas spp.	EstA, PspA, PspB	Esterase or proteases
Rickettsiales	rOmpA, rOmpB	Adhesins
Salmonella typhimurium	ApeE	Esterase
Serratia marcescens	PrtS, PrtT, Ssp-H1, Ssp-H2	Proteases
Shigella flexneri	SepA, Pic, SigA, IcsA	Proteolytic toxins, proteases or mucinase
Xenorhabdus luminescens	PlaA	Lipase

2. STRUCTURAL FEATURES AND SECRETION OF AUTOTRANSPORTERS IN GENERAL

The autotransporter pathway exports proteins with diverse functionalities, including proteases, toxins, adhesins and invasins (Table 1). The most common feature of the autotransporter proteins is that their genes encode large polyproteins that are organized in three functional domains: a N-terminally located signal peptide, an internal passenger or functional domain, and a C-terminal translocation unit (TU). The TU consists of a linking region and a β-barrel, which are together required for autodisplay (Fig. 1). It is believed that these three domains confer all the features required to mediate the distinctive steps in transport across the IM, periplasm and OM (Jose *et al.*, 1995; Pohlner *et al.*, 1987; Klauser *et al.*, 1993). The autotransporter polyprotein is exported into the periplasm by action of the general secretory pathway (Pugsley, 1993). The leader sequence is cleaved by the signal peptidase, releasing the mature polyprotein into the periplasm. Once in the periplasm, the β-barrel portion of the protein inserts into the OM to form an aqueous β-barrel pore channel across the membrane. After formation of the β-barrel pore, the linker presumably folds through the pore, pulling the passenger domain through the channel (Klauser *et al.*, 1990) (Fig. 1). Autodisplayed passenger proteins are often cleaved from the autotransporter and released into the environment (Benjellountouimi *et al.*, 1995; Brunder *et al.*, 1997; Djafari *et al.*, 1997; Eslava *et al.*, 1998; Guyer *et al.*, 2000; Halter *et al.*, 1984; Henderson *et al.*, 1999; Otto *et al.*, 1998; Provence and Curtiss III, 1994; Stein *et al.*, 1996). Another possibility is that an intact polyprotein remains at the cell surface, with a carboxyl-terminal membrane-bound domain and an amino-terminal domain extending into the environment (O'Toole *et al.*, 1994; St Geme and Cutter, 2000). Alternatively, the passenger protein is cleaved but remains in contact with the cell surface via a strong interaction with the β-domain (Suhr *et al.*, 1996). No energy coupling or accessory factors are known to be required for the secretion across the OM.

3.   PHYLOGENY OF AUTOTRANSPORTERS

Autotransporters can be classified based on function and phylogeny of the passenger domain. Phylogenetic classification is based upon nucleotide and amino acid homologies, as well as on homologous positions of active site motifs. Members of the same family are presumed to share a common ancestor; subfamilies are proposed, based upon a relatively greater degree of homology (Henderson *et al.*, 1998). An example of horizontal gene transfer is illustrated by the immunoglobulin A1 proteases (IgA1). Members of these proteases can be found in *Neisseria* and *Haemophilus* (Henderson and Nataro, 2001). The *Escherichia coli* proteins Pet (Eslava *et al.*, 1998), Tsh (Hbp) (Otto *et al.*, 1998; Provence and Curtiss III, 1994), EspP (Brunder *et al.*, 1997), EspC (Stein *et al.*, 1996), and the *Shigella* proteins SepA (Benjellountouimi *et al.*, 1995), Pic (Henderson *et al.*, 1999) and SigA (Al-Hasani *et al.*, 2000) possess a serine protease motif (consensus GD*S*GSP) at similar positions suggesting that these autotransporters are related serine proteases. The proteins belong to the family of Serine Protease autotransporters of Enterobacteriaceae (SPATE-family) (Henderson *et al.*, 1998). Members of this protein subfamily have evolved specific functions that are adaptive for the particular niche occupied by the pathogen. Other subfamilies of autotransporters are the nonprotease autotransporters of *Enterobacteriaceae* like AIDA-I, *Bordetella-*, *Moraxella-*, *Haemophilus-*, *Helicobacter-* and *Rickettsia*-autotransporters (Table 1).

194

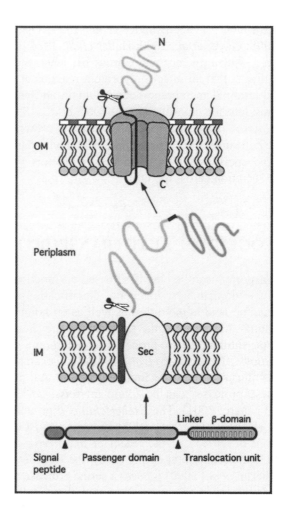

Fig. 1. Model of autotransporter secretion pathway (Pohlner *et al.*, 1987). Proteins are translated as a polyprotein with three domains; the signal sequence, the passenger domain and the translocation unit (TU). The C-terminal TU is required for autodisplay of the mature protein and consists of a linker region and a β-domain. The proteins cross the IM via the Sec system, followed by cleavage of their amino-terminal signal sequence at the periplasmic side of the IM by signal peptidase. Once in the periplasm, the carboxyl-terminal or β-domain of the protein inserts into the OM to form a β-barrel structure. The passenger domain is translocated to the cell surface through the pore. The passenger domain is released into the extracellular environment by proteolysis.

4. MULTIDOMAIN STRUCTURE AND SECRETION MECHANISM OF AUTOTRANSPORTERS

As stated above autotransporter proproteins consist of three domains. These distinct domains, an amino-terminal leader peptide, a surface-exposed mature protein and a carboxyl-terminal domain, are present in nearly all autotransporter molecules. Each domain will be described separately.

4.1 The N-terminal leader peptide; inner membrane translocation

The N-terminus of the autotransporter protein harbors characteristic features of signal sequences that mediate translocation across the cytoplasmic membrane. These features are characteristic of the prototypical Sec-dependent signal sequence; a polar stretch of positively charged amino acids (N-domain), a hydrophobic core region (H-domain) of neutral amino acids, and a less-hydrophobic region (C-domain) that contains a signal peptidase recognition site (Nielsen *et al.*, 1997). The proteins of the SPATE subfamily exhibit unusually long signal peptides, consisting of at least 50 amino acid residues, like EspP: 50 amino acid residues (Brunder *et al.*, 1997), EspC: 53 (Stein *et al.*, 1996), PssA: 55 (Djafari *et al.*, 1997), SepA: 56 (Benjellountouimi *et al.*, 1995), Pet: 52 (Eslava *et al.*, 1998), Tsh: 52 (Stathopoulos, 1999), Hbp: 52 (Otto *et al.*, 1998), Pic: 55 (Henderson *et al.*, 1999) and SigA: 54 (Al-Hasani *et al.*, 2000). These signal peptides are cleaved off during passage of the inner membrane by a signal peptidase. The extra stretch of amino acid residues is located between the first methionine and the charged N-domain. These atypical signal sequences begin with M_1N (R/K), followed closely by a motif (Y/F, X, I/L/V, X, Y/W) containing conserved aromatic and hydrophobic residues. Immediately preceding the positively charged N-domain is an eight-amino-acid motif ($I_{18}AVSELAR$), which varies only slightly in all of these unusual signal sequences (Henderson *et al.*, 1998). The meaning of this extended signal sequence is still not known, but recent data (B. R. Otto) suggest that this extended signal sequence has a relatively high affinity for SRP (see below).

There are several pathways that can be involved in the targeting of the autotransporter preprotein to the inner membrane and in the translocation of proteins across the inner membrane. Transport of the autotransporter precursor across the inner membrane has always been assumed to occur via the SecB pathway (Pohlner *et al.*, 1987). The SecB pathway of *E. coli* involves the tetrameric chaperone SecB. SecB binds post-translationally or at a late co-translational state to the mature region of a presecretory protein (Kumamoto and Francetic, 1993). The SecB-preprotein complex is targeted to the membrane, where SecA is activated for high-affinity recognition of SecB and of the preprotein by binding to the membrane-embedded SecYEG translocon (Driessen, 1998). After translocation the signal peptidase LepB cleaves off the signal peptide and the protein is released into the periplasm. Another pathway involved in targeting and translocation of proteins makes use of the signal recognition particle (SRP).

In *E. coli* a SRP pathway exists that is homologous to the SRP targeting pathway of the eukaryotic ER membrane (Luirink *et al.*, 1994). SRP co-translationally binds particular hydropohobic signal sequences when they emerge from the ribosome. The complex of SRP, preprotein and ribosome is targeted via FtsY to the SecYEG translocon (Valent *et al.*, 1998). Most inner membrane proteins follow this SRP route, although, certain secreted proteins with relatively hydrophobic signal sequences, such as β-lactamase, may use of this pathway preferentially (Luirink *et al.*, 1994). The signal peptides of autotransporters are considerably different from the signal peptides of proteins targeted by the SecB pathway. Therefore, it is likely that the SRP pathway is involved in the targeting of the pre-autotransporter proteins, although, the signal sequences of autotransporters possess a moderated hydrophobicity (Fig. 2). The co-translational translocation can prevent aggregation, degradation and accumulation in the cytoplasm. Recently, an interaction between nascent polypeptides of the autotransporter Hbp and SRP, trigger factor, SecA as well as with SecY has been found by cross-linking analysis (unpublished results). In addition, steady-state experiments with a SecB-knockout strain showed no apparent effect on the amount of Hbp present in the culture supernatant of this strain. These results suggest that autotransporters, in particularly Hbp might follow the SRP targeting route

A third pathway translocates proteins across the inner membrane in their native folded state (Santini, 1998). Proteins with a so-called twin-arginine motif [(S/T)RRxFLK] present in their signal peptide specifically use this pathway. Transport is mainly driven by the proton motive force and is independent of SecY, SecE, SecA and SecB. Autotransporters may also use this TAT-pathway. A weak twin-arginine motif present in some of the signal peptides of autotransporters, including Hbp (Otto *et al.*, 1998), may indicate the use of the TAT-pathway. However, experiments with several mutant strains depleted in components of the TAT translocase did not show apparent defects in Hbp transport (unpublished data).

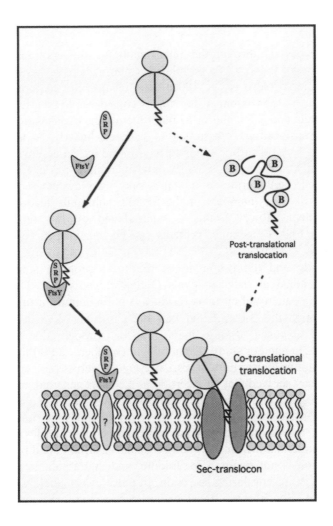

Fig. *2.* Model of autotransporter protein targeting to the *E. coli* inner membrane. In the SRP-mediated pathway the cytoplasmic protein SRP co-translationally binds the signal peptide of the autotransporter protein as soon as it emerges from the ribosome (Valent *et al.*, 1998). FtsY picks up the complex of SRP, nascent polypeptide chain and ribosome in the cytoplasm. After docking of FtsY at the membrane, the ribosome-nascent polypeptide chain complex is released from SRP and the presecretory protein will insert into the Sec-translocon. This way of co-translational translocation probably prevents aggregation, degradation and accumulation of autotransporters in the cytoplasm. In the SecB-mediated pathway, the protein SecB will post-translationally bind to the mature region of the protein and subsequently targets the presecretory protein to the Sec-translocon.

4.2   Transport through the periplasm

After passage through the inner membrane, the autotransporter pro-protein is thought to exist as a periplasmic intermediate as has been shown for IgA1 protease (Jose *et al.*, 1996). The incapability to form extensive disulfide loops in the passenger domains is considered to be a feature of the IgA polyprotein and is crucial for its outer membrane translocation. As outer membrane translocation is assumed to proceed linearly as an unfolded peptide chain, large disulfide loops formed in the oxidative environment of the periplasm would possibly hamper outer membrane passage (Jose *et al.*, 1996). Indeed, IgA1 protease contains only two cysteines and related polyproteins like the members of the SPATE subfamily also have a low cysteine content. However, this is contradicted by two recent studies showing that folded passenger proteins can be translocated (Brandon and Goldberg, 2001; Veiga *et al.*, 1999). Hybrid proteins consisting of a single-chain antibody and the β-domain of the IgA1 protease were properly translocated, albeit at reduced levels (Veiga *et al.*, 1999). These results suggests that a paucity of cysteine residues is not required for proteins to be secreted through the β-barrel and that the passenger domains of hybrid proteins are exposed during their transfer through the periplasmic compartment. In another study (Brandon and Goldberg, 2001) the transit of the native autotransporter protein IcsA of *Shigella* across the periplasm was examined. They showed that an insertion in the *dsbB* gene did not affect the surface expression or unipolar targeting of IcsA. The *dsbB* gene product mediates disulfide bond formation of many periplasmic intermediates. Furthermore, they indicated that the passenger domain of IcsA is folded during secretion, that it forms an intramolecular disulfide bond but disulfide bond formation is not required for folding, and that a soluble periplasmic state of IcsA is present during secretion. In pulse-chase labeling studies the processing of Hbp during translocation across the cell envelope was followed. One minute after the chase, the pre-proform (including the signal sequence, the passenger domain and the β-barrel domain) of Hbp was not detectable anymore, the proform (passenger domain and β-barrel domain) disappeared three minutes later, while the mature Hbp did appear in the culture supernatant (unpublished data, Otto *et al.*). These results indicate that Hbp, just like IcsA, exist as a (probably periplasmic) intermediate for a relatively long time. It is unknown how the autotransporter proproteins survive the passage through the periplasmic space without extensive degradation by periplasmic proteases. The putative periplasmic intermediates probably form a complex with periplasmic chaperones or they possess autochaperone characteristics. However, there are no experimental data to further support one of these possibilities. A novel model for the secretion of IcsA that may be applicable to other autotransporter proteins was proposed (Brandon and Goldberg, 2001). After passage of the inner membrane, the proteins will be folded into a structure that is resistant to periplasmic proteases. The β-domain becomes associated with the outer membrane, and the N-terminal part of the mature protein passes through the pore of the β-barrel. In conjunction with these translocation events across the cell envelope, IcsA forms one disulfide bond in the periplasm that probably stabilizes the folded structure and mediates its resistance to proteolysis. It is unknown how this structured polypeptide can be translocated through the hydrophyllic channel created by the β-domain.

4.3 The C-terminal translocation unit (TU); outer membrane translocation

The autotransporter proteins show homology throughout the entire sequences, but a high degree of homology has been found between the C-terminal parts, whereas a lower degree of conservation has been detected in the N-terminal parts of the molecules (Eslava *et al.*, 1998). These conserved domains of the autotransporter polyproteins are assumed to form a C-terminal translocation unit (TU) (Maurer *et al.*, 1999; Pohlner *et al.*, 1987), consisting of a linking region and a membrane-supported β-barrel (Jahnig, 1990) with certain similarities to the well-known porin structure (de Cock, 1997). Deletions can not be made in the β-core region without complete loss of the translocation function (Klauser *et al.*, 1993; Otto *et al.*, 1998). A β-barrel is a complex protein structure composed of multiple amphipathic antiparallel β-sheets. The first and last β-sheets spontaneously form hydrogen bonds to close the ring conformation. The amphipathic primary structure favors the forming of an aqueous pore channel through the hydrophobic lipid layer of the outer membrane. Computer modeling and gel-mobility studies support the formation of a β-barrel structure from the C-terminal translocator domain (Henderson *et al.*, 1998; Maurer *et al.*, 1999; Loveless, 1997). In addition, the β-domain of the *Bordetella pertussis* autotransporter BrkA was recently shown to form pores in black lipid bilayer membrane experiments (Shannon and Fernandez, 1999). Purified recombinant BrkA C-terminal protein formed channels in lipid bilayers, whereas the BrkA N-terminal protein did not form channels.

It is unknown whether a single β-domain forms the channel or whether oligomerization is required. Recently, Veiga *et al.*, (2002) found oligomeric rings with a central cavity of ~2 nm shaped by the carboxyl-terminal domains of the IgA protease. However, unpublished results from Tommassen and van Putten (Utrecht, NL) based on the crystal structure of a β-barrel indicated that one β–barrel may form a pore on its own.

A consensus configuration for the structure of the β-barrels of autotransporters has been proposed (Henderson *et al.*, 1998). The carboxyl-terminal part of the protein forms a β-barrel consisting of approximately 14 amphipathic, antiparallel β-sheets. This generates the β-core: a hydrophobic coat which contacts the membrane, and a closely hydrophilic interior (Fig. 1). In addition, the β-domains of the polyprotein precursors share a consensus amino acid motif at the extreme carboxyl terminus (Jose *et al.*, 1995; Loveless, 1997). The terminal amino acid residue is always phenylalanine or tryptophan, preceded by alternating charged and aromatic residues. A similar motif is found in the *E. coli* porin proteins OmpF and PhoE. This motif is important for an efficient and correct assembly of the protein into the outer membrane. Deletion of amino acid residues from this motif in PhoE abolishes outer membrane localization and trimer formation (de Cock, 1997). Henderson (Henderson *et al.*, 1998) have predicted a signature sequence for the β-domains of the autotransporters:

[SATQRGHF]-[AKLTQNFGLIV]-[FGATLISN]-[GPDSNKIATW]-
[KRHNDQS]-X₃-[DPAQHKG]-[QNDWKYFITLGH]-[ASTQNG]-
[LIVFGASY]-[NQHEGAIFTS]-[LIVA]-[NGSKT]-[LIVMFYG]-
[RKQSTGN]-[LIFVYW]-[STAKRNVQEM]-[FW].

This signature may be useful in structure-function analysis and in the identification of new autotransporters. The computer-predicted models of the β-barrel of autotransporters provide only a general idea of their structure, but as indicated by the crystal structures of OmpA (Pautsch and Schulz, 1998) and FepA (Buchanan *et al.*, 1999), they can be wrong. The most important structural features of the β-barrels of autotransporters that need to be elucidated are the exact number of strands, whether the extreme N-terminal strand of the barrel faces in or out, and whether or not the pore is plugged after export of the passenger domain.

The linker of the TU is required for autodisplay of the mature protein. The linker attached to the periplasmic side of the β-barrel presumably folds through the aqueous pore, pulling the probably partly folded translocation-competent passenger domain through the pore beginning with its C-terminal part (Klauser *et al.*, 1990). In the case of the SPATE subfamily, the protein is cleaved of from the TU once at the surface (Fig. 3). After export and cleavage of the passenger domain, the linker region probably remains connected to the β-barrel plugging in this way the aqueous pore. In the case of the autotransporter protein AIDA-I (Maurer *et al.*, 1999) the C-terminal portion of the linking region is assumed to remain within the β-barrel pore, while the N-terminal portion of the linker anchors the passenger domain to the cell surface (Fig. 3). The minimal essential linking region of AIDA-I has been characterized and spans 28 to 48 amino acids (Maurer *et al.*, 1999), while the size of the linking region of other autotransporter proteins remains to determined. In addition, it is unknown if

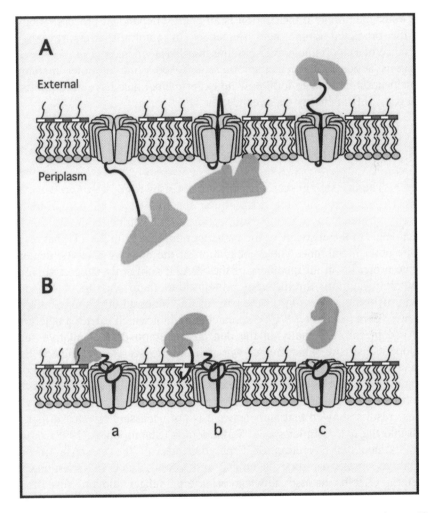

Fig. 3. Model for the secretion of autotransporter proteins across the outer membrane. A), Translocation of the passenger domain through the β-barrel pore. The linker region of the TU folds into the β-barrel channel tightening the partly folded passenger domain against the channel and plugging in this way the pore at the periplasmic side. Further insertion of the linker into the channel may triggers the unfolding of the passenger domain and the linker will

pull the protein through the channel. As soon as the passenger domain is surface exposed, the protein will fold in its mature conformation. B), Processing of the cell-surface exposed passenger domain. Once at the cell surface several possibilities may occur: (a) the protein remains intact as a large polyprotein, (b) the linker is cleaved half away, the C-terminal part of the linking region remains within the β-barrel pore, while the N-terminal portion of the linker anchors the passenger domain to the cell surface, (c) the passenger domain is cleaved off from the linker and is released into the environment. The linker probably plugs the channel preventing leakage through the pore in this way.

the outer membrane translocation is an energy-dependent process. Analysis of the predicted amino acid sequences of autotransporters reveals the presence of P-loop nucleotide-binding motifs (Henderson *et al.*, 1998). This suggests a possible mechanism for energy coupling of outer membrane translocation. However, until now no experimental data have been published that can confirm energy dependence.

4.4 Release of the passenger domain

The last step in secretion of IgA proteases from *N. gonorrhoeae* and from *H. influenzae* into the extracellular space involves the autoproteolytic activity of the catalytic serine endopeptidase site (Pohlner *et al.*, 1987). This event leads to the cleavage of the passenger domain from the TU that resides in the outer membrane. The conservation of the putative catalytic center of serine proteases in all members of the SPATE subfamily suggested similar proteolytic activity for different proteins, but there was no evidence for autoproteolytic processing in the case of any other autotransporter proteins known. Stein (Stein *et al.*, 1996) created a mutant which carries a deletion in the *espC* gene, especially in the consensus region of the putative serine endopeptidase site. This mutant was still able to secrete a truncated EspC product. Similar experiments were carried out with the autotransporter protein Tsh and also with this protein the extracellular secretion was not abolished by a mutation in the serine protease motif (Stathopoulos, 1999). These results showed that autocleavage is not necessary to release EspC or Tsh into the extracellular space. Stathopoulos (Stathopoulos, 1998) showed that extracellular secretion of EspC depends on the presence of outer membrane proteases such as OmpT and OmpP found in Gram-negative bacteria. OmpT was also shown to release passenger domains fused to the helper domain of IgA protease of *N. gonorrhoeae* into the extracellular space (Klauser *et al.*, 1993). Previous studies (Otto *et al.*, 1998; Stathopoulos, 1999) have indicated, that in *E. coli* strains defective in *ompT*, Hbp and the identical protein Tsh (Provence and Curtiss III, 1994) were still properly processed. This suggests the possible involvement of OmpP in the processing of proteins of the SPATE subfamily. Nevertheless, the processing site Asn-Asn$_{1101}$ that is present in all autotransporters of the SPATE subfamily is not a known cleavage site for the proteases of the OmpT family (Stathopoulos, 1999). It is likely that an unknown asparagine-specific endopeptidase or autoproteolysis is accountable for the processing of all these proteins.

5. CONCLUDING REMARKS AND FUTURE PERSPECTIVE

In spite the fact of the many studies on the secretion mechanism of autotransporter proteins, it is still not known how translocation across the cell envelope occurs. There are a lot of questions to be answered. Which targeting pathway to the inner membrane is used by autotransporters? Autotransporters possess cleavable N-terminal signal peptides and are thus likely to employ the Sec-machinery. Many of these signal peptides are unusually long, consisting of at least 47 amino acids. The function of this extension is unknown, however, our unpublished results about the co-translational translocation of Hbp (Fig. 2) suggest that proteins with these unique signal peptides are targeted to the Sec-translocon via the SRP-pathway. This probably prevents folding of the pre-proprotein in the cytoplasm, especially of the β-domain.

Another important issue is if the proproteins need accessory factors during their transfer through the periplasm and if they are folded or not in the periplasmic space. Brendon and Goldberg showed that the proteins are folded to a certain extend to prevent proteolytic degradation and therefore probably the proteins do not need chaperones (Brandon and Goldberg, 2001). A problem is how these structured polypeptides can be translocated through the hydrophyllic channel created by the β-domain. Presumably, the linker of the TU folds into the formed channel tightening the partly folded passenger domain against the channel. In this way the channel will be plugged at the periplasmic side. Further insertion of the linker into the channel may triggers the unfolding of the C-terminal part of the passenger domain. The concerted actions of pulling by the linker and pushing by the stepwise unfolding of the protein will transfer the protein through the channel. As soon as the passenger domain is surface exposed, the linker will plug the channel and the protein will fold in its mature conformation. In this push and pull model additional factors are likely needed on both sides of the outer membrane. However, there is no evidence for such factors until now. Another mechanism for the translocation of the passenger domain could be a mechanism like the twin-arginine translocation (TAT) pathway (Robinson and Bolhuis, 2001). In that case, the passenger domain is translocated across

204

the outer membrane as a fully folded protein. In this model, several β-barrels might form a translocation pore that can adapt to the size of the substrate and provide a tight fit to avoid leakage during translocation. Another question for further research concerns the processing of the mature protein. Is an outer membrane protease responsible for the release of the passenger domain from the β-barrel or is it due to autoproteolytic activity?

The challenge for the future is to generate more detailed biochemical, genetic and structural information to reveal this fascinating secretion pathway. Especially, experimental data of four, poorly characterized, stages in the secretion of autotransporters are needed. Firstly, the targeting of these proteins to the inner membrane. Secondly, the kinetics of transfer of Hbp in relation to the subcellular localization of intermediates of Hbp during the secretion process, and the folding status of Hbp during the translocation across the cell envelop. Thirdly, the interaction of Hbp with putative periplasmic chaperones, and finally the possible formation of a higher order structure of the β-barrel in the outer membrane. A major achievement would be the resolution of the crystal structures of the passenger domain and the TU-domain, but also of a stable translocation intermediate. The latter will obtain a translocator frozen at the moment a polypeptide is translocated through it. Furthermore, an *in vitro* translocation system, composed of artificial lipid vesicles and purified passenger-fusion proteins as proposed (Klauser *et al.*, 1993), could be used to study outer membrane assembly and folding-sensitive translocation in more detail.

6.   REFERENCES

Al-Hasani, K., Henderson, I.R., Sakellaris, H., Rajakumar, K., Grant, T., Nataro, J.P., Robins-Browne, R. and Adler, B. (2000) Infect Immun 68: 2457-2463.
Benjellountouimi, Z., Sansonetti, P.J. and Parsot, C. (1995) Mol Microbiol 17: 123-135.
Brandon, L.D. and Goldberg, M.B. (2001) J Bacteriol 183: 951-958.
Brunder, W., Schmidt, H. and Karch, H. (1997) Mol Microbiol 24: 767-778.
Buchanan, S.K., Smith, B.S., Venkatramani, L., Xia, D., Esser, L., Palnitkar, M., Chakraborty, R., van der Helm, D. and Deisenhofer, J. (1999) Nat Struct Biol 6: 56-63.
De Cock, H., Struyvé, M., Kleerebezem, M., van der Krift, T. Tommassen, J. (1997) J Mol Biol 269: 473-478.
Djafari, S., Ebel, F., Deibel, C., Kramer, S., Hudel, M. and Chakraborty, T. (1997) Mol Microbiol 25: 771-784.
Driessen, A.J.M., Fekkes, P. and van der Wolk, J.P.W. (1998) Curr Opin Microbiol 1: 216-222.
Eslava, C., NavarroGarcia, F., Czeczulin, J.R., Henderson, I.R., Cravioto, A. and Nataro, J.P. (1998) Infect Immun 66: 3155-3163.
Guyer, D.M., Henderson, I.R., Nataro, J.P. and Mobley, H.L. (2000) Mol Microbiol 38: 53-66.
Halter, R., Pohlner, J. and Meyer, T.F. (1984) EMBO J 3: 1595-1601.

Henderson, I.R., Czeczulin, J., Eslava, C., Noriega, F. and Nataro, J.P. (1999) Infect Immun 67: 5587-5596.

Henderson, I.R. and Nataro, J.P. (2001) Infect Immun 69: 1231-1243.

Henderson, I.R., Nataro, J.P., Kaper, J.B., Meyer, T.F., Farrand, S.K., Burns, D.L., Finlay, B.B. and St Geme, J.W., 3rd (2000) Trends Microbiol 8: 352.

Henderson, I.R., Navarro-Garcia, F. and Nataro, J.P. (1998) Trends Microbiol 6: 370-378.

Jahnig, F. (1990) Trends Biochem Sci 15: 93-95.

Jose, J., Jahnig, F. and Meyer, T.F. (1995) Mol Microbiol 18: 378-380.

Jose, J., Krämer, J., Klauser, T., Pohlner, J. and Meyer, T.F. (1996) Gene 178: 107-110.

Klauser, T., Pohlner, J. and Meyer, T.F. (1990) EMBO J 9: 1991-1999.

Klauser, T., Pohlner, J. and Meyer, T.F. (1993) Bioassays 15: 799-805.

Kumamoto, C.A. and Francetic, O. (1993) J Bacteriol 175: 2184-2188.

Loveless, B.J. and Saier, M.H. (1997) Mol Membr Biol 14: 113-123.

Luirink, J., ten Hagen-Jongman, C.M., van der Weijden, C.C., Oudega, B., High, S., Dobberstein, B. and Kusters, R. (1994) EMBO J 13: 2289-2296.

Maurer, J., Jose, J. and Meyer, T.F. (1999) J Bacteriol 181: 7014-7020.

Nielsen, H., Engelbrecht, J., Brunak, S. and von Heijne, G. (1997) Protein Eng 10: 1-6.

O'Toole, P.W., Austin, J.W. and Trust, T.J. (1994) Mol Microbiol 11: 349-361.

Otto, B.R., van Dooren, S.J.M., Nuijens, J.H., Luirink, J. and Oudega, B. (1998) J Exp Med 188: 1091-1103.

Pautsch, A. and Schulz, G.E. (1998) Nat Struct Biol 5: 1013-1017.

Pohlner, J., Halter, R., Beyreuther, K. and Myer, T.F. (1987) Nature 325: 458-462.

Provence, D.L. and Curtiss III, R. (1994) Infect Immun 62: 1369-1380.

Pugsley, A.P. (1993) Microbiol Rev 57: 50-108.

Robinson, C. and Bolhuis, A. (2001) Nat Rev Mol Cell Biol 2: 350-356.

Santini, C.L., Ize, B., Chanal, A., Müller, M., Giordano, G. and Wu, L.F. (1998) EMBO J 17: 101-112.

Shannon, J.L. and Fernandez, R.C. (1999) J Bacteriol 181: 5838-5842.

St Geme, J.W., 3rd, and Cutter, D. (2000) J Bacteriol 182: 6005-6013.

Stathopoulos, C. (1998) Membr Cell Biol 12: 1-8.

Stathopoulos, C., Provence, D.L. and Curtiss III, R. (1999) Infect Immun 67: 772-781.

Stein, M., Kenny, B., Stein, M.A. and Finlay, B.B. (1996) J Bacteriol 178: 6546-6554.

Suhr, M., Benz, I. and Schmidt, M.A. (1996) Mol Microbiol 22: 31-42.

Valent, Q.A., Scotti, P.A., High, S., de Gier, J.W.L., von Heijne, G., Lentzen, G., Wintermeyer, W., Oudega, B. and Luirink, J. (1998) EMBO J 17: 2504-2512.

Veiga, E., de Lorenzo, V. and Fernandez, L.A. (1999) Mol Microbiol 33: 1232-1243.

Veiga, E.,Sugawara, E., Nikaido, H., de Lorenzo, V. and Fernandez, L.A. (2002) EMBO J 21: 2122-2131.

Chapter 11

ASSEMBLY OF ADHESIVE ORGANELLES ON GRAM-NEGATIVE BACTERIA

Sheryl S. Justice, Karen W. Dodson, Matthew R. Chapman, Michelle M. Barnhart and Scott J. Hultgren

Department of Molecular Microbiology
Washington University School of Medicine
660 S. Euclid Ave., Box 8230
St. Louis, MO 63110, USA

1. INTRODUCTION

Bacterial pathogens interact with their hosts in a multitude of ways in order to persist, survive and multiply. The molecular signals triggered by these interactions, ultimately determines the fate of the microbe and the consequences to the host. The interaction between a microbe and its host often depends on the presentation of adhesins on the microbial surface. Adhesins are often assembled into hair-like fibers called pili (Fig. 1). Bacterial adhesins bind with stereochemical specificity to host receptors, thus targeting the bacteria to a defined niche within the environment. For example, pili facilitate intimate contact with eukaryotic cells resulting in the colonization of the cell surface, or invasion to gain access to a nutritionally rich environment within the eukaryotic cell allowing evasion of the host defenses. In host-pathogen interactions pili often activate a cascade of events that will ultimately determine whether the microbe will remain extracellular or trigger processes that will allow its uptake into host cells and/or activation of host defense mechanisms.

B. Oudega (ed.), Protein Secretion Pathways in Bacteria, 207–232.

2. CLASSES OF ADHESIVE ORGANELLES

There are many classes of adhesive organelles that have been identified for Gram-negative bacteria. This chapter will focus primarily on representative systems of: (*i*) pili of the chaperone-usher system such as P and type 1 pili, (*ii*) type IV pili that are assembled by the general secretion pathway, (*iii*) CS1 pili assembled by the alternate chaperone/usher pathway, and (iv) Curli, assembled by the nucleation/precipitation pathway. We will discuss how these organelles are distinguished by their molecular architecture and mechanism of assembly. Aspects of the regulation of some of the systems will also be discussed.

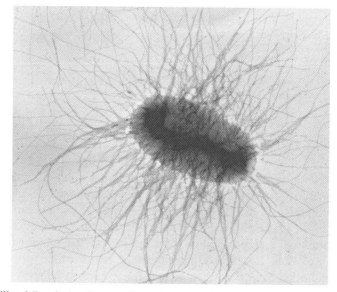

Fig. 1. Piliated *E. coli*. An electron micrograph of an *E. coli* cell with numerous P pili on the cell surface.

2.1 Structure, function and assembly of bacterial fibers

2.1.1 P and type 1 pili (chaperone/usher)

P and type 1 pili are produced by uropathogenic strains of *Escherichia coli*. The ability of *E. coli* to cause pyelonephritis and cystitis depends on the function of P and type 1 pili in binding to the kidney and bladder epithelium, respectively (Langerman *et al.,* 1997; Mulvey *et al.,* 1998; Roberts *et al.,* 1984). These structures are the prototype organelles assembled by the chaperone/usher pathway for which there are over 30 members (Hung *et al.*, 1996). P pili are rod shaped structures with a diameter of approximately 7 nm. The length of a P pilus varies from 5-7 μm. P pili are highly stable and remain intact through a variety of harsh environments and in most cases require the presence of urea or acidic conditions to promote disassembly in the presence of detergents.

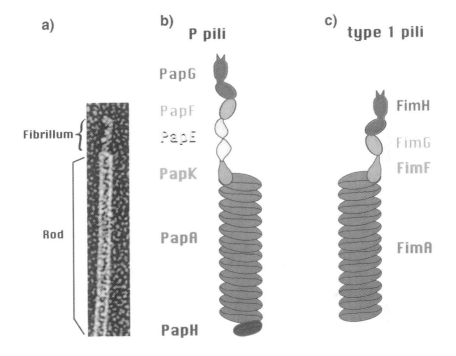

Fig. 2. P and type I Pili. a) A high resolution electron micrograph of a P pilus of *E. coli* showing the pilus rod and the tip fibrillum. A model of the subunit organization within P pili (b) and type 1 pili (c) is depicted. The subunit names are indicated in the same color as the subunit within the pilus.

The assembly of P pili requires eleven genes encoded by the *pap* (pyelonephritis associated pili) gene cluster. The *pap* gene cluster is located in specific regions of the chromosome containing genes implicated in the pathogenicity of the organism. These specific regions have been termed pathogenicity islands (Blum *et al.*, 1995; Guyer *et al.*, 1998; Hacker *et al.*, 1990). Using electron microscopy, it was shown that P pili have a composite structure consisting of a thin tip fibrillum joined to a thicker pilus rod (Fig. 2a). The tip fibrillum contains the PapG adhesin joined to its distal end by the PapF adapter (Jacob-Dubuisson *et al.*, 1993). Most of the fibrillum consists of repeating monomers of PapE arranged in an open helical configuration (Kuehn *et al.*, 1992) that provides flexibility to the fibrillum tip. The fibrillum is joined to the rod by the PapK adaptor (Jacob-Dubuisson *et al.*, 1993). The rod is comprised of repeating PapA subunits arranged in a right-handed helical cylinder with 3.28 subunits per turn (Bullitt and Makowski, 1995). The pilus rod is thought to be anchored to the bacterium via the PapH protein (Baga *et al.*, 1987). The protein at the tip of the pilus, PapG, provides the adhesive attributes of the pilus (Fig. 2b). A critical event in pyelonephritis is the ability of the PapG adhesin to bind to the globoside receptor present in the human kidney. The PapG adhesin consists of an N-terminal receptor binding domain and a C-terminal pilin domain. The pilin domain shares homology with the other subunits of the pilus and serves to attach the receptor domain to the tip fibrillum via interactions with PapF (described below). The receptor domain binds to Galα(1-4)Gal containing glycolipids present on the host epithelia. The PapG receptor domain (Fig. 3) has been shown by crystallography to be rich in β-strands.

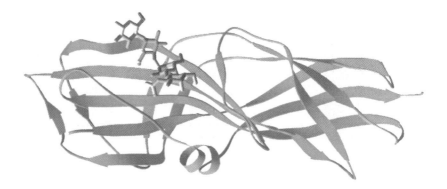

Fig. 3. Receptor binding site for the adhesin domain of PapG. The structure of PapG bound with GBO4 receptor (globoside) was used to determine the binding site for the Galα(1,4) Gal moieties of the glycolipids on uroepithelial cells (Dodson *et al.*, 2001). The structure of PapG is depicted as a ribbon diagram with the backbone of the sugar moiety of the globoside (pink) bound to the side of the PapG adhesin domain.

The receptor GalNAcb1-3Gala1-4Galb1-4Glc(GbO4) binds in a groove on the side of the molecule created by three β-strands (g, k and j) and the loop which connects the o β-strand to the A helix (oA loop). There are extensive interactions between the protein and each of the four individual sugars of the receptor. Hydrophobic interactions occur between the carbons in the sugar rings of the receptor and hydrophobic residues in the protein including an invariant tryptophan. Almost all of the hydroxyls of the receptor are involved in either direct or water mediated hydrogen bonding interactions with protein residues. Comparisons of the bound and unbound state of the receptor binding domain showed that while there are no conformational changes in the protein upon binding, waters must be displaced from the binding site of the unbound state in order for the receptor to bind. PapG binds to Galα(1,4)Gal moieties present on the glycolipids of uroepithelial cells and erythrocytes (Leffler and Svanborg-Eden, 1980; Striker *et al.,* 1995).

Type 1 pili are produced by most members of the *Enterobacteriaceae* family and have been shown to be required for the establishment of bladder infections by uropathogenic *E. coli* (Hultgren *et al.,* 1985; Mulvey *et al.,*1998; Schilling *et al.,* 2001). There are at least nine *fim* genes that are required for the assembly of type 1 pili. The molecular architecture of the type 1 pilus greatly resembles that of the P pilus (Fig. 2c) (Brinton, 1965). The lack of a PapE homologue results in a pilus with a short tip fibrillum (Jones *et al.,* 1992). The penultimate subunit, the FimH adhesin, specifically binds (Fig. 2c) to mannose-oligosaccharides (Krogfelt *et al.,* 1990; Ofek *et al.,* 1977; Sharon, 1987) and FimH is joined to the FimA pilus rod via FimG.

The assembly of P and type 1 pili can be divided into 4 distinct stages. These include: translocation of pilin subunits across the inner membrane; chaperone-assisted folding of the subunits; targeting of the chaperone-subunit complexes to the outer membrane usher; and assembly into the final filamentous structure. All of the pilus components required for assembly contain a typical signal sequence that directs the protein to the *sec*-dependent pathway for translocation across the inner membrane.

After the pilus subunits are translocated into the periplasmic space, the specific chaperone (PapD for P pili and FimC for type 1 pili) forms a heterodimeric complex with each pilus subunit. This interaction plays three specific roles in pilus biogenesis (Fig. 4). First, the chaperone aids in the release of the pilin subunit from the inner membrane as pilin subunits remain "tethered" to the inner membrane in the absence of the chaperone (Jones *et al.,* 1997). Second, pilin subunits are rapidly degraded in the periplasm when the chaperone is limiting or absent, suggesting a role for the chaperone in the folding and stability of the pilin subunits. Lastly, the chaperone prevents premature pilin subunit-subunit interactions from occurring in the periplasm. These three functions are all part of the same process that occurs simultaneously and the structural basis for its mechanism of action is explained below.

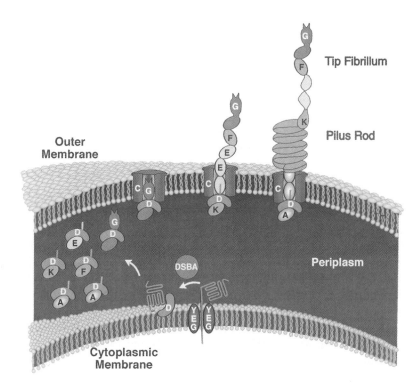

Fig. 4. Model for assembly of P and type 1 pili (chaperone/usher Pathway). A cartoon representation of the chaperone usher pathway used in type 1 and P pilus biogenesis. The nomenclature in this figure is that of the P pili, but the process can be directly applied to the type 1 system.

The structure of PapD has been solved by x-ray crystallography to a resolution of 2.0 Å (Holmgren and Brändén, 1989). PapD has two immunoglobulin (Ig) like domains. An Ig domain is characterized by 7 antiparallel β-sheets (Fig. 5). The two Ig-like domains are oriented to give the chaperone a boomerang appearance (Fig. 6).

Crystallization of the FimC-FimH chaperone-adhesin complex and of the PapD-PapK chaperone-pilin complex revealed insights into the mechanism of action of PapD and the basis for pilus assembly. The PapK pilin is comprised of an Ig-like fold except that it terminates after the sixth β-strand and thus is missing the C-terminal "G" β-strand. Due to the absence of the 7th "G" β-strand, pilins do not have all of the steric information necessary to fold independently. One function of the chaperone is to transiently donate its "G" β-strand to complete the Ig fold of the pilin in a process termed donor strand complementation, which facilitates the folding of the chaperone directly on the chaperone template.

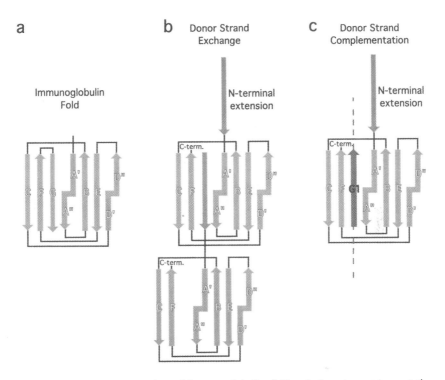

Fig. 5. Greek Key representation of immunoglobulin folds. a) A cartoon representation depicting the β-strand configuration of the complete Ig fold. For the canonical Ig fold the β-strands are in antiparallel configuration. b) Model for the molecular interactions between pilin subunits in the intact pilus. In this model, the amino-terminus of the distal subunit (blue strand) is in the antiparallel configuration. c) In de donor strand complementation, the G1 strand of the chaperone (pink strand) complements for the absence of the 7th strand of the pilin subunit.

The FimH adhesin has two domains, a mannose binding domain (adhesin) and a pilin (assembly) domain. The pilin domain participates in the same donor strand complementation reaction as that described for the PapD-PapK complex. The absence of the 7th "G" β-strand in the pilin results in a deep groove on the surface of the mature pilin that exposes its hydrophobic core. This groove comprises the critical surface on a subunit that participates in subunit-subunit interactions during pilus assembly (Barnhart *et al.*, 2000). Thus, as a direct consequence of chaperone binding, the interactive surface of the pilin subunit is concealed and aggregation and/or degradation of the subunit is prevented. In addition, the alternating hydrophobic residues in the

214

"G" β-strand of the chaperone become part of the hydrophobic core of the pilin.

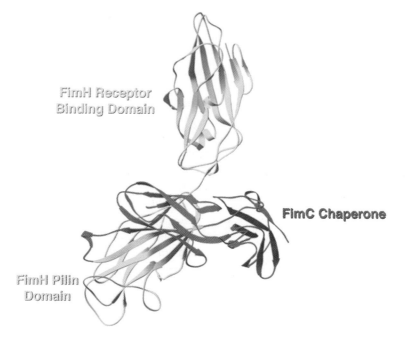

Fig. 6. Ribbon structure of FimC-FimH complex. The structure of the FimC-FimH complex was resolved by x-ray crystallography (Choudhury *et al.*, 1999). The FimC chaperone adopts a boomerang appearance and is depicted in teal. The G1 strand of the chaperone that complements the Ig fold in the pilus subunit is indicated in red.

This explains the mechanism by which the folding of a subunit is simultaneously coupled with the capping of the interactive surface and the stabilization of the hydrophobic core.

Chaperone-subunit complexes are targeted to the usher in the outer membrane where the chaperone is dissociated from the subunit, exposing its interactive groove allowing it to be assembled into the pilus. Outer membrane ushers are donut shaped oligomeric structures with internal pores of ~2–3 nm (Saulino *et al.*, 2000; Thanassi *et al.*, 1998). This pore is large enough to accommodate a folded pilin subunit. However, the pilus rod is a right-handed helical cylinder comprised of 3.28 subunits per turn. This helical cylinder is 70 Å in diameter, and is too thick to fit through the usher pore. Thus, it was postulated that subunits comprising the rod are translocated through the usher as a linear fiber and subsequently coil into a right-handed helical cylinder on the outside surface of the cell. This may be in part what drives the outward growth of the organelle, or at least prevents its retrograde movement back into the cell. Evidence supporting this model came from biochemical and electron microscopic characterization of oligomeric assembly intermediates demonstrating that the growing pilus emerges through the usher channel as a linear fiber (Saulino *et al.*, 1998). The usher is also involved in determining the order in which the pilin subunits are assembled into the growing pilus. The PapC and FimD ushers have both been shown to bind to the chaperone/pilin subunit complexes with affinities that directly correlate with the order in which they are assembled into the pilus (Dodson *et al.*, 1993; Saulino *et al.*, 1998).

Pilus assembly has been proposed to occur via a mechanism termed donor strand exchange (Barnhart *et al.*, 2000; Knight *et al.*, 2000., Sauer *et al.*, 1999). This model was envisioned based on the crystal structures of the PapD-PapK and FimC-FimH complexes. These structures revealed that the highly conserved N-terminus of the pilin subunits are exposed in the complexes. This N-terminal extension contains a motif that is analogous to the "G" β-strand of the chaperone. Thus, during pilus assembly, it was proposed that the "G" β-strand of the chaperone is exchanged for an N-terminal extension from the subunit of an incoming chaperone-subunit complex. This model also may explain why subunit-subunit interactions are more stable than chaperone-subunit complexes. For example, in donor strand exchange the N-terminus of each subunit completes the Ig fold of its neighbor in the antiparallel conformation (Fig. 5). If the subunits are modeled in this fashion, the dimensions of the modeled pilus are very similar to those assembled *in vivo* (Sauer *et al.*, 1999). Thus, the pilus consists of perfectly canonical Ig domains, each of which contributes a strand to the fold of the neighboring subunit. In contrast, donor strand complementation by the chaperone results in the positioning of the "G" β-strand of the chaperone parallel with the "F" strand of the pilus subunit.

The donor strand exchange hypothesis was supported from the experiments of Barnhart *et al.* (2000). They hypothesized that if the donor strand complementation and donor strand exchange hypotheses were correct, then completion of the Ig fold of a subunit with a seventh β-strand provided in "cis", should allow proper folding of the subunit in the absence of the chaperone. To test this, the N-terminal extension of the FimG adaptor was genetically fused to FimH, since FimG joins FimH to the tip fibrillum (Jones *et al.*, 1992). Thus, if pilus assembly occurs via donor strand exchange then the N-terminal extension of FimG presumably would complete the Ig fold of the pilin domain of FimH. Indeed, Barnhart *et al.* demonstrated that fusing the region encoding the N-terminal extension of FimG to the 3' end of *fimH* alleviated the requirement for the chaperone for FimH folding and stability. These experiments are provocative because they demonstrated that pilins do not contain all of the information necessary for

216

folding but instead receive the missing steric information from the chaperone during donor strand complementation (Barnhart *et al.*, 2000).

2.1.2 Type 4 pili

As with the P and type 1 pili, type 4 pili are also involved in the adhesion of bacteria to host cell surfaces, as indicated by the critical role they play in the virulence of enteropathogenic *E. coli* (EPEC) (Bieber *et al.*, 1998) *Neisseria gonorrhea* (Koomey *et al.*, 1987), *Vibrio cholera* (Tayler *et al.*, 1987), *Pseudomonas* (Ramphal *et al.*, 1991) and many other bacterial species. In addition, type 4 pili have also been implicated in twitching motility, in Myxococcus fruiting body formation and in the adsorption of bacteriophages (Wall and Kaiser, 1999). The external diameter of a type 4 pilus is ~60Å. The type 4 pilus of *Neisseria gonorrhea* is thought to have a specialized adhesin at its tip called PilC (Rudel *et al.*, 1995) (Fig. 7).

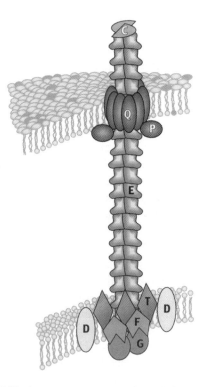

Fig. 7. Type 4 Pili. A cartoon representation of the subunit organization within the prototypical type 4 pilus of *N. gonorrhoeae*.

Specialized adhesins for other type 4 pili have not been described. The *Neisseria* type 4 pilus rod is composed of PilE. Models for the assembly of type 4 pili and regions of antigenic variation arose from the solution of the PilE by X-ray crystallography (Parge *et al.*, 1995). The 2.6 Å resolution structure revealed an α–β roll fold with a long hydrophobic N-terminal α_1-helical extension (residues 2-54) that forms the handle of an over-all ladle-shaped molecule (Fig. 8). Other elements of the structure include: (*i*) an extended disaccharide-bound sugar loop (residues 55-77), with N-acetylglucoseamine–α(1,3)-galactose O-linked at position Ser-63, (*ii*) two β–hairpins forming a four-stranded anti-parallel β-sheet (residues 78-93 and 103-122), and (*iii*) a disulfide containing region (residues 121-158), which, despite the hypervariable nature, appears to be a regular β-hairpin (β_5–β_6) followed by a loop connection. Parge *et al.* (1995) modeled the assembly of the monomers into the pilus rod using the constraints of the dimensions for a type 4 pilus, 5 subunits per turn, a 41 Å pitch, and an outer diameter of 62 Å. Using these parameters the authors speculated that the β-sheet and sugar loop region of each pilin monomer form a continuous 25-stranded tubular β-wrap around a core of packed α_1 helices. An interesting feature of this model is that essentially only the hypervariable and sugar-binding domains of each pilin monomer are exposed in the final assembled pilus structure. Since these regions are involved in the antigenic variation, it is logical that they would be on the exposed surface of the assembled pilus.

In contrast to the phase variation of type 1 pili (discussed below), phase variation of type 4 pili in *N. gonorrhoeae* alters the amino acid sequence of the major pilus subunit PilE. Antigenic variation in this system utilizes the RecA-dependent homologous recombination system (Koomey *et al.*, 1987; Mehr and Seifert, 1997). Multiple copies of the untranscribed pilin gene, *pilS*, are present on the chromosome. Antigenic variation occurs when the DNA within a *pilS* gene is used to replace a variable region of the DNA within the *pilE* gene (Haas and Meyer, 1987; Hagblom *et al.*, 1985; Seifert *et al.*, 1988). Unlike the phase variation of type 1 pili, the transfer of the sequences from the *pilS* to the *pilE* gene is irreversible.

The signal peptides of type 4 pilins are unusual in that they contain an extended positively charged amino terminus prior to the long stretch of hydrophobicity. Type 4 pilins utilize a dedicated signal peptidase that functions by cleaving only the short positively charged N-terminus. Thus, the hydrophobic region becomes the N-terminus of the mature peptide and forms a long α–helix as seen in the crystal structure (Fig. 8). Cleavage of the signal peptide is typically coupled with methylation of the N-terminal amino acid, typically, a phenylalanine (Strom and Lory, 1992). These signal peptidases have been extensively characterized in *Pseudomonas* (PilB)

218

(Lory and Strom, 1997), *Neisseria* (PilD) (Strom *et al.*, 1993) and *Vibrio* (TcpJ) (Kaufman *et al.*, 1991). Interestingly, the type 4 pre-pilin peptidases are also necessary for the cleavage of signal peptides from type 4 pilin-like proteins that are involved in type II secretion.

Fig. 8.   Ribbon structure of PilE. The structure of PilE was resolved to 2.6 Å by X-ray crystallography (Parge *et al.*, 1995).

It is also interesting to note that some of the type 4 pilin components have homologous in the type II secretion machinery. In fact, in some organisms, such *Pseudomonas aeruginosa*, the type 4 pilus and the type II secretion machinery share the PilA component (Lu *et al.*, 1997). The significant homology of these systems suggests that these two pathways arose from divergent evolution of a common organelle.

In *Nesseria*, PilT is implicated in the dissassembly and/or degradation of PilE subunits (Wolfgang *et al.*, 2000). PilQ has homology to the pIV/PulD family of proteins that are proposed to form oligomeric channels in the outer membrane (Russel *et al.*, 1997). PilF and PilG are non-structural components that appear to act in the assembly and stabilization of the pilus, respectively (Wolfgang *et al.*, 1998). Most of the type 4 biogenesis systems contain homologues that carry out these functions. In addition, there are numerous other proteins that play complicated roles in assembly and the reader is advised to consult many of the excellent reviews on the subject for further information (Tonjum and Koomey, 1997; Wu and Fives-Taylor, 2001).

The assembly of type 4 pili can be broken down into three distinct stages, *i*) polymerization of PilE to form the pilus rod, *ii*) stabilization of the pilus rod, and *iii*) protrusion of the intact pilus through the outer membrane (Fig. 9).

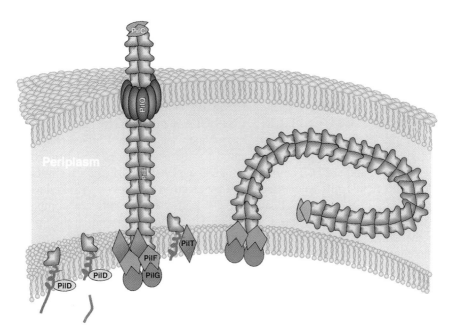

Fig. 9. Model for assembly of type 4 pili (General Secretion Pathway). A cartoon representation of the proposed model for assembly of type IV pilus using the nomenclature from *N. gonorrhoeae*.

In contrast to the assembly of P and type 1 pili, the assembly of type 4 pili does not require a dedicated chaperone to promote assembly of the pilus, and polymerization of the pilus rod occurs within the periplasmic compartment presumably at the inner membrane. Genetic inactivation of most of the components results in the failure to produce pilus rods as a direct consequence of the instability of PilE subunits. Following translocation of PilE across the inner membrane, PilD cleaves PilE to form the stable oligomerization competent protein for assembly into the filamentous pilus rod. Using strains with multiple mutations, Koomey and colleagues discovered that in presence of a *pilQ/pilT* double mutant, intact pili were observed in the periplasmic space, but not on the cell surface (Fig. 9) (Wolfgang *et al.,* 2000). This experiment suggests that the pilus is formed in the periplasmic space, and implicates PilQ as the channel through which the intact pilus rod is protruded through the outer membrane (Wolfgang *et*

220

al., 2000). Moreover, PilQ of *P. aeruginosa* was observed as an oligomeric ring with an inner diameter large enough to support protrusion of the intact pilus (Bitter *et al.,* 1998). Taken together, these observations, in addition with the homology to known oligomeric gated channels, suggest that PilQ forms the pore through which the assembled pilus extrudes.

2.1.3   CS1 Pili

The CS1 class of pili is thought to be involved in the intestinal colonization of ETEC   (Sakellaris *et al.,* 1996). The CS1 pili family includes: CS2, CS4, CS14, CS17, CS19 CFA/I, pili of ETEC, as well as the type 2 pili of *Burkholdeia cepacia* associated with cystic fibrosis (Sajjan *et al.,* 1995). Four *coo* genes are essential for the assembly of CS1 pili, *cooA*, *cooB*, *cooC*, and *cooD*. The morphological aspects of CS1 pili are similar to the P and type 1 pili when visualized by electron microscopy (Sakellaris *et al.,* 1996). In contrast to P and type 1 pili, the CS1 pili are composed primarily of a single subunit, CooA (Fig. 10). CooD is present at the tip and is the adhesin required for attachment of the bacterium to host tissues (Sakellaris *et al.,* 1999). Although the mechanism is unknown, CooC and CooB aid in the assembly of the pilus. This assembly machinery is referred to as the alternate chaperone/usher pathway since work by Scott and colleagues demonstrated that although CooB and CooC share no similarity to PapD-like chaperones or PapC-like ushers, they have similar functions (Voegele *et al.,* 1997).

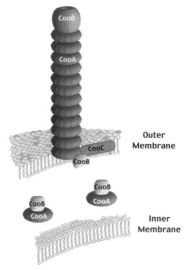

Fig. 10. CS1 Pili. A cartoon representation of the subunit organization within the CS1 pilus.

The CooD and CooA subunits are transported into the periplasm using the *sec*-dependent pathway. Once in the periplasm, CooA and CooD form complexes with the chaperone, CooB (Fig. 10). CooC is located in the outer membrane and is believed to be involved in the translocation of the subunits across the outer membrane. However, the exact role of CooC in the assembly pathway is not understood. The isolation of CooB-CooC complexes suggests that CooB may also play an important role in the stabilization and activity of the CooC outer membrane complex (Voegele *et al.*, 1997).

2.1.4 Curli

Curli are 4 to 7 nm wide extracellular organelles assembled by *E. coli* and *S. enterididis* (Collison *et al.*, 1991; Olsen *et al.*, 1989) (Fig. 11).

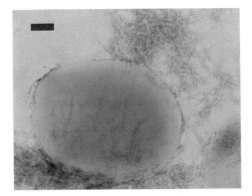

Fig. 11.Curli. An electron micrograph of a negatively stained *E. coli* producing curli. The bar indicates 200 nm.

The adhesive nature of curli allows binding to fibronectin (Olsen *et al.*, 1989), plasminogen (Sjobring *et al.*, 1994), and human contact phase proteins (Ben Nasr *et al.*, 1996). Curli are distinguished by their fibrous appearance, ability to resist protease digestion, ability to bind the diazo dye Congo Red, and to remain insoluble when boiled in 1% SDS (Collinson *et al.*, 1991; Collinson *et al.*, 1999). Interestingly, these are properties shared with eukaryotic amyloid fibers, which are hallmarks of many human aliments including Alzheimer's disease, Parkinson;s disease, the prion diseases as well as systematic amyloidosis. Curli biogenesis is directed by the nucleation/precipitation assembly machinery and represents a new paradigm for understanding amyloid formation. Two *csg* operons, *csgBA* and *csgDEFG*, direct curli formation. The major structural component of *E.*

222

coli curli is the 15.3 kDa CsgA protein. Unpolymerized CsgA is predicted to be secreted into the extracellular milieu where it is subsequently primed for nucleation by the minor curli subunit, CsgB (Bian and Normark, 1997) (Fig. 12). CsgA secretion is dependent on the outer membrane-localized lipoprotein CsgG (Chapman *et al.,* 2002; Loferer *et al.,* 1997). In the absence of CsgE, CsgA is unstable and very few curli are assembled, suggesting CsgE is a chaperone for CsgA. CsgF mutants secrete wild type levels of CsgA subunits, although CsgA is not efficiently assembled into curli (Chapman *et al.,* 2002). CsgF may work independently or in concert with CsgB to guide *in vivo* extracellular nucleation of CsgA.

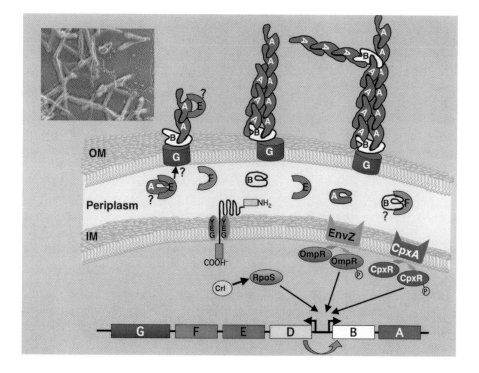

Fig. 12. Model for assembly of Curli (nucleation precipitation pathway). A high resolution electron micrograph of curli on the surface of *E. coli* (inset). A cartoon representation of the model for the nucleation precipitation pathway. Putative roles for the Csg protiens is indicated. Elements involved in regulation of expression of the *csg* genes are indicated.

3. REGULATION OF PILUS ASSEMBLY

The adhesive organelles discussed in this chapter are highly specialized structures that are required for defined situations. In some cases, the presence of these adhesive organelles may be detrimental if present constantly throughout the lifetime of the organism. For example, a bacterium uses pili to attach to a host organism resulting in the instigation of a disease. If pili are produced under circumstances where the receptor for the corresponding adhesin is not present, the bacteria would be wasting metabolic energy as well as providing the host with a target for the immune system. Furthermore, if the disease progresses to the point of host mortality, the bacterium would need to exit the host in order to continue to survive. Highly complex regulatory systems are present in bacteria to ensure that adhesive structures are produced at the appropriate time in the right environment. In support of this idea, cross-talk between adhesin operons has been demonstrated in *E. coli*. PapB, a transcriptional regulator of the *pap* operon, affects transcription of the *fim* operon (Xia *et al.*, 2000) (discussed below).

Again, this section is not intended to be a comprehensive review of all the regulatory systems known for the biogenesis of adhesive organelles. For continuity, the systems chosen for discussion fall into one of the four classes of pili described in the previous section. Although not described here, there have been great insights into the regulation of other pilus systems such as the Tcp (toxin co-regulated pilus) of *Vibrio cholerae*. The regulation of *tcp* expression is subject to a complex regulatory cascade responding to multiple signals including environmental cues such as temperature and pH (Kovacikova and Skorupski, 2001; Skorupski and Taylor, 1999). It is interesting to note that as the name implies, the regulation of pili is coordinated with other virulence factors (Hase and Mekalanos, 1998).

3.1 Regulation of Pilus Biogenesis

3.1.1 P pili

Expression of the *pap* gene cluster is regulated by catabolite activator protein (CAP) (Forsman *et al.*, 1992), PapB and PapI (Baga *et al.*, 1985; Forsman *et al.*, 1989; Kaltenbach *et al.*, 1998), the leucine-response regulatory protein (Lrp) (Kaltenbach *et al.*, 1998; Van der Woude *et al.*, 1995; Weyand and Low, 2000), deoxyadenosine methylase (DAM) (Blyn *et al.*, 1990; Braaten *et al.*, 1991; Nou *et al.*, 1995) and the histone-like protein (HNS) (Van der Woude *et al.*, 1995; White-Ziegler *et al.*, 1998). Expression from the P_{BA} promoter is subject to regulation by phase variation (Blyn *et al.*, 1989). The position of the sites for these effectors on the *pap* gene promoters is depicted in Fig. 13. Under catabolite repression, expression of *papB* and *papI* is inhibited resulting in the lack of pilus biogenesis. When the GATC-I site is methylated, the Lrp is bound to sites 1-3, which overlap the promoter for *papB* resulting in the lack of pilus biogenesis. Methylation of GATC-II abolishes the binding of Lrp to sites 1-3 and the Lrp binds to sites 4 and 5 as a complex with PapI. This then allows transcription of the *pap* gene cluster, and pilus biogenesis proceeds.

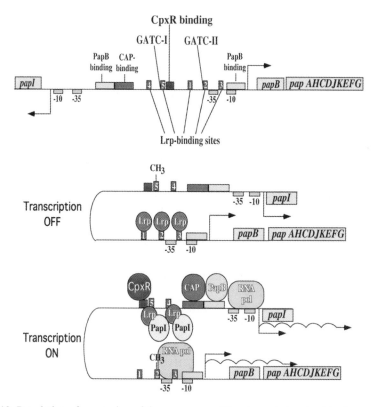

Fig. 13. Regulation of expression of the *pap* operon. The elements involved in the regulation of expression of the *pap* gene cluster are indicated.

Two component regulatory systems play important roles in the regulation of transcription in bacteria. Of these, the Cpx system consisting of an inner membrane histidine kinase, CpxA, and a cytoplasmic response regulator, CpxR, responds to the accumulation of misfolded proteins in the periplasm as well as to stresses affecting the bacterial cell envelope (Snyder *et al.,* 1995) by increasing the production of proteins which promote folding in the periplasm such as the prolyl isomerase PpiA, DsbA which promotes disulfide bond formation, as well as the DegP protease which degrades misfolded proteins (Jones *et al.,* 1997). CpxA functions as an autokinase, and the phosphorylated form of CpxA acts as a phosphatase (Raivio and Silhavy, 1997). Phosphorylation of CpxR by CpxA enhances the ability of CpxR to bind to the regulatory sequences of target genes (Pogliano *et al.,* 1997; Raivio and Silhavy, 1997), ultimately leading to an increase in transcription of the target genes. Specifically during pilus biogenesis, the Cpx pathway responds to the presence of OFF pathway pilus subunits that do not correctly enter into the biogenesis pathway. Furthermore, the *pap* operon is one of the targets of CpxR regulation (Hung *et al.,* 2001). A DNA-binding site for CpxR was discovered in the promoter region of the *pap* gene cluster (Hung *et al.,* 2001) (Fig. 13). The induction of *pap* expression by CpxR was capable of overcoming catabolite repression of the *pap* gene cluster (Hung *et al.,* 2001). Pilus assembly leads to the phosphorylation of CpxR, presumably due to the presence of misfolded non-functional pilin subunits in the periplasm, which consequently results in an increase in the synthesis of the periplasmic accessory factors which promote folding of pilins further committing the bacterium to synthesize pili. This would in turn lead to an increase in the ability of the bacterium to adhere to host cells.

The type 1 gene cluster encodes for two recombinases that are required for phase variation, FimB and FimE (Gally *et al.,* 1996; Klemm, 1986). The main promoter of the type 1 operon lies within a 314 bp DNA element that serves as a switch for fim expression (Abraham *et al.,* 19985; Blomfield *et al.,* 1991;Dorman and Higgens, 1987; McClain *et al.,* 1991). FimE only acts to flip the promoter from on-to-off (Blomfield *et al.,* 1991; Gally *et al.,* 1996). FimB is capable of flipping the promoter in both directions, off-to-on and on-to-off (Gally *et al.,* 1996). The P pili regulator PapB acts to inhibit FimB activity and acts to increase expression of FimE to keep the promoter in the "off" orientation (Holden *et al.,* 2001). The cross-talk between systems and the redundancy in activity of the regulators supports the theory that the timing of synthesis of adhesive organelles is critical in the survival of bacteria.

3.1.2   Type 4 pili

Synechocystis is a freshwater cyanobacterium that exhibits three movements in response to light. These movements include: phototaxis to orient the cell with respect to the direction of the light, photophobic movement, and photokinesis. Phototaxis is dependent upon the presence of type 4 pili. TaxD contains a domain that has homology to the phy domains of cyanobacterial and vascular plant phytochromes which sense light. The phy domains bind linear tetrapyrrole chromophores resulting in the absorption of light. Inactivation of *taxD* leads to the abolishment of type 4 synthesis, but does not affect the level of *pilA* mRNA. This suggests that the regulatory effect of TaxD on pilus biogenesis was at the posttranscriptional level (Bhaya, 2001). The authors propose the following model for the regulation of pilus biogenesis by light. The TaxD protein absorbs light from the chromophore that would then interact with TaxAY perhaps through the scaffold molecule TaxW. This light-activated complex would trigger autophosphorylation of TaxAY, which would then act on PilT ultimately leading to positive orientation towards the light. As in other type 4 pilus systems, inactivation of *pilT* leads to hyperpiliation, suggesting that PilT plays a role in regulating the length of pili during biogenesis.

A regulatory system that influences expression of pili has also been described in *Caulobacter crescentus*. Although the major pilin subunit PilA has homology with the type 4 pilus, the remainder of the genes required for assembly of the pilus are similar to the pilus system of *Actinobacillus actinomycetemcomitans* (Haase *et al.*, 1999; Inoue *et al.*, 1998). As a consequence of an asymmetric cell division in *Caulobacter*, a stationary stalk cell releases a swarmer cell that is incapable of entering the cell cycle until differentiation into a stalk cell. Among the most dramatic advances that have come from studies in *Caulobacter*, was the discovery of CtrA, a protein that plays a major role in regulating transcription of a number of cell cycle-regulated genes (Domian *et al.*, 1997) in addition to flagellar biosynthesis (Quon *et al.*, 1996) (Fig. 14). After completion of cell division, CtrA is present only in the released swarmer cells (Domian *et al.*, 1997). The major role for CtrA is to bind to chromosomal sites surrounding the origin, thereby preventing premature reinitiation of chromosome replication until the appropriate time in the cell cycle (Quon *et al.*, 1998). In addition to a single polar flagellum on the swarmer cell, several pili are observed on this flagellated pole. Conversely, pili are not observed on the stalk cell. Skerker and Shapiro (2000) recently investigated the regulation of pili expression in *C. crescentus*. The presence of CtrA in postdivisional swarmer cells is responsible for the expression and subsequent assembly of pili on the flagellated pole of the swarmer cell (Skerker and Berg, 2001).

Another important aspect in the synthesis of pili includes regulating the number of organelles per cell. Thus far, very little is known about how the bacterial cell determines the number of organelles to place on the cell surface. One organism for which a model can be invoked based upon experimental data is the type 4 pili system in *Neisseria gonorrhoeae*.

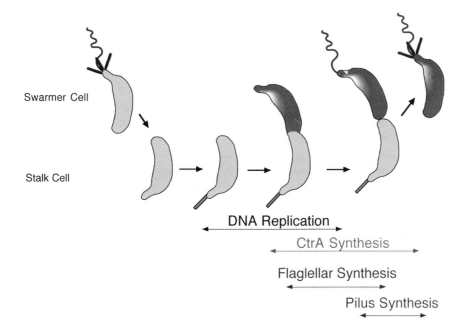

Fig. 14. Regulation of type IV pilus gens by CtrA. A cartoon representation of the effects of CtrA expression of morphology of *C. crescentus*. Production of CtrA (green cells) is required for biogenesis of flagella and pili. The relationships between differentiation and initiation of cell division with pilus biogenesis are indicated.

The Gram-negative diplococcus *N. gonorrhoeae* is the etiologic agent responsible for the sexually transmitted disease gonorrhea. The presentation of type 4 pili on the surface of *N. gonorrhoeae* is essential for natural competence (Sparling *et al.,* 1986) and for adherence and invasion of host tissues (Mosleh *et al.,* 1997). Long and colleagues (2001) placed the *N. gonorrhoeae pilE* gene under control of the regulatable promoter, P_{lac}. They discovered that the number of pili per cell was a direct consequence of the level of *pilE* induction. For example, at low levels of induction, one pilus of wild type length was observed on a few cells in the population. At very high levels of induction, multiple and bundled pili were observed on all cells in the population. The authors proposed a model suggesting that at low levels of expression, PilE is targeted to the inner membrane where only one assembly site can be utilized. Only upon detection of increased levels of PilE are other assembly sites activated. This hypothesis suggests that the assembly sites for pilus biogenesis are regulated, at least, in response to PilE concentration. In contrast to P and type 1 pili, genes required for type 4 biogenesis typically are not encoded in operons. The presence of genes in monocistronic messages would allow careful modulation of the individual components of the system and would allow fine-tuning of the expression in response to different growth environments.

In many bacterial systems, retraction of the pilus would provide a mechanism to facilitate specific bacterial actions. In *E. coli*, it appears that the pili in contact with uroepithelial cells are significantly shorter than the other pili on the same bacterial cell. If the pili were capable of retracting, the shortened pili would then "pull" the bacterium closer to the host cell. This could increase the possibility of invasion into the host cell. In other bacterial species where motility against a fixed surface is a normal mode of transportation, pilus retraction is a gratifying hypothesis for the mechanism of movement via pilus organelles. Evidence is mounting that suggests that pilus retraction may, in fact, be a normal activity of pilus function.

It has been suggested that adsorption of bacteriophage was propagated by retraction of the pilus (Marvin and Hohn, 1969). The role of pilus retraction in translocation of bacteriophage across the cell membrane was strengthened by studies in *E. coli*. (Jacobson, 1972; Novotny and Fivestaylor, 1974), *P. aeruginosa* (Bradley, 1972a, 1972b, 1974), and in *C. crescentus* (Skerker and Shapiro, 2000; Skerker and Berg, 2001). Additionally, pilus retraction was implicated in promoting cell contact during conjugation. Recently, the application of fluorescent microscopy with live cells provided a means to define pilus extension and retraction in *P. aeruginosa* (Skerker and Berg, 2001). In these experiments, the pili were covalently labeled with a fluorescent dye. The extension and retraction of pili were clearly visible. The rate of retraction was similar to the rate of motility, suggesting that pilus retraction provided the force which drives motility. Furthermore, it was determined that the pilus was too flexible for extension to contribute to the motility of the bacterium. This work lends further support to the observations of pilus retraction in *N. gonorrhoeae* (Merz *et al.*, 2000) and *Myxococcus xanthus* (Sun *et al.*, 2000).

3.1.3   Curli

In most laboratory strains of *E. coli*, curli expression is maximal during stationary phase, after growth at temperatures below 32 °C (Arnqvist *et al.*, 1994) and under conditions of low osmolarity (Olsen *et al.*, 1989). The gene products of two divergently transcribed operons are required for assembly of Curli (Hammar *et al.*, 1995) (Fig. 12). Expression of *csgB* and *csgA* occurs in cells that have entered stationary phase after growth at temperatures below 32 °C (Arnqvist *et al.*, 1994) under conditions of low osmolarity (Olsen *et al.,* 1989). Transcription from P_{csgBA} also requires the activity of a curli specific protein, CsgD (Hammar *et al.,* 1995), which has homology with the helix-turn-helix motif of the FixJ family of transcriptional regulators (Zogaj *et al.,* 2001). Transcription from a single promoter upstream of *csgBA* (P_{csgBA}) requires the activity of the stationary phase specific sigma factor (σ^s) (*rpoS*) (Olsen *et al.,* 1993). The histone-like protein, H-NS, is responsible for repression of transcription of *csgA* until the repression activity of H-NS is relieved by the activity of σ^s (Hammar *et al.,* 1995).

4. CONCLUSION

Thus far, the investigation of the adhesive organelles of bacteria has provided new insights into the molecular details in the assembly of macromolecular complexes. In the case of the type 1 and P pili, novel protein-protein interactions have been unveiled. In addition, the molecular role of chaperones in the proper folding of individual proteins has been elucidated. It is quite likely that understanding the protein-protein interactions from the adhesive organelles will serve as paradigms for assembly of protein complexes in other systems and species. Moreover, since the majority of the organelles discussed here are critical in the lifestyles and/or pathogenesis of the organism, the interference with the assembly or activity of these organelles may dramatically alter the environments in which they can persist and/or survive.

5. REFERENCES

Abraham, J.M., Freitag, C.S., Clements, J.R. and Eisenstein, B.I. (1985) Proc Natl Acad Sci USA 82: 5724-5727.

Arnqvist, A., Olsen, A. and Normark, S. (1994) Mol Microbiol 13: 1021-1032.

3. Baga, M., Goransson, M., Normark, S. and Uhlin, B.E. (1985) EMBO J 4: 3887-3893.

Baga, M., Norgren, M. and Normark, S. (1987) Cell 49: 241-251.

Barnhart, M.M., Pinkner, J., Soto, G., Sauer, F.G., Langermann, S., Waksman, G., Frieden, C. and Hultgren, S.J. (2000) Proc Nat Acad Sci USA 97: 7709-7714.

Ben Nasr, A., Olsen, A., Sjobring, U., Muller-Esterl, W. and Bjorck, L. (1996) Mol Microbiol 20: 927-935.

Bhaya, D., Takahashi, A., Shahi, P. and Grossman, A.R. (2001) J Bacteriol 183: 6140-3.

Bian, Z. and Normark, S. (1997) EMBO J 16: 5827-5836.

Bieber, D., Ramer, S.W., Wu, C.Y., Murray, W.J., Tobe, T., Fernandez, R. and Schoolnik, G.K. (1998) Science 280: 2114-2118.

Bitter, W., Koster, M., Latijnhouwers, M., de Cock, H. and Tommassen, J. (1998) Mol Microbiol 27: 209-219.

230

Blomfield, I.C., McClain, M.S., Princ, J.A., Calie, P.J. and Eisenstein, B.I. (1991) J Bacteriol 173: 5298-5307.

Blum, G., Falbo, V., Caprioli, A. and Hacker, J. (1995) FEMS Microbiol Lett 126: 189-195.

Blyn, L.B., Braaten, B.A., White, Z.C., Rolfson, D.H. and Low, D.A. (1989) EMBO Journal 8: 613-620.

Blyn, L.B., Braaten, B.A. and Low, D.A. (1990) EMBO J 9: 4045-4054.

Braaten, B., Blyn, L.B., Skinner, B.S. and Low, D.A. (1991) J Bacteriol 173: 1789-1800.

Bradley, D.E. (1972a) J Gen Microbiol 72: 303-319.

Bradley, D.E. (1972b) Biochem Biophys Res Commun 47: 142-149.

Bradley, D.E. (1974) Virol 58: 149-163.

Brinton Jr., C.C. (1965) Trans NY Acad Sci 27: 1003-1165.

Bullitt, E. and Makowski, L. (1995) Nature 373: 164-167.

Chapman, M., Robinson, l., Pinkner, J., Roth, R., Heuser, J., Hammar, M., Normark, S., Hultgren, S.J. (2002) Science 295: 851-855.

Choudhury, D., Thompson, A., Stojanoff, V., Langermann, S., Pinkner, J., Hultgren, S.J. and Knight, S.D. (1999) Science 285, 1061-1066.

Collinson, S.K., Emody, L., Muller, K.-H., Trust, T. and Kay, W.M. (1991) J Bacteriol 173: 4773-4781.

Collinson, S.K., Parker, J.M., Hodges, R.S. and Kay, W.W. (1999) J Mol Biol 290: 741-756.

Dodson, K.W., Jacob-Dubuisson, F., Striker, R.T. and Hultgren, S.J. (1993) Proc Natl Acad Sci USA 90: 3670-3674.

Dodson, K.W., Pinkner, J.S., Rose, T., Magnusson, G., Hultgren, S.J. and Waksman, G. (2001) Cell 105: 733-743.

Domian, I.J., Quon, K. C. and Shapiro, L. (1997) Cell 90: 415-424.

Dorman, C.J. and Higgins, C.F. (1987) J Bacteriol 169: 3840-3843.

Forsman, K., Goransson, M. and Uhlin, B.E. (1989) EMBO J 8: 1271-1277.

Forsman, K., Sonden, B., Goransson, M. and Uhlin, B.E. (1992) Proc Natl Acad Sci USA 89: 9880-9884.

Gally, D.L., Leathart, J. and Blomfield, I.C. (1996) Mol Microbiol 21: 725-738.

Guyer, D.M., Kao, J.S. and Mobley, H.L. (1998) Infect Immun 66: 4411-4417.

Haas, R. and Meyer, T. (1986) Cell 44: 107-115.

Haase, E.M., Zmuda, J.L. and Scannapieco, F.A. (1999) Infect Immun 67: 2901-2908.

Hacker, J., Bender, L., Ott, M., Wingender, J., Lund, B., Marre, R. and Goebel, W. (1990) Microbial Pathog 8: 213-225.

Hagblom, P., Segal, E., Billyard, E. and So, M. (1985) Nature 315: 156-158.

Hammar, M., Arnqvist, A., Bian, Z., Olsen, A. and Normark, S. (1995) Mol Microbiol 18: 661-670.

Hase, C.C. and Mekalanos, J.J. (1998) Proc Natl Acad Sci USA 95: 730-734.

Holden, N.J., Uhlin, B.E. and Gally, D.L. (2001) Mol Microbiol 42: 319-330.

Holmgren, A. and Brändén, C. (1989) Nature 342: 248-251.

Hultgren, S.J., Porter, T.N., Schaeffer, A.J. and Duncan, J.L. (1985) Infect Immun 50: 370-377.

Hung, D.L., Knight, S.D., Woods, R.M., Pinkner, J.S. and Hultgren, S.J. (1996) EMBO J 15: 3792-3805.

Hung, D.L., Raivio, T.L., Jones, C.H., Silhavy, T.J. and Hultgren, S.J. (2001) EMBO J 20, 1508-1518.

Inoue, T., Tanimoto, I., Ohta, H., Kato, K., Murayama, Y. and Fukui, K. (1998) Microbiol Immunol 42: 253-258.

Jacob-Dubuisson, F., Heuser, J., Dodson, K., Normark, S. and Hultgren, S.J. (1993) EMBO J. 12: 837-847.

Jacobson, A. (1972) J Virol 10: 835-843.

Jones, C.H., Jacob, D.F., Dodson, K., Kuehn, M., Slonim, L., Striker, R. and Hultgren, S.J. (1992) Infect Immun 60: 4445-4451.

Jones, C.H., Danese, P.N., Pinkner, J.S., Silhavy, T.J. and Hultgren, S.J. (1997) EMBO J 16: 6394-6406.

Kaltenbach, L., Braaten, B., Tucker, J., Krabbe, M. and Low, D. (1998) J Bacteriol 180: 1224-1231.

Kaltenbach, L.S., Braaten, B.A. and Low, D.A. (1995) J Bacteriol 177: 6449-6455.

Kaufman, M.R., Seyer, J.M. and Taylor, R.K. (1991) Genes Dev 5: 1834-1846.

Klemm, P. (1986) EMBO J 5: 1389-1393.

Knight, S.D., Berglund, J. and Choudhury, D. (2000) Curr Opin Chem Biol 4: 653-660.

Koomey, J.M., Gotschlich, E.C., Robbins, K., Bergstrom, S. and Swanson, J. (1987) Genetics 117: 391-398.

Kovacikova, G. and Skorupski, K. (2001) Mol Microbiol 41: 393-407.

Krogfelt, K.A., Bergmans, H. and Klemm, P. (1990) Infect Immun 58: 1995-1999.

Kuehn, M.J., Heuser, J., Normark, S. and Hultgren, S.J. (1992) Nature 356: 252-255.

Langermann, S., Palaszynski, S., Barnhart, M., Auguste, G., Pinkner, J.S., Burlein, J., Barren, P., Koenig, S., Leath, S., Jones, C.H. and Hultgren, S.J. (1997) Science 276: 607-611.

Leffler, H. and Svanborg-Eden, C. (1980) FEMS Microbiol Lett 8: 127-134.

Loferer, H., Hammar, M. and Normark, S. (1997) Mol Microbiol 26: 11-23.

Long, C.D., Hayes, S.F., van Putten, J.P., Harvey, H.A., Apicella, M.A. and Seifert, H.S. (2001) J Bacteriol 183: 1600-1609.

Lory, S. and Strom, M.S. (1997) Gene 192: 117-121.

Lu, H.M., Motley, S.T. and Lory, S. (1997) Mol Microbiol 25: 247-259.

Marvin, D.A. and Hohn, B. (1969) Bacteriol Rev 33: 172-209.

McClain, M.S., Blomfield, I.C. and Eisenstein, B.I. (1991) J Bacteriol 173, 5308-5314.

Mehr, I.J. and Seifert, H.S. (1997) Mol Microbiol 23: 1121-1231.

Merz, A.J., So, M. and Sheetz, M.P. (2000) Nature 407: 98-102.

Mosleh, I.M., Boxberger, H.J., Sessler, M.J. and Meyer, T.F. (1997) Infect Immun 65: 3391-3398.

Mulvey, M.A., Lopez-Boado, Y.S., Wilson, C.L., Roth, R., Parks, W.C., Heuser, J. and Hultgren, S.J. (1998) Science 282: 1494-1497.

Nou, X., Braaten, B., Kaltenbach, L. and Low, D.A. (1995) EMBO J 14: 5785-5797.

Novotny, C.P. and Fives-Taylor, P. (1974) J Bacteriol 117: 1306-1311.

Ofek, I., Mirelman, D. and Sharon, N. (1977) Nature 265: 623-625.

Olsen, A., Jonsson, A. and Normark, S. (1989) Nature 338: 652-655.

Olsen, A., Arnqvist, A., Hammar, M., Sukupoli, S. and Normark, S. (1993) Mol Microbiol 7: 523-536.

Parge, H.E., Forest, K.T., Hickey, M.J., Christensen, D.A., Getzoff, E.D. and Tainer, J.A. (1995) Nature 378: 32-38.

Pogliano, J., Lynch, A.S., Belin, D., LIn, E.C.C. and Beckwith, J. (1997) Genes Dev. 11: 1169-1182.

Quon, K.C., Marczynski, G.T. and Shapiro, L. (1996) Cell 84: 83-93.

Quon, K.C., Yang, B., Domian, I.J., Shapiro, L. and Marczynski, G.T. (1998) Proc Natl Acad Sci USA 95: 120-125.

Raivio, T.L. and Silhavy, T.J. (1997) J Bacteriol 179: 7724-7733.

Ramphal, R., Koo, L., Ishimoto, K.S., Totten, P.A., Lara, J.C. and Lory, S. (1991) Infect Immun 59: 1307-1311.

Roberts, J.A., Hardaway, K., Kaack, B., Fussell, E.N. and Baskin, G. (1984) J Urol 131: 602-607.

Rudel, T., Scheuerpflug, I. and Meyer, T.F. (1995) Nature 373: 357-359.

Russel, M., Linderoth, N.A. and Sali, A. (1997) Gene 192: 23-32.

Sajjan, U.S., Sun, L., Goldstein, R. and Forstner, J.F. (1995) J Bacteriol 177: 1030-1038.

232

Sakellaris, H., Balding, D.P. and Scott, J.R. (1996) Mol Microbiol 21: 529-241.

Sakellaris, H., Penumalli, V.R. and Scott, J.R. (1999) J Bacteriol 181: 1694-1697.

Sauer, F.G., Futterer, K., Pinkner, J.S., Dodson, K.W., Hultgren, S.J. and Waksman, G. (1999) Science 285: 1058-1061.

Saulino, E.T., Thanassi, D.G., Pinkner, J. and Hultgren, S.J. (1998) The EMBO J 17: 2177-2185.

Saulino, E.T., Bullitt, E. and Hultgren, S.J. (2000) Proc Natl Acad Sci USA 97: 9240-9245.

Schilling, J.D., Mulvey, M.A. and Hultgren, S.J. (2001) Urology 57: 56-61.

Seifert, H.S., Ajioka, R.S., Marchal, C., Sparling, P.F. and So, M.A. (1988) Nature 336: 392-395.

Sharon, N. (1987) FEBS Lett 217: 145-157.

Sjobring, U., Pohl, G. and Olsen, A. (1994) Mol Microbiol 14: 443-452.

Skerker, J.M. and Shapiro, L. (2000) EMBO J 19, 3223-3234.

Skerker, J.M. and Berg, H.C. (2001) Proc Natl Acad Sci USA 98: 6901-6904.

Skorupski, K. and Taylor, R.K. (1999) Mol Microbiol 31: 763-771.

Snyder, W.B., Davis, L.J., Danese, P.N., Cosma, C.L. and Silhavy, T.J. (1995) J Bacteriol 177: 4216-4223.

Sommer, J.M. and Newton, A. (1988) J Bacteriol 170: 409-415.

Sparling, P.F., Cannon, J.G. and So, M. (1986) J Infect Dis 153: 196-201.

Striker, R., Nilsson, U., Stonecipher, A., Magnusson, G. and Hultgren, S.J. (1995) Mol Microbiol 16: 1021-1030.

Strom, M.S. and Lory, S. (1992) J Bacteriol 174: 7345-7351.

Strom, M.S., Nunn, D.N. and Lory, S. (1993) Proc Natl Acad Sci USA 90: 2404.

Sun, H., Zusman, D.R. and Shi, W. (2000) Curr Biol 10: 1143-1146.

Taylor, R.K., Miller, V.L., Furlong, D.B. and Mekalanos, J.J. (1987) Proc Natl Acad Sci USA 84: 2833-2837.

Thanassi, D.G., Saulino, E.T., Lombardo, M.-J., Roth, R., Heuser, J. and Hultgren, S.J. (1998) Proc Natl Acad Sci 95: 3146-3151.

Tonjum, T. and Koomey, M. (1997) Gene 192: 155-163.

Van der Woude, M.W., Kaltenbach, L.S. and Low, D.A. (1995) Mol Mibrobiol 17: 303-312.

Voegele, K., Sakellaris, H. and Scott, J.R. (1997) Proc Natl Acad Sci USA 94: 13257-13261.

Wall, D. and Kaiser, D. (1999) Mol Microbiol 32: 1-10.

Weyand, N.J. and Low, D.A. (2000) J Biol Chem 275: 3192-3200.

White-Ziegler, C.A., Angus Hill, M.L., Braaten, B.A., van der Woude, M.W. and Low, D.A. (1998) Mol Microbiol 28: 1121-1137.

Wolfgang, M., Park, H.S., Hayes, S.F., van Putten, J.P. and Koomey, M. (1998) Proc Natl Acad Sci USA 95: 14973-14978.

Wolfgang, M., van Putten, J.P., Hayes, S.F., Dorward, D. and Koomey, M. (2000) EMBO J 19: 6408-6418.

Wu, H. and Fives-Taylor, P.M. (2001) Crit Rev Oral Biol Med 12: 101-115.

Xia, Y., Gally, D., Forsman-Semb, K. and Uhlin, B.E. (2000) EMBO J 19: 1450-1457.

Zogaj, X., Nimtz, M., Rohde, M., Bokranz, W. and Romling, U. (2001) Mol Microbiol 39: 1452-1463.

Chapter 12

EXPORT OF BACTERIOCINS

Bauke Oudega

Department of Molecular Microbiology, IMBW/BioCentrum Amsterdam
Faculty of Biology, Vrije Universiteit Amsterdam
De Boelelaan 1087
1081 HV Amsterdam, NL

1.   INTRODUCTION

Bacteriocins are bactericidal proteins that are produced by bacteria. They usually kill only bacterial cells that are closely related to the producing bacteria. The narrow host range is determined by the sophisticated uptake mechanisms that these proteins use to enter susceptible bacterial cells. Bacteriocins range in size from rather small toxic peptides to relatively large protein structures of 60-70 kDa. The relatively small bactericidal peptides include microcins and lantibiotics (De Vos *et al.,* 1995), most of which are apparently exported by the producing cell via ABC transporters. This chapter will focus on the larger bacteriocins, especially on colicins and cloacin DF13, and on their mechanism of export. These bacteriocins are the best studied bacteriocins with respect to production, export, uptake and mode of action. These relatively large proteins are produced by *Escherichia coli* strains and by related gram-negative organisms. The mechanism by which these colicins are translocated across both the inner membrane and the outer membrane to reach their extra-cellular destination is totally different from all other mechanisms studied so far. This process appears to be semi-specific and is often referred to as quasi-lysis. The exact mechanism is not known, but it has been applied for the production of heterologous proteins by *E. coli*. In this chapter, the mechanism of export of colicins and of cloacin DF13, and the biotechnological applications of this mechanism of export will be described.

B. Oudega (ed.), Protein Secretion Pathways in Bacteria, 233–247.
© 2003 *Kluwer Academic Publishers. Printed in the Netherlands.*

2. BIOLOGICAL ACTIVITY AND STRUCTURE OF COLICINS

Colicins are plasmid-encoded bacteriocins, produced by *E. coli* strains and capable of killing other *E. coli* cells and cells of closely related species (de Graaf *et al.,* 1986).

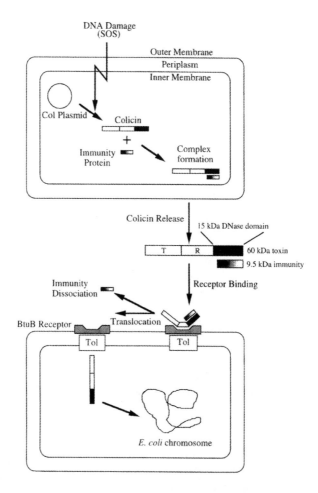

Fig. 1. The formation, secretion and uptake pathway of a DNase type colicin-immunity protein complex (E2). During uptake, the bound immunity protein must dissociate from the DNase, but when and where this occurs is not known. T, colicin translocation domain. R, colicin receptor binding domain. Tol, represents the four Tol proteins that are involved in uptake of this particular colicin (Kleanthous *et al.,* 1998).

Colicins are mostly 50-70 kDa in size and they are released into the extra-cellular environment of the producing cells in relatively large amounts; up to 100-150 mg per liter of culture supernatant has been found. They kill susceptible cells by penetrating the cell envelop and exercising their biological activity. Four distinct types of colicin activity have been identified. The first type forms pores in the cytoplasmic membrane (colicin E1, A, B, Ia, Ib, K, and N), the second type has DNase activity (colicin E2, E7, E8, and E9, and a third type has RNase activity (colicin E3, E4, E6). Finally, the fourth type causes inhibition of synthesis of peptidoglycan and of LPS O-antigen by inhibition of lipid-carrier regeneration (colicin M) (Gross *et al.*, 1996; Lazdunski *et al.*, 1998). These four distinct types of activity are often referred to as type E1, E2, E3 or M.

In order to reach their targets, the bacteriocins first have to cross the outer membrane of a susceptible cell (Fig. 1). Colicins bind to high-affinity receptor proteins in the outer membrane. These receptors often function in the uptake of nutrients or vitamins. After binding to the receptor, the bacteriocin molecules are translocated across the cell envelope via uptake systems which are coupled to the receptor proteins. Two distinct uptake systems are used, the TolA route and the TonB route, by group A and group B bacteriocins, respectively. Each of these pathways involves several periplasmic and inner membrane proteins. For the killing of a single cell, one or a few bacteriocin molecules are in principle enough.

Despite their different modes of action, the various colicins have in principle a similar domain structure. The central region of the bacteriocin protein is involved in binding to the receptor molecules of the target cells, whereas the N-terminal part is important for the uptake and translocation of the bacteriocin. The C-terminal part possesses ionophore activity or the catalytic activity.

Early biophysical studies on the structure of colicins showed that the ratio between the length and the width is relatively large (Konisky, 1982), which indicated that these proteins have an unusual shape (see also Fig. 1). The crystal structure of colicin Ia (colicin E1 type) has been determined recently (Fig. 2; Wiener *et al.,* 1997). This colicin contains 626 amino acid residues, is about 21 nm long and 1-4 nm wide. This unusual structure is long enough to span the complete cell envelope of *E. coli*. The protein consists of three functional domains that are separated by a pair of relative long helices of about 16 nm. A central domain at the bend of the hairpin-like structure (formed by the two helices) forms the outer membrane receptor binding segment. The N-terminal domain involved in translocation is located at the other end of the hairpin. It consists of three 8 nm long helices arranged in a helical sheet. This domain can bind to the receptor after which the bacteriocin can cross the outer membrane and periplasm through the

TonB pathway. The so-called TonB box recognition motif of colicin Ia (a stretch of amino acid residues that can bind TonB) is located on one side of this sheet structure. The third domain, the C-terminal domain containing the biological killing activity (in this case channel formation), is located close to the translocation domain. It is made up of 10 helices, eight amphipatic helices burying two hydrophobic helices. Other structural studies indicate that the structure of colicin Ia will be most likely found in every colicin, although there may be some variation in length of the two central helices. The pore-forming domains of colicin A, E1 and of an N-terminally truncated form of colicin N have also been crystallized and their 3D structure has been determined (Elkins *et al.*, 1997; Parker *et al.*, 1992; Lazdunski *et al.*, 1998). These pore-forming domains have a similar type of structure as the comparable domain of colicin Ia.

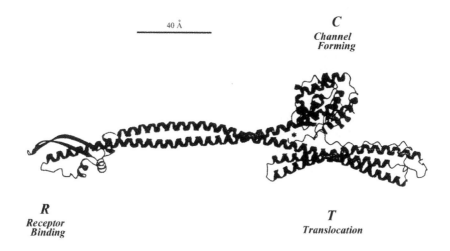

Fig. 2. Ribbon representation of the structure of colicin Ia. The molecule is ± 21 nm long and consists of three functional domains separated by a pair of helices, each 16 nm long. The translocation (T), receptor binding (R) , and channel forming (C) domains are indicated. The location of the N-terminal peptide which includes the TonB box is marked by an asterisk between the T and C domains (Wiener *et al.*, 1997).

3. PRODUCTION AND EXPORT OF COLICINS

The genetic information for the production and export of colicins is mostly located on plasmids (Pugsley, 1984). Two types of colicinogenic plasmids have been identified. Small, multicopy-number plasmids, that are not self-transmissible, encode colicins like A, E1-E9, K, and N. Large, low-copy-number plasmids, that are transmissible, encode colicins like colicins B, Ia, Ib, and M. In addition to the colicin protein itself, colicinogenic cells synthesise a protein that protects the producing cell against the biological activity of the colicin it produces. This protein is called the immunity protein. Most of these immunity proteins are much smaller then their homologous colicin, often about 10 kDa. Immunity against the nuclease-type bacteriocins, i.e. colicin E2 and E3 and related colicins, is provided by binding of the immunity protein in the cytoplasm to the region of the bacteriocin possessing the catalytic (RNase or DNase) activity. These bacteriocins are secreted into the extra-cellular environment as an equimolar complex (heterodimer) with their corresponding immunity protein (see also Fig. 1). In contrast, pore-forming bacteriocins are secreted as monomers and their immunity proteins are localised in the cytoplasmic membrane. These immunity proteins protect the producing cells from exogenous bacteriocin molecules.

The genes encoding the bacteriocin and the immunity protein are located in a gene cluster. Except for colicin Ia, Ib, B, and M, colicin gene clusters contain a third gene which codes for a protein required for the secretion of the bacteriocins, the so-called bacteriocin release protein (BRP, also referred to as 'kil protein' or 'lysis protein').

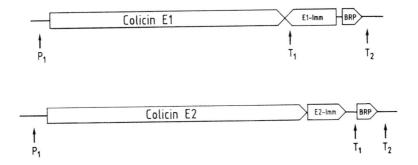

Fig. 3. Structural organization of colicin operons. Two types of operons are indicated, the colicin E1 (pore forming) type and the colicin E2/E3 (Dnase/RNase) type. Genes are indicated by boxes. P_1, (mitomycin C) inducible promoter for the expression of bacteriocin gene, BRP gene and in some cases the immunity protein gene. P_2, promoter for expression of immunity protein. T_1 and T_2, terminators of transcription.

238

Expression of the colicin gene is controlled by a promoter that can be activated by the SOS response, for instance, upon irradiation with UV or by exposure to DNA damaging agents, like mitomycin C. In the case of the nuclease-type of colicins, the genes encoding the immunity protein and the BRP are transcribed from the same promoter as the bacteriocin gene and in the same direction, whereas the immunity protein gene of pore-forming colicins is transcribed from a constitutive promoter in the opposite direction with respect to the colicin and BRP gene (Fig. 3).

4. BACTERIOCIN RELEASE PROTEINS

4.1 Effects of BRP expression

Most colicins are secreted into the extra-cellular medium after expression of their homologous bacteriocin release protein. The colicin itself as well as the immunity protein are not synthesized with an N-terminal signal peptide that could direct these proteins to the Sec-translocase. Colicin are produced as soluble proteins in the cytoplasm. A role for BRP in the secretion of bacteriocins was first shown by analysis of insertion and deletion mutants of the BRP gene of the cloacin DF13 gene cluster (van Tiel-Menkveld *et al.,* 1981; Oudega *et al.,* 1982). Cloacin DF13 is a colicin E3 type of bacteriocin that can also be produced by *E. coli*. The requirement for BRP in bacteriocin release was confirmed by complementation experiments in which the expression of the bacteriocin and its BRP were separately controlled (Luirink *et al.,* 1987, 1988). With respect to the secretion of colicins, the colicin gene cluster encoded BRP's are functionally interchangeable. This suggests at least a common mode of action for the BRP's (Pugsley *et al.,* 1983).

BRP-mediated secretion of bacteriocins appears to be semi-specific. BRP expression leads to bacteriocin release, but also to the release of some cytoplasmic proteins (elongation factor G and Tu, and chloramphenicol acetyltransferase) and many periplasmic proteins (β-lactamase, alkaline and acid phosphatase and RNase I) (van der Wal *et al.,* 1995). Full induction of a colicin gene cluster causes a concomitant full induction of the BRP gene. This results in a decline in culture turbidity which coincides with bacteriocin secretion and the release of other proteins. This process is often called quasi-lysis. Eventually, loss of viability is observed, especially after over-expression of subcloned BRP's. Since strong expression of BRP's causes quasi-lysis and eventually lethality, BRP's are often referred to as 'lysis proteins' or 'kil proteins'. However, these latter names do not justify other reports. It has been shown that moderate and relatively low level expression of (subcloned) BRP under the proper culture conditions has no or almost no significant effect on the culture turbidity of the producing cells, whereas the cells do secrete large amounts of bacteriocin molecules and low levels of other proteins (de Graaf *et al.*, 1986; van der Wal, 1995).

4.2 Requirement of OMPLA

Some of the effects of induction of a colicin gene cluster on the cell envelope resemble the effects of bacteriophage infection or treatment with EDTA, which are known to be caused by activation of the detergent-resistant outer membrane phospholipase A (PldA or OMPLA) of *E. coli*. The activity of OMPLA in *E. coli* cells expressing the colicin E2 gene cluster increases 30- to 40-fold prior to quasi-lysis and bacteriocin release, indicating that these phenomena are the consequences of induced PldA activity (Pugsley *et al.*, 1984). Upon BRP expression, the major outer membrane phospholipid phosphatidylethanolamine (PE) is degraded, and the amounts of free fatty acids and lysoPE, the breakdown products of PE, increase (Pugsley *et al.*, 1984; Luirink *et al.*, 1986; Cavard *et al.*, 1987; Suit *et al.*, 1988). The membrane perturbing properties of the accumulated lysophospholipids might be required for the release of bacteriocins and other proteins (Cullis *et al.*, 1979). Strains defective in OMPLA (*pldA*) are also defective in quasi-lysis and bacteriocin release (Pugsley *et al.*, 1984; Luirink *et al.*, 1986; Cavard *et al.*, 1987). The requirement for OMPLA can be bypassed by other ways of membrane permeabilization. Release of bacteriocins by *pldA* strains, for instance, can be achieved by treatment of cells with the detergent Triton X-100 (Pugsley *et al.*, 1983, 1984; Cavard *et al.*, 1989). Furthermore, colicin A was found to be released by *tolQ* strains, which are known to be leaky for periplasmic proteins. However, in all these cases, functional BRP remains necessary for bacteriocin release, suggesting that BRP causes additional modifications of the cell envelope to provoke bacteriocin release.

Recently, the crystal structure of OMPLA has been solved (Snijder *et al.*, 1999). The protein forms a β-barrel in the outer membrane, in which the barrel consists of 12 antiparallel β-strands. In its inactive form the protein is a monomer. Induction of its activity results in dimerization. Expression of BRP results in activation and dimerization of OMPLA (Dekker *et al.*, 1999). It remains to be investigated whether BRP directly affects OMPLA or whether OMPLA is activated as a results of an effect of BRP on the outer membrane integrity.

4.3   Nature of BRP's

BRP's are synthesized as precursors polypeptides (preBRP's) ranging in size from 45 to 52 amino acid residues. BRP's are very similar in their primary structure. All preBRP's contain a Leu-X-Y-Cys sequence around the signal peptide cleavage site (Fig. 4). This so-called lipobox is required for lipid-modification and processing by signal peptidase II of bacterial lipoproteins (Wu *et al.*, 1986). The lipoprotein nature of various BRP's has been established (Pugsley *et al.*, 1987; Luirink *et al.*, 1987).

BRP's are predominantly found in the outer membrane of *E. coli* and other bacteriocinogenic gram-negative bacteria (van der Wal, 1995). The lipid-modification of BRP's was found to be important for the sub-cellular localization (Luirink *et al.*, 1988) and for functioning, since unmodified mutant BRP's do not function in quasi-lysis and bacteriocin release (Pugsley *et al.*, 1987; Cavard *et al.*, 1987; Luirink *et al.*, 1988; Kanoh *et al.*, 1991).

	Signal peptide	Mature protein (BRP)
CloDF13	MKKAKAIFLFILIVSGFLLVA	CQANYIRDVQGGTVAPSSSSELTGIAVQ
ColE2	MKKITGIILLLLAVIILSA	CQANYIRDVQGGTVSPSSTAEVTGLATQ
ColE3	MKKITGIILLLLAVIILSA	CQANYIRDVQGGTVSPSSTAEVTGLATQ
ColE8	MKKITGIILLLLAVIILAA	CQANYIRDVQGGTVSPSSTAEVTGLATQ
ColE9b	MKKILTGIILLLLAAIILAA	CQANTIRDVQGGTVSPSSSAELTGLATQ
ColE1	MRKKFFVGIFAINLLVG	CQANYIRDVQGGTIAPSSSSKLTGIAVQ
ColA	MKKIIICVILLAIMLLAA	CQVNNVRDTGGGSVSPSSIVTGVSMGSEGVGNP
ColN	MCGKILLILFFIMTLSA	CQVNHIRDVKGGTVAPSSSSRLTGLKLSKRSKDPL

Fig. 4. Primary structure of a number of different BRP's and their signal peptides (van der Wal *et al.*, 1995).

4.4   BRP signal peptides

An unusual feature of most but not all BRP's is that their signal peptide is not proteolytically degraded after processing of the BRP precursor by signal peptidase II. These cleaved off stable signal peptides accumulate exclusively in the cytoplasmic membrane (Howard *et al.*, 1992; van der Wal *et al.*, 1992). Normally, signal peptides are rapidly degraded after cleavage from the precursor (Hussain *et al.*, 1982) by signal peptide peptidases. The stability of the signal peptides is an intrinsic property, since the colicin E2 and cloacin DF13 BRP signal peptides can be visualised on a gel after radio-labelling when they are expressed as separate entities (Kanoh *et al.*, 1991; van der Wal *et al.*, 1992). By constructing hybrid signal peptides, consisting of a part of the unstable signal peptide of the murein lipoprotein of *E. coli* and a part of the stable cloacin DF13 BRP signal peptide, it was found that the N-terminal part of the BRP signal peptide together with the C-terminal alanine residue are important for the observed stability (van der Wal *et al.*, 1994).

Expression of the stable BRP signal peptides, as separate entities, results in quasi-lysis and lethality (Kanoh *et al.*, 1991; van der Wal *et al.*, 1992). The molecular mechanism of how these stable signal peptide cause quasi-lysis and lethality is not clear. The role of stable signal peptides in cloacin DF13 release has been further investigated (Luirink *et al.*, 1991). A BRP derivative with an unstable signal peptide was found to cause quasi-lysis, lethality and the leakage of periplasmic proteins, but it was not capable of provoking the release of cloacin DF13 anymore. This suggests that the stable BRP signal peptide is involved in bacteriocin release (Luirink *et al.*, 1991). These and other experiments (van der Wal *et al.*, 1995) showed that, in addition to the mature lipid-modified BRP, the stable signal peptide is required for efficient bacteriocin release.

5.   COLICIN EXPORT

As described above, the release of colicins and cloacin DF13 into the culture medium is accompanied by the release of a subset of host proteins. The stable BRP signal peptide, the mature BRP itself and OMPLA play a role in this process. Colicins and cloacin DF13 accumulate in the cytoplasm of cells that are defective in release due to a OMPLA or BRP mutation (Cavard *et al.*, 1987; van der Wal *et al.*, 1992). This suggests that the stable BRP signal peptides, the mature BRP and OMPLA together form trans-envelope pores or export-regions in the cell envelope through which the bacteriocins and a number of cytoplasmic proteins leave the cells. The release of periplasmic proteins concomitant with bacteriocin release might be explained by the activation of OMPLA in regions of the outer membrane that are not part of the trans-envelope pores.

Attempts to visualise these putative trans-envelope pores by electron microscopy have not been successful, however, some indirect evidence for their existence has been obtained. Membranes of BRP producing cells are difficult to separate by isopycnic sucrose density gradient centrifugation experiments, and proteins of the two membranes are detected in fractions of intermediate density (Howard *et al.,* 1991; Howard *et al.,* 1992; unpublished observations). This suggests that upon expression of a BRP, the density of the various membrane fractions is changed. Membranes of *pldA* strains producing the colicin A BRP are also difficult to separate, indicating that BRP's cause alterations in the cell envelope independent of PldA (Howard *et al.,* 1991). As discussed above, neither BRP nor OMPLA alone are able to provoke the release of bacteriocins, which supports the idea that the simultaneous effects of the BRP and PldA are required for the formation of the putative trans-envelope pores or translocation sites. The stable signal peptides might contribute to the formation of trans-envelope pores at the level of the cytoplasmic membrane. The mature BRP might cause damage to the cytoplasmic and especially to the outer membranes upon which PldA is activated, possibly by the membrane perturbing activity of the BRP, analogous to OMPLA activation by the membrane perturbants (Horrevoets *et al.,* 1991). The bacteriocins might leave the cells through these proposed pores or translocation sites because that have that special long-shaped structure and they are soluble proteins in the cytoplasm not interacting with other components. In short, they have the right shape and they are available. A raison why colicins do not accumulate in the periplasm might be that their special shape prevents this; they are so long that they contact at least one membrane during the translocation process. The other few cytoplasmic proteins that also leave the cell upon BRP expression might fulfil the same requirements. How the bacteriocin molecules and other cytoplasmic proteins are targeted to these translocation sites and how the translocation process is initiated or energized is not known.

6. APPLICATION OF BRP'S IN EXPORT OF HETEROLOGOUS PROTEINS

A lot of effort has been put in the adaptation of the *E. coli* bacteriocin secretion systems for the secretion of heterologous proteins into the medium. Attempts to apply the bacteriocin secretion system for the release of proteins fused to colicin molecules have not been successful (Cavard *et al.,* 1986; Pugsley *et al.,* 1986; Géli *et al.,* 1989; Luirink, 1989). With respect to the proposed mechanism of secretion, it is now clear that colicins do not possess secretion signals and that their release is dependent on their special structure. This may be the most important reason why fusion proteins are not released by cells expressing BRP (Baty *et al.,* 1987). Nevertheless, the observation that BRP's can promote the leakage of periplasmic proteins has led to the use of subcloned BRP's in achieving the release of heterologous proteins into the culture medium from the *E. coli* periplasm and in one case from the cytoplasm (Lloubès *et al.,* 1993).

Proteins of different origin and size have been released into the culture medium by simultaneously expressing the target protein and a BRP. In most cases a cleavable N-terminal signal peptide was used to direct the target protein into the periplasm. The BRP's were expressed in moderate amounts in order to achieve the release of the target protein and at the same time to avoid quasi lysis and lethality. Several proteins of prokaryotic and human origin have been released by using a one-plasmid system encoding both the colicin E1 BRP and the target protein. In this system, the BRP is constitutively expressed at a low level. The target protein is expressed from a *Bacillus* penicillinase promoter and is directed into the periplasm by its natural signal peptide or the *Bacillus* penicillinase signal peptide (see review van der Wal *et al.,* 1995).

To optimise BRP-mediated protein release several expression vectors have been constructed, encoding a BRP controlled by an inducible promoter. The cloacin DF13 BRP and a target protein, the *E. coli* periplasmic molecular chaperone FaeE, were placed under control of separately inducible promoters. Using this system it was found that the efficiency of BRP-mediated protein release increased when a one-plasmid system is used instead of a binary vector system (van der Wal *et al.,* 1995).

Proteins released by the action of a BRP can be easily isolated from the culture medium, as has been shown for human growth hormone (hGH), released by concomitant expression of the cloacin DF13 BRP (Fig. 5; Hsiung *et al.,* 1989). Furthermore, disulfide bridges in hGH released by the action of BRP have been shown to be correctly formed, and the protein showed equal biological activity as compared to that of authentic hGH. Correct formation of disulfide bridges has also been shown for the recombinant human IgG-Fc. This protein was released as a 55-60 kDa dimer with correct interchain as well as intrachain disulfide bridges. Furthermore, a human tumor necrosis was released as a dimer or a trimer, and the biological activity of the released form was higher than that of the cytoplasmic protein, possibly due to differences in the tertiary structure.

244

Secretion of proteins might prevent the formation of inclusion bodies either in the cytoplasm or in the periplasm.

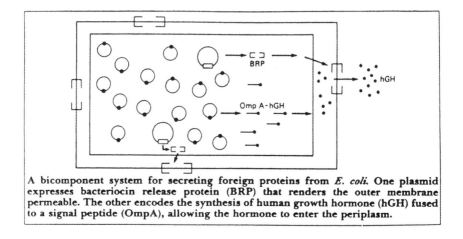

A bicomponent system for secreting foreign proteins from *E. coli*. One plasmid expresses bacteriocin release protein (BRP) that renders the outer membrane permeable. The other encodes the synthesis of human growth hormone (hGH) fused to a signal peptide (OmpA), allowing the hormone to enter the periplasm.

Fig. 5. Schematic representation of the use of the cloacin DF13 BRP for the release of human growth hormone.

For instance, overproduction of hGH has been shown to result in the formation of protein aggregates (Schoner *et al.*, 1985; Lloubès *et al.*, 1993), which can be prevented by simultaneously expressing a BRP. The colicin A BRP was used to obtain extra-cellular correctly folded and biological active hGH (Lloubès *et al.*, 1993). In this study, the hormone peptide, N-terminally extended with a methionine residue (Met-hGH), was directly released from the cytoplasm. The amount of accumulated Met-hGH was about 20 times higher than that of mature hGH, targeted by its natural signal peptide to the periplasm, since the precursor was slowly processed and appeared to be sensitive to proteolysis in the cytoplasm. This also shows that efficient secretion vectors are indispensable for optimal protein release by a BRP.

Other studies, directed towards the applicability of BRP's on a semi-large scale (Aono 1988; Yu *et al.*, 1992), in a 20 L fermentor (Steidler *et al.*, 1994), and in a cell recycle system (Yu *et al.*, 1993) have shown that it is feasible to use BRP's in continuous cultures in order to isolate large amounts of a protein of interest.

As described above, there are many examples of the application of the BRP gene for the release of (heterologous) proteins by *E. coli*. However, high expression levels of BRP cause quasi-lysis and lethality. These phenomena are in part caused by accumulation of the stable cleaved off BRP signal peptides in the cytoplasmic membrane, and in part by the mature lipid-modified BRP's. BRP has been further optimised for protein secretion, especially for secretion of proteins out of the periplasm, by using less stable signal peptide for its targeting (van der Wal *et al.*, 1998). Furthermore, mutant BRP derivatives were constructed that are strongly reduced in their capacity to cause quasi-lysis and lethality. These optimal BRP's are still capable of inducing the release of β-lactamase from the periplasm (van der Wal *et al.*, 1998). The use of these optimized, mutated BRP's will improve the further application of BRP in the release of heterologous proteins from the *E. coli* periplasm into the culture medium.

7.   FUTURE PERSPECTIVES

The mechanism of export of bacteriocins is not completely clear, especially the role of the BRP and its signal sequence have to be further analyzed. But this type of research now depends on the development of highly sophisticated techniques that can be used to visualize in great detail molecular interactions in a single cell. Nevertheless the BRP and derivatives have been used for the export of proteins into the culture medium. The use of this system, will, however, be restricted to the production of proteins for scientific and/or pharmaceutical interest. The amounts secreted are just not high enough for large scale (bulk) production of proteins of commercial interest.

8.   REFERENCES

Aono, R. (1988) Appl Microbiol Biotechnol 28: 414-418.
Baty, D., Lloubès R., Géli, V., Lazdunski, C. and Howard, S.P. (1987) EMBO J 6: 2463-2468.
Cavard, D., Crozel, V., Gorvel, J.P., Pattus, F., Baty, D. and Lazdunski, C. (1986) J Mol Biol 187: 449-459.
Cavard, D., Baty, D., Howard, S.P., Verhey, H.M. and Lazdunski, C. (1987) J Bacteriol 169: 2187-2194.
Cavard, D., Howard, S.P. and Lazdunski, C. (1989) J Gen Microbiol 135: 1715-1726.

246

Cavard, D. (1991) J Bacteriol 173: 191-196.

Cullis, P.R. and De Kruijff, B. (1979) Biochim Biophys Acta 559: 399-420.

De Graaf, F. K. and Oudega, B. (1986) Curr Top Microbiol Immunol 125: 183-205.

Dekker, N., Tommassen, J. and Verhey, H.M. (1999) J Bacteriol 181: 3281-3283.

De Vos, W.M., Kuipers, O.P., Van der Meer, J.R. and Siezen, R.J. (1995.) Mol Microbiol 17: 427-437.

Elkins, P., Bunker, A., Cramer, W.A. and Stauffacher, V.V. (1997) Structure 5: 443-458.

Géli, V., Baty, D., Knibiehler, M., Lloubès, R., Pessegue, B., Shire, D. and Lazdunski, C. (1989) Gene 80: 129-136.

Gross, P. and Braun V. 1996) Mol Gen Genet 251: 388-396.

Horrevoets, A.J.G., Francke, C., Verheij, H.M. and De Haas, G.H. (1991) Eur J Biochem 198: 255-261.

Howard, S.P., Leduc, M., van Heijenoort, J. and Lazdunski, C. (1987) FEMS Microbiol Lett 42: 147-151.

Howard, S.P., Cavard, D. and Lazdunski, C. (1991) J Gen Microbiol 137: 81-89.

Howard, S.P. and Lindsay, L. (1992) In: Bacteriocins, microcins and lantibiotics. (James R., Lazdunski C., Pattus F.). Springer-Verlag, Berlin 317-329.

Hsiung, H.M., Mayne, N.G. and Becker, G.W. (1986) Bio/Technol 4: 991-995.

Hsiung, H.M., Cantrell,, A., Luirink, J., Oudega, B., Veros, A.J. and Becker, G.W. (1989) Bio/Technol 7: 267-271.

Hussain, M., Ozawa, Y., Ichihara, S. and Mizushima, S. (1982) Eur J Biochem 129: 223-239.

Kanoh, S., Masaki, H., Yajima, S., Ohta, T. and Uozumi, T. (1991) Agric Biol Chem 55: 1607-1614.

Kleanthous, C., Hemmings, A.M., Moore, G.R. and James, R. (1998) Mol Microbiol 28: 227-233.

Konisky, J. (1982) Ann Rev Microbiol 36: 125-1 44.

Lazdunski, C.J., Bouveret, E., Rigal, A., Journet, L., Lloubes, R. and Benedetti, H. (1998) J Bacteriol 180: 4993-5002.

Lloubès, R., Vita, N., Bernadac, A., Shie D., Leplatois, P., Géli V., Frenette, M., Knibiehler M., Lazdunski, C. and Baty, D. (1993) Biochimie 75: 451-458.

Luirink, J., van der Sande, C., Tommassen, J., Veltkamp, E., de Graaf, F.K. and Oudega, B. (1986) J Gen Microbiol 132: 825-834.

Luirink, J., de Graaf F.K. and Oudega, B. (1987) Mol Gen Genet 206: 126-132.

Luirink, J., Hayashi, S., Wu, H.C., Kater, M.M., de Graaf, F.K. and Oudega, B. (1988) J Bacteriol 170: 4153-4160.

Luirink, J., Watanabe, T., Wu, H.C., Stegehuis, F., de Graaf, F.K. and Oudega, B. (1987) J Bacteriol 169: 2245-2250.

Luirink, J. (1989) Thesis, Amsterdam.

Luirink, J., Duim, B., de Gier, J.W.L. and Oudega, B. (1991) Mol Microbiol 5: 393-399.

Oudega, B., Stegehuis, F., van Tiel-Menkveld G.J. and de Graaf, F.K. (1982) J Bacteriol 150: 1115-1121.

Parker, M.W., Postma, J.P., Pattus, F., Tucker, A.D. and Tsernoglou D. (1992) J Mol Biol 224: 639-657.

Pugsley, A.P. and Schwarz, M. (1983) J Bacteriol 156: 109-114.

Pugsley A.P. and Schwarz, M. (1983) Mol Gen Genet 190: 366-372.

Pugsley, A.P. (1984) Micro Sci 1: 168-175.

Pugsley, A.P. and Schwartz, M. (1984) EMBO J 3: 2393-2397.

Pugsley, A.P. and Cole, S.T. (1986) J Gen Microbiol 132: 2297-2307.

Pugsley, A.P. and Cole, S.T. (1987) J Gen Microbiol 133: 2411-2420.

Schoner, R.G., Ellis, L.F. and Schoner, B.E. (1985) Bio/Technol 2:151-154.

Steidler, L., Fiers, W. and Remaut, E. (1984) Biotechnol Bioeng 44: 1074-1082.

Snijder, H.J., Ubarretxena-Belandia, I., Blaauw, M., Kalk, K.H., Verhey, H.M., Egmond, M.R., Dekker N. and Dijkstra, B.W. (1999) Nature 401: 717-721.

Suit, J.L. and Luria, S.E. (1988) J Bacteriol 170: 4963-4966.

Wiener, M., Freymann, D., Ghosh, P. and Strouid R.M. (1997) Nature 385: 461-464.

Van der Wal F.J., Oudega B., Kater, M.M., ten Hagen-Jongman, C.M., de Graaf, F.K. and Luirink, J. (1992) Mol Microbiol 6: 2309-2318.

Van der Wal, F.J., Valent, Q.A., ten Hagen-Jongman, C.M., de Graaf, F.K., Oudega, B. and Luirink, J. (1994) Microbiology 140: 369-378.

Van der Wal F.J., Luirink J. and Oudega, B. (1995) FEMS Microbiol Rev 17: 381-399.

Van der Wal, F.J., Koningstein, G., ten Hagen, C.M., Oudega B. and Luirink J. (1998) Appl Env Microb 64: 392-398.

Van Tiel-Menkveld, G.J., Veltkamp, E. and de Graaf, F.K. (1981) J Bacteriol 146: 41-48.

Wu, H.C. and Tokunaga, M. (1986) Curr Top Microbiol Immunol 125:127-157.

Yu, P. and San, K.Y. (1992) Biotech Progress 8: 25-29.

Yu, P. and San, K.Y. (1993) Biotech Progress 9: 587-593.

Chapter 13

BIOGENESIS OF FLAGELLA: EXPORT OF FLAGELLAR PROTEINS VIA THE FLAGELLAR MACHINE

Tohru Minamino[1] and Shin-Ichi Aizawa[2]

[1] Protonic NanoMachine Project, ERATO, JST, 1-7 Hikaridai, Seika, Kyoto, 619-0237, Japan
[2] Department of Biosciences, Teikyo University, 1-1 Toyosatodai, Utsunomiya 320-8551, Japan

1.   INTRODUCTION

The flagellum is a motility apparatus that many bacteria use for responding to various environmental stimuli. The major part of the flagellum is a helical filament, which can thrust the cell body through liquid when it rotates. The rotary force (torque) is generated by the basal structure (therefore called the flagellar motor), which is anchored both in the outer and inner membranes. The basal structure is made of more than 20 different proteins. The bacterial flagellum, a complicated complex made of more than 10 different proteins, is constructed through both the membranes and cell wall towards the outside of the cell. Since most of the component proteins do not retain signal peptides recognized by the general secretion pathway, they have to be secreted by their own export system. The export apparatus consists of a channel and a gate. Flagellar proteins to be exported are transported to the gate with the help of chaperones and pushed into the channel by a flagella-specific ATPase. There are two modes in substrate specificity at the gate: one for rod/hook-type proteins and one for filament-type proteins. The switching of the two modes is triggered by hook completion, which seems to be necessary for maintaining an ordered and thus economic secretion.

In this chapter the flagellar growth, the flagellar structure and its components will be described, as well as the mechanism of export of the

B. Oudega (ed.), Protein Secretion Pathways in Bacteria, 249–270.

flagellar proteins from the cytoplasm and their integration into the flagellar structure outside the cell.

2. FLAGELLAR GROWTH

2.1 Distal Growth

One of the prominent features of flagella is its manner of growth. The flagellum grows distally, adding new born subunits onto the growing end at the tip. In contrast, many other filaments such as the eukaryotic flagellum or pili grow proximally, adding new subunits at the base. These two patterns can be distinguished simply by pulse-labelling of nascent proteins with a radioactively-labelled amino acid. The radioactivity will accumulate at the tip of growing filaments in distal growth, whereas it will stay at the base in proximal growth. However, in the case of flagellar growth, radioactive materials have not been employed. Instead, the unusual amino-acid p-fluoro-phenylalanine, that changes left-handed normal filaments into right-handed curly filaments when incorporated in the molecule, has been added to the medium. Consequently, curly filaments were found attached to the distal end of normal filaments, indicating that new filaments grew distally (Iino, 1969).

This distal growth was also confirmed by *in vitro* experiments using purified filaments. A filament had asymmetrical ends as observed by electron microscopy. The proximal end is pointed like an arrowhead, whereas the distal end lookes like a fishtail. The reconstitution of filaments from monomers showed unidirectional growth from the fishtail-end, proving that filaments grow only from their distal ends (Asakura *et al.*, 1964).

If the proteins were secreted randomly through the surface of the bacterial cells, they would diffuse away in the medium and never reach the tips of the growing filaments. Therefore, distal growth demands that the cell should elaborate a method to transfer the nascent proteins to the distal tip of flagella, suggesting the existence of an export system specific for flagellar proteins (Fig. 1).

2.2 Semi self-assembly

The assembly mode of the flagellum has been regarded as self-assembly as seen in bacteriophages, because these two supramolecular structures are of the similar size, are composed of many components, and

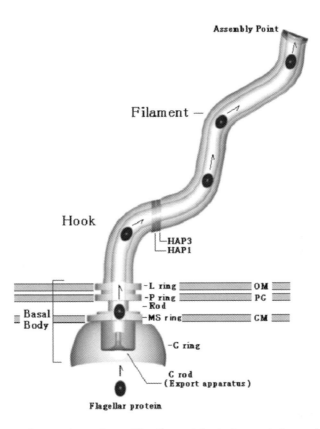

Fig. 1. Flagellar protein secretion pathway. Flagellar proteins to be secreted are selectively delivered into the gate of the flagellum-specific export apparatus and travel down the central channel of the flagellum, which runs through the rod, hook, and filament. OM, outer membrane; PG, peptidoglycan layer; CM, cytoplasmic membrane.

have rotational symmetry. However, the flagellar assembly is not so simple because of the assembling site. Flagella grow outside the cell, whereas bacteriophages are reproduced inside the cell.

Flagellar filaments self-assemble *in vitro* but not *in vivo*. Also, hooks self-assemble *in vitro*, but not *in vivo*. Both structures require helper proteins to elongate *in vivo*, a capping protein FliD for the filament (Ikeda *et al.*, 1989) and another capping protein FlgD for the hook (Ohnishi *et al.*, 1994). Optimal conditions for polymerization of filaments *in vitro* and *in vivo* are different. Filament elongation *in vivo* must be as fast as cell division, whereas *in vitro* it takes overnight for the filaments to grow as long as a few microns. The two capping proteins, FliD and FlgD, are regarded as a kind of chaperone that elaborates folding of flagellin or the hook protein molecules at efficient rates, so that flagellar elongation coheres with the speed of the cell cycle.

2.3 Flagellar morphogenesis

The morphological pathway of flagellar construction has been revealed by analyzing flagellar mutants that cannot produce complete flagella, but that produce intermediate structures only. The first extensive work has been carried out by Suzuki *et al.* (1978), which has been revised 14 years later by Kubori *et al.* (1992).

The first structure of flagellar construction recognized by electron microscopy is the MS ring, which is made of a single protein FliF (Kubori *et al.*, 1992). The MS ring acts as the construction base, together with the export apparatus built upon it. Once the whole export apparatus, the C rod and C ring, is completed on the cytoplasmic surface of the MS ring, the export of flagellar proteins into the periplasmic space starts to built the rod and following structures in an ordered way (Fig. 2) (Aizawa, 1996).

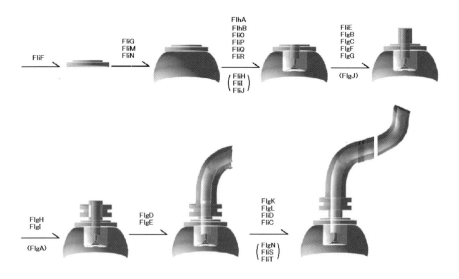

Morphological assembly pathway

Fig. 2. Morphological pathway of flagellar assembly. Proteins of the flagellum are indicated above the arrows and those not integrated in the structure are in parenthesis.

2.4   Flagellar growth and cell division

The manner of flagellar growth is different between peritrichous flagella and monotrichous flagella. On cells having peritrichous flagella (many flagella around the cell body), the number of flagella is kept constant after each cell division, securing the steady movement over generations. Filaments can elongate over several generations; faster in the beginning and slower as it becomes longer (Aizawa and Kubori, 1998). Interestingly, the number of flagella per cell follows the cell division, but filament elongation takes more than two generations (Karlinsey *et al.*, 2000b).

On the other hand, on cells with a monotrichous flagellum (single flagellum at the pole, or at the side of the cell body), flagellar construction has to be complete within each generation. In *Caulobacter crescentus* which shows distinguishable cell cycles, the flagellum is actively disassembled at its base and released into the media during the transition from a swarmer cell to a stalk cell.

Taken together, each flagellum has its own export apparatus at its base and grows independently from the other flagella (for peritrichous flagella) or accordantly with the cell cycle (for monotrichous flagellum).

2.5   Secretion of flagellar proteins

Bacterial cells secrete many molecules during growth, for instance alcohols as metabolic products, lactones for quorum sensing, toxins for pathogenicity, and flagellar proteins. In *S. typhimurium*, flagellar proteins and virulence factors are the major proteins secreted into the media as detected by the Coomassie Brilliant Blue (CBB) staining of SDS gels (Komoriya *et al.*, 1999). The secretion of virulence factors is not affected by flagellar mutations, indicating that these two secretion systems are independent of each other. The amount of flagellar proteins secreted into media varies from one mutant to another, suggesting that the secretion is not deliberate, but a consequence of spill-over during construction.

The secretion of flagellar proteins has been extensively examined by Minamino and Macnab (1999, 2000a, b). They employed either flagellin FliC or the hook-capping protein FlgD as an indicator of secretion events, and measured the amounts of those proteins in the cytoplasm, the periplasm, and the media. Both FliC and FlgD are secreted into the periplasm of a *flgJ* mutant but not into the periplasm of *flhA, flhB, fliH, fliI, fliJ, fliO, fliP, fliQ* and *fliR* mutants. In conclusion, FlhA, FlhB, FliH, FliI, FliJ, FliO, FliP, FliQ and FliR are required for the translocation of flagellar proteins across the cytoplasmic membrane. FlgJ, however, is not directly involved in the export event. It is the flagellum-specific muramidase that hydrolyzes the peptidoglycan layer to allow the rod to penetrate it (Nambu *et al.,* 1999).

3.   FLAGELLAR EXPORT APPARATUS

The existence of an export apparatus specific for assembly of flagella has been suggested from various observations as mentioned above, which can be summarized as follows.
1. Most flagellar proteins are not synthesized with a signal sequence commonly found in proteins secreted through the general secretion pathway. The flagellar protein secretion system is called type III, and has been typically found as a secretion system for virulence factors in pathogenic bacteria (Macnab, 1999).
2. Each flagellum must have its own secretion machinery at its base (Aizawa and Kubori, 1998).
3. In the morphological pathway, there are more than ten genes products involved in the early steps of the flagellar construction, suggesting a supra-molecular structure at the base (Kubori *et al.,* 1992).
4. More then ten gene products are responsible for flagellar secretion (Minamino and Macnab, 1999, 2000c).

The early gene products are: FlgJ, FlgN, FlhA, FlhB, FliF, FliG, FliH, FliI, FliJ, FliM, FliN, FliO, FliP, FliQ, FliR, FliS, and FliT. Among those proteins, FliF, FliG, FliM, and FliN have been identified as constituents of a relatively large complex of the MS and C ring (see below). The other proteins are further divided into two groups, integral-membrane components (FlhA, FlhB, FliO, FliP, FliQ, and FliR) and cytoplasmic components (FlgN, FliH, FliI, FliJ, FliS, and FliT).

3.1   Integral membrane components FlhA, FlhB, FliO, FliP, FliQ and FliR

Sequence analysis programs showed that FlhA, FlhB, FliO, FliP, FliQ, and FliR are integral-membrane proteins. FlhA contains 692 amino acid residues and consists of two domains: a hydrophobic N-terminal domain with 8 transmembrane segments and a hydrophilic C-terminal cytoplasmic domain (Minamino *et al.,* 1994). Since mutations within the N-terminal region suppress a specific *fliF* mutation (the in-frame deletion of

Ala-174 and Ser-175 of FliF), the transmembrane regions of FlhA may interact directly with the MS ring (Kihara et al., 2001). The C-terminal cytoplasmic domain binds to the cytoplasmic components FliH, FliI and FliJ, as well as to the C-terminal cytoplasmic domain of FlhB (Minamino and Macnab, 2000c).

FlhB consists of 383 amino acid residues and has two domains, a hydrophobic N-terminal domain with 4 transmembrane segments and a C-terminal cytoplasmic domain (Minamino et al., 1994). The C-terminal cytoplasmic domain is responsible for the switching of substrate-specificity from hook protein to flagellin (Kutsukake et al., 1994; Williams et al., 1996). It binds to FliH, FliI, FliJ and to the C-terminal cytoplasmic domain of FlhA. This indicates its important role, together with FlhA, in the gating of secretion (Minamino and Macnab, 2000c).

FliO is a 13 kDa membrane protein with a small N-terminal cytoplasmic domain, a single transmembrane helix and a large periplasmic domain (Ohnishi et al., 1997). FliO is a putative component of the C rod, which lies at the center of the cytoplasmic face of the MS-ring/C-ring structure (Katayama et al., 1996).

FliP is synthesized with a signal peptide and exists in two forms, a 25 kDa precursor form and a 23 kDa mature form (Ohnishi et al., 1997). Site-directed mutagenesis at the cleavage site reduces, but does not eliminate, complementation of a *fliP* mutant, suggesting that a signal peptide of FliP is important for kinetically efficient insertion of FliP into the membrane (Ohnishi et al., 1997). FliP probably possesses 4 transmembrane segments and a substantial periplasmic domain between segments 3 and 4 (Ohnishi et al., 1997). This protein is often co-purified with the hook-basal structure, indicating an interaction between the two (Fan et al., 1997).

FliQ is a 9.6 kDa protein with 2 transmembrane segments. It is the smallest component of the complex (Ohnishi et al., 1997). FliQ is also a putative component of the C rod (Katayama et al., 1996).

FliR is a 29 kDa hydrophobic protein, consisting of 5 or 6 transmembrane segments connected with short loops (Ohnishi et al., 1997). FliR is also co-purified with the hook-basal body (Fan et al., 1997).

FlhA, FlhB, FliO, FliP, FliQ and FliR are probably the integral-membrane components of the flagellar export apparatus, forming a protein-conducting channel within the cytoplasmic membrane (Minamino and Macnab, 1999). It has become evident in the last decade that FlhA, FlhB, FliP, FliQ, FliR (but not FliO) show a strong similarity to proteins involved in the export of virulence effector proteins of pathogenic bacteria, indicating a common origin between the two systems (Hueck, 1998; Macnab, 1999; Aizawa, 2001).

3.2   The cytoplasmic components FliH, FliI and FliJ

Vogler *et al.* (1992) carried out filament regrowth experiments to determine whether certain temperature-sensitive mutants have the ability to regrow sheared filaments at the restrictive temperature. They found that three genes, *flhA*, *fliH*, and *fliI*, might be involved in the export process.

FliI is an ATPase. Mutations within its ATPase domain inhibited the export of any flagellar protein (Dreyfus *et al.*, 1993; Fan and Macnab, 1996). The FliI ATPase is a 456 amino acid protein, consisting of at least two domains, a large N-terminal flagellum-specific domain of about 110 residues and a C-terminal domain comprising the ATPase catalytic site (Fan and Macnab, 1996). The ATPase region shows a significant sequence similarity to the catalytic β-subunit of the proton-translocating F_0F_1 ATPase, retaining well-conserved Walker boxes (Vogler *et al.*, 1991).

A molecule of FliI forms a stable complex with two molecules of FliH in the cytoplasm (Minamino and Macnab, 2000b, c). Purified FliI exists as monomers (Fan and Macnab, 1996; Minamino and Macnab, 2000b), whereas FliH forms homodimers (Minamino and Macnab, 2000b). When mixed together, a FliH-FliI heterotrimer is formed (Minamino and Macnab, 2000b).

FliI with a R7C/L12P double mutation fails to form a complex with FliH (Minamino and Macnab, 2000b). Consistently, the truncation of the first 7 amino acid residues from the N-terminus of FliI is sufficient to prevent the formation of a complex with FliH, indicating that the N-terminus of FliI associates with FliH (T. Minamino, J. Tame, K. Namba, R. M. Macnab, unpublished results). Interestingly, the ATPase activity of the FliH/FliI complex is 10-fold lower than that of FliI alone *in vitro*, indicating that FliH severely inhibits the ATPase activity of FliI (Minamino and Macnab, 2000b). Therefore, it is likely that FliH functions as a negative regulator of FliI ATPase activity before the flagellar export apparatus is complete. This ATP hydrolysis must be linked with the translocation of its export substrates across the cytoplasmic membrane into the channel of the growing flagella (Minamino and Macnab, 2000b).

How does FliI convert ATP to the energy driving protein translocation? The binding of adenosine nucleotides to the FliH/FliI complex causes a conformational change in the N-terminal flagellum-specific region of FliI, which can be measured by protease-sensitivity depending on its nucleotide-bound state (T. Minamino, J. Tame, K. Namba, R. M. Macnab, unpublished results). As a result a conformational change of the N-terminal region of FliI might occur upon ATP binding, and the ATP hydrolysis might results in translocation of flagellar proteins across the cytoplasmic membrane by interaction with the integral-membrane components of the type III flagellar export apparatus.

It should be noted that both FliH and FliI physically bind to immobilized FliJ, to the cytoplasmic domains of FlhA and FlhB, and to their substrates *in vitro* (see below; Silva-Herzog and Dreyfus, 1999; Minamino and Macnab, 2000c).

3.3   The C ring

The genes *fliG, fliM,* and *fliN* are called the "switch genes", emphasizing their roles in the switching of flagellar motor rotation. However, genetic analysis has shown that they are not only for torque generation and for the switching of rotational direction, but they also function in flagellar assembly, and hence have a more multifunctional function (Yamaguchi *et al.,* 1986a, b; Kubori *et al.,* 1992). Recent secretion experiments showed that the rod/hook-type proteins were not secreted into the periplasm of *fliG, fliM,* and *fliN* mutants, confirming their involvement in protein export (Minamino and Macnab, 1999).

It has been shown that FliG, FliM, and FliN form a cup-shaped structure on the cytoplasmic face of the MS ring complex, called the C ring (Khan *et al.,* 1992; Francis *et al.,* 1992, 1994). The main emphasis for the role of the C ring has been torque generation (Lloyd *et al.,* 1996), but its role in the export of flagellar proteins has been shown by a study on the mechanism of the hook length control. In wild-type cells, the length of the hook is fairly tightly controlled at about 55 nm (Hirano *et al.,* 1994). Mutants with a defect in FliK have abnormally elongated hooks, called polyhooks, without the filaments. For that reason, FliK has been called a hook-length control protein (Patterson-Delafield *et al.,* 1973; Suzuki and Iino, 1981). However, *fliK* mutants show a peak in hook-length distribution at the wild-type value of 55 nm, indicating that FliK does not directly determine the hook-length (Koroyasu *et al.,* 1998). Makishima *et al.* (2001) have found short hooks in *fliG, fliM,* and *fliN* mutants. They claimed that the hook length is determined by the amount of hook subunits determined by the C ring and exported by the export apparatus. Probably, the C ring provides the structural environment necessary for the export event to occur.

3.4   Secretin

Hook initiation can occur in mutants defective in *flgI* (the structural gene for the P ring),in *flgH* (structural gene for the L ring) and in *flgA* gene (coding for the periplasmic chaperone for FlgI to assemble into the P-ring). However, the hook can not elongate beyond the outer membrane in these mutants (Kubori *et al.,* 1992). Furthermore, hook-type proteins pass through the cytoplasmic membrane, but not through the outer membrane in these mutants (Minamino and Macnab, 1999). Therefore, the LP ring complex is necessary for the export of hook-type proteins across the outer membrane, similar to the outer membrane secretin ring structure found in pathogenic bacteria. In the type III secretion pathway for virulence effector proteins, the secretin is a large homomultimeric annular complex in the outer membrane that allows virulence factors to cross the outer membrane (Pugsley, 1993; Hueck, 1998).

4.   FLAGELLAR CHAPERONES

Many chaperones have been identified in the type III export systems for virulence effector proteins (Hueck, 1998). These type III chaperones display substrate specificity and only bind to one or two of their specific substrate protein(s). Although primary sequence similarity among these type III chaperones is low or not evident, they have common structural properties. They are small proteins (ca. 15 to 20 kDa in mass) with many charged residues. They have a putative C-terminal amphipathic helix that might mediate interaction with their cognate substrate protein(s).

The physical path of flagellar proteins travelling down to their assembly point is the 3 nm central channel within the flagellar structure (Namba *et al.,* 1989; Morgan *et al.,* 1993), suggesting that flagellar proteins can be transported in a partially unfolded and stretched conformation. All flagellar axial proteins are indeed unfolded in their monomeric form in solution, but tend to form aggregates at high concentrations (Namba, 2001). Therefore, premature aggregation of these disordered proteins should be prevented in the cytoplasm before their export begins. Since four flagellar cytoplasmic proteins, FliJ, FliS, FlgN and FliT have several features in common with the type III cytoplasmic chaperone family, it has been hypothesized that they act as the flagellum-specific chaperones (Fraser *et al.,* 1999; Minamino and Macnab, 1999; Stephens *et al.,* 1997; Yokoseki *et al.,* 1995).

4.1 General chaperone

FliJ is required for export of all flagellar proteins via the flagellar export pathway (Minamino and Macnab, 1999, 2000c; Minamino *et al.*, 2000a). All spontaneous *fliJ* mutants, including the ones with null mutations, have the ability to form swarms on motility agar plates after prolonged incubation at 30 °C, indicating that they display a leaky motile phenotype (Minamino *et al.*, 2000a).

FliJ is a 147 amino acid cytoplasmic protein (Vogler *et al.*, 1991). One particular *fliJ* mutant which forms significantly bigger swarms than null mutants encodes the N-terminal 73 amino acid residues of FliJ, about one-half of the protein (Minamino *et al.*, 2000a). Overproduction of this truncated protein stimulates both motility and export of flagellar proteins to wild-type levels, indicating that the N-terminal 73 amino acids can confer full function (Minamino *et al.*, 2000a). The N-terminal region of FliJ containing a potential α-helical coiled-coil structure is essential for FliJ function, whereas its C-terminal region may contribute to maintaining the proper conformation and stability of the N-terminal region (Minamino *et al.*, 2000a).

FliJ binds to all flagellar proteins exported via the flagellar export pathway as determined by affinity blotting (Minamino and Macnab, 2000c). Several export substrates such as FliE (a rod protein) and FlgG (distal rod protein), when overproduced, formed inclusion bodies in the cytoplasm (Minamino and Macnab, 2000c; Minamino *et al.*, 2000b). Co-overproduction of FliJ and either of the two export substrates, FliE or FlgG, hindered their aggregation in the cytoplasm. This strongly suggests that FliJ has a chaperone-like activity, and prevents export substrates from premature aggregation in the cytoplasm (Minamino *et al.*, 2000a).

4.2 Substrate-specific chaperones

FliS is a cytoplasmic protein of 15 kDa (Kawagishi *et al.*, 1992). A *fliS* mutant produces much shorter filaments than the wild type. Since this mutant produces the complete basal body structures, FliS must be responsible for the elongation step rather than the initiation step of filament assembly. The *fliS* mutation does not affect FliC (flagellin) synthesis, but decreases the amount of FliC exported out of the cells (Yokoseki *et al.*, 1995). Therefore, the short filaments of this mutant could be a consequence of impaired FliC export. Purified FliS binds to FliC, in agreement wit the hypothesis that FliS functions as an export chaperone to facilitate the export of FliC (Auvray *et al.,* 2001; Yokoseki and Kutsukake, 1995).

A *fliS* mutation enhances the expression of flagellar genes. This results in a 2 to 3 fold increase in the number of flagella and in the amount of FlgM (anti-sigma factor) exported out of the cells (Yokoseki *et al.,* 1996). FliS binds to FlgM as well, suggesting that FliS negatively regulates FlgM export to avoid unnecessary overproduction of flagellar structures (Yokoseki and Kutsukake, 1995).

A *flgN* mutant produces only a few flagella. Electron microscopy of the isolated flagellar structures have shown that this mutant produces many hook-basal bodies without filaments, but occasionally ones with filaments, indicating that FlgN is required for the efficient initiation of filament assembly (Kutsukake *et al.,* 1994). The *flgN* mutation specifically reduces the export of FlgK and FlgL (the hook-filament junction proteins) (Fraser *et al.,* 1999; Bennett *et al.,* 2001). FlgN is a cytoplasmic, 16.5 kDa protein, containing a putative C-terminal amphipathic helix. FlgN binds to immobilized FlgK and FlgL, specifically to their C-terminal amphipathic helical domains *in vitro* (Fraser *et al.,* 1999), and prevents *in vitro* aggregation of FlgK monomers (Bennett *et al.,* 2001). FlgN might acts as a chaperone specific for both FlgK and FlgL.

In addition to the role as a substrate-specific chaperone, genetic evidence indicates that FlgN may also act as a translational regulator of FlgM before completion of hook assembly (Karlinsey *et al.,* 2000a). If that is the case, FlgN should have a dual role in flagellar biogenesis, before and after hook completion. Since loss of FlgN especially impairs the initiation of filament formation, FlgN might primary function as an export chaperone.

FliT is a cytoplasmic protein of 14 kDa (Kawagishi *et al.,* 1992). Mutants defective in FliT retain the ability to swim at almost wild-type levels, indicating that FliT is not absolutely required for flagellar formation (Yokoseki *et al.,* 1995). FliT binds to FliD (filament cap protein), specifically to its C-terminal helical domain *in vitro* (Fraser *et al.,* 1999), and the loss of FliT in the cells reduces the amount of FliD exported into the media. This suggests that FliT functions as an export chaperone specific for FliD (Bennett *et al.,* 2001).

The N-terminally truncated FlgN or FliT polypeptides containing the C-terminal amphipathic domain cannot complement the chaperone deficiency of the *flgN* or *fliT* mutants, respectively, whereas translational fusions of these domains to glutathione S-transferase can. This indicates that their C-terminal domains, probably their amphipathic helices, are responsible for the substrate binding (Bennett *et al.,* 2001).

4.3   Hypothetical model for the flagellar protein export

Based on available evidence concerning the flagellar export pathway, a tentative model for the export process is presented here (Fig. 3).

1. An export substrate, with the aid of cytoplasmic chaperones such as FliJ, is brought into the FliH/FliI complex to form a FliH/FliI/FliJ/substrate complex in the cytoplasm.

2. This complex diffuses to the cytoplasmic domains of FlhA (FlhA$_C$) and FlhB (FlhB$_C$) and forms a FlhA$_C$/FlhB$_C$/FliH/FliI/FliJ/substrate complex at the central pore of the MS ring.

3. At this stage, FliI hydrolyzes ATP and drives the translocation of the substrate across the plane of the cytoplasmic membrane through the transmembrane region of the export apparatus (FliO, FliP, FliQ and FliR and the transmembrane domains of FlhA and FlhB), accompanied by dissociation of the FliH/FliI complex and of FliJ from the export apparatus.

4. Finally, the translocated substrates diffuse down the channel in the nascent structure and assemble at its distal end.

5.   MODES OF EXPORT

The flagellar export apparatus strictly recognizes flagellar proteins to be secreted in an ordered way. How can this specificity be explained? Expression of flagellar genes is divided into three hierarchical levels (Kutsukake, 1990). The FlhDC complex of the first level activates flagellar operons of the second level, which contains most of the component genes of the basal body. In the second level, there are two regulatory proteins, FliA and FlgM. FliA is a flagella specific sigma factor that initiates the transcription of the third level, including flagellin FliC. FlgM is an anti-FliA protein, that binds to FliA to inhibit its function. Therefore, between the level 2 and 3, there is a big gap in terms of translation and secretion of flagellar proteins.

Hook completion is a morphological checkpoint of flagellar assembly. Until hook assembly is complete, FlgM remains in the cytoplasm and inhibits expression of the *flgK*, *flgL*, *fliD* and *fliC* genes, which are responsible for filament formation (Kutsukake and Iino, 1994). Upon completion of hook structure, FlgM goes outside of the cells through the hook-basal body structure, causing relief in this inhibition. This suggests that the flagellar export apparatus has the ability to sense completion of the hook and to turn FlgM export on (Hughes *et al.*, 1993; Kutsukake, 1994). What monitors the state of hook assembly and signals the flagellar export

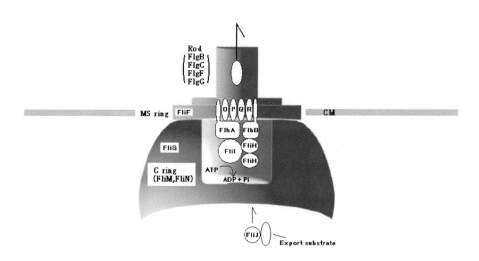

Fig. 3. Schematic model of the flagellar export apparatus. The geometry of each component is not known except whether or not they are membrane proteins. The integral-membrane components of the export apparatus consisting of FlhA, FlhB, FliO, FliP, FliQ and FliR, are located within the central pore of the MS ring. The substantial cytoplasmic domains of FlhA and FlhB project into the cavity of the C ring. The flagellum-specific ATPase, FliI/FliH complex, is soluble but should interact with the export channel. FliJ serves as a chaperone to prevent export substrates from premature aggregation in the cytoplasm.

apparatus to stop export of hook protein and to start export of FlgM? Kutsukake *et al.*, (1994) provided the first evidence to suggest that two flagellar proteins FliK and FlhB may be responsible for this process.

5.1   Export signal of flagellar proteins

Since flagellar proteins pass through the flagellar export pathway to their final destinations, they must contain a flagellum-specific export signal recognized by the flagellar export apparatus. N-terminal segments of flagellar proteins retain the ability to be exported through this pathway (Iyoda and Kutsukake, 1995; Kornacker and Newton, 1994; Kuwajima *et al.*, 1989; Minamino *et al.*, 1999b). Consistently, single amino acid substitutions within the N-terminal region of FlgM severely impair its export (Chilicot and Hughes, 1998). It therefore appears that their N-terminal regions must contain the putative export signal.

There is a consensus sequence (ANNLAN) in the N-terminal region of some flagellar proteins (Homma *et al.*, 1990a, b). However, these amino acids reflect their similarities as axial components of flagellum (rod proteins, hook protein, HAPs, and flagellin) rather than signals for export. They are robably required for flagellar assembly but not for export (Kornacker and Newton, 1994). Therefore, it seems likely that the export apparatus recognizes a higher ordered structural motif in these proteins, rather than a primary sequence.

Since rod/hook-type proteins belong to a different export class than the filament-type proteins (See above), the export signal of rod/hook-type proteins must differ from that of the filament-type proteins (Minamino *et al.*, 1999a). However, the basis for this remains unknown.

5.2   Switching

Even though the hooks have already reached mature length in *fliK* mutants, FlgM cannot go outside the cells through the polyhook-basal bodies produced by these mutants. So expression of genes responsible for filament formation remains inhibited in these mutants (Hughes *et al.*, 1993; Kutsukake, 1994). Therefore, FliK has the dual function of controlling the length of the hook and facilitating FlgM export. Extragenic *fliK* suppressor mutants possessing second-site mutations in the *flhB* gene, whose product is an integral-membrane component of the flagellar export apparatus, can assemble filaments at the tip of the polyhooks (Kutsukake *et al.*, 1994; Williams *et al*, 1996). This suggests that mutant FlhB proteins have the ability to switch the substrate specificity from hook protein to FlgM even in the absence of FliK. Therefore, mutations in *fliK* prevent FlhB from switching the substrate specificity of the export and so polyhooks are produced.

Other flagellar proteins are also subjected to the FliK-FlhB regulatory system. It has been shown that rod-type (FliE, FlgB, FlgC, FlgF, FlgG) and hook-type (FlgD, FlgE) proteins are exported most efficiently prior to hook completion, whereas the export of filament-type proteins (FlgK, FlgL, FliD, FliC) occurs more efficiently only after hook completion (Komoriya et al., 1999; Minamino et al., 1999a). It appears that the export properties of rod/hook-type proteins are different from those of filament-type proteins and that their export order exactly parallels the assembly order of the rod-hook-filament structure (Komoriya et al., 1999; Minamino et al., 1999a; Minamino and Macnab, 1999).

FliK is a 405 amino acid soluble protein consisting of three regions, an N-terminal region, a proline-rich central region, and a highly conserved C-terminal region (Kawagishi et al., 1996; Williams et al., 1996). Spontaneous polyhook mutations are either frame shift or nonsense mutations, resulting in truncation of the FliK C-terminus (Suzuki and Iino, 1981; Williams et al., 1996). Intragenic suppressor mutations, whose second-site mutations invariably restore this C-terminal region, regain only the ability to switch substrate specificity of the flagellar export apparatus and to assemble filaments at the tip of the polyhooks (Williams et al., 1996). Therefore, it has been proposed that the C-terminal region of FliK is directly involved in switching substrate specificity. This C-terminal region is highly conserved among the FliK proteins of other bacteria (Gonzalez-Pedrajo et al., 1997).

FliK had been believed to reside permanently in the cytoplasm because it is soluble and has not been identified as a component of the wild-type flagellar structure or of flagellar precursors (Kawagishi et al., 1996). However, it has been shown that FliK itself is an export substrate recognized by the flagellar export apparatus and that its export occurs most efficiently during hook formation (Minamino et al., 1999b). Export is impaired by deletion within the N-terminal region of FliK, but not by C-terminal truncations, indicating that the N-terminal region is important for the export of FliK (Minamino et al., 1999b). Export deficient fliK mutants produce polyhooks sometimes with the filaments, suggesting that the export of FliK during hook assembly is important for both hook-length control and the switching of substrate specificity (Minamino et al., 1999b).

Although FliK has not yet been found in the complete flagellar structure or flagellar precursor structures as mentioned above, several lines of evidence suggest that FliK probably associates with the growing flagellum to sense the state of hook assembly and then to activate the switching together with FlhB (Minamino et al., 1999b). Consistently, switching of substrate specificity does not occur prior to the hook completion, even though FliK is being exported in the *flgE rflH* mutant (Kutsukake, 1997; Minamino *et al.,* 1999b).

FlhB is an integral-membrane protein with a substantial C-terminal cytoplasmic domain (FlhB$_C$) (Minamino *et al.,* 1994). Since all *flhB* mutations, which allow the *fliK* mutant cells to assemble filaments at the tip of polyhooks, are located within FlhB$_C$, the latter protein must be involved in switching from the rod/hook-type substrates to filament-type substrates, indicating that FlhB$_C$ may exist in two substrate specificity states (Kutsukake *et al.,* 1994; Williams *et al.,* 1996). Wild-type FlhB$_C$ is not stable (half-life of ± 5 min). It is specifically cleaved at Pro270 into two polypeptides, FlhB$_{CN}$ and FlhB$_{CC}$,. These two parts still retain the ability to interact with each other after cleavage. On the other hand, mutant FlhB proteins that can undergo switching of substrate specificity even in the absence of FliK, are much more resistant to cleavage (half-lives of 20 to 60 min). It seems likely that the suppressor mutations affect the state of the putative hinge region between the FlhB$_{CN}$ and FlhB$_{CC}$ sub-domains and, therefore, that their conformation must be different from that of the wild type (Minamino and Macnab, 2000a).

The cleavage products of wild-type FlhB$_C$, existing as a FlhB$_{CN}$/FlhB$_{CC}$ complex on an affinity blot membrane, bind stronger to rod/hook-type substrates than to filament-type substrates (Minamino and Macnab, 2000a, c). In contrast, the intact form of FlhB$_C$ (mutant or wild type) or the FlhB$_{CC}$ polypeptide alone binds both of them to about the same degree. FlhB$_{CN}$ itself does not bind substrates appreciably (Minamino and Macnab, 2000a). This suggests that FlhB$_C$ acts as an export switch to have two substrate specificity states and that its conformational change, mediated by the interaction between FlhB$_{CN}$ and FlhB$_{CC}$, is responsible for the specificity of the switching process. The fact that all *flhB* mutations that suppress *fliK* mutations are found in FlhB$_{CC}$ supports this idea (Kutsukake *et al.,* 1994; Williams *et al.,* 1996).

FliK binds to FlhB$_C$ *in vitro* (Minamino and Macnab, 2000a). However, FliK itself is an export substrate (Minamino *et al.,* 1999b) and its binding properties for FlhB$_C$ does not appear to be different from those of other export substrates (Minamino and Macnab, 2000a,c). Therefore, whether FlhB$_C$ communicates with FliK to mediate the switching of substrate specificity is still an open question.

266

5.3 A model for the substrate specificity switching of the flagellar export apparatus

Based on the available information, a model for the substrate specificity switching of the flagellar export apparatus is presented here (Fig. 4) (Kutsukake, 1997; Kutsukake *et al.*, 1994; Minamino *et al.*, 1999a; Minamino and Macnab, 1999; Muramoto *et al.*, 1998; Williams *et al.*, 1996).

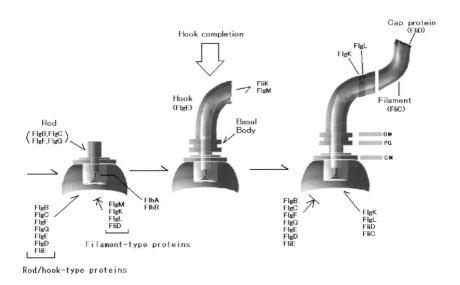

Hypothetical Model for the substrate specificity switching

Fig. 4. Switching of substrate specificity of the export apparatus. Switching between two modes in substrate specificity is triggered by hook completion, and a concomitant secretion of FliK and FlgM. In the first mode, the rod /hook-type proteins are secreted to construct the hook-basal body. During this period the small amounts of FlgK, FlgL and FliD, which are the filament-type proteins, are already expressed but not secreted. Upon completion of the hook, FliK and FlgM are secreted into the medium. The secretion of FlgM allows the filament-type proteins to be fully expressed and secreted in the second mode, triggered by FliK secretion.

1. During rod and hook assembly, FlhB supports the export of the rod/hook-type proteins, but not that of the filament-type proteins. At this stage, a regulatory protein, RflH (Flk), inhibits the export of filament-type proteins (Kutsukake, 1997). RflH (Flk) has also been postulated to

regulate *flgM* translation in response to completion of the L and P rings and the elongation of the hook (Karlinsey *et al.*, 1997).

  2. Upon completion of hook assembly, the hook-length signal is somehow transmitted to FliK and/or FlhB, shutting off rod/hook-type protein export and turning on filament-type protein export.

5.4 Other systems

  As stated above the mechanism of flagella assembly and export resembles that of type III secretion systems (see also chapter on type III secretion). Type III secretion systems have a similar machinery for length control and substrate specificity switching. The *Salmonella* needle complex for instance is an export apparatus that allows virulence effector proteins to cross both the cytoplasmic and the outer membrane (Kubori *et al.*, 1998). The length of the needle portion is fairly well controlled at 80 nm (Kubori *et al.*, 1998). Loss of the type III secretion associated protein InvJ causes elongated needles (Kubori *et al.*, 2000). The *invJ* mutation severely impairs the secretion of virulence effector proteins. InvJ itself is also a secretion protein (Collazo and Galan, 1997). There is sequence homology between InvJ and FliK (Komoriya *et al.*, 1998). Since the *Salmonella* type III virulence factor secretion system also has a FlhB homolog, SpaS. InvJ, presumably along with SpaS, might mediate the switching of substrate specificity from a component of the needle structure to virulence effector proteins.

6. CONCLUSIONS AND FUTURE PERSPECTIVES

  The mechanism of flagellar construction has been elucidated at a molecular level. The roles of almost all flagellar proteins are known by now. The flagellum works as a protein secretion system before it works as a motile system. The physical principle underlying the unique rotary motion of the flagellar motor has to be uncovered. Flagellar biogenesis apparently synchronizes with the cell division, indicating that it is under control of the global regulation to maintain the life cycle of a cell. The link between flagellar genes and house-keeping genes is still missing. The flagellar secretion system shares characteristics with the so-called type III secretion system for virulence factors, suggesting a common origin. Analysis of evolutional paths of the flagellum may reveal the relationship between the flagellum and other export systems, and when and how the flagellum started rotating.

7.   ACKNOWLEDGEMENTS

We thank Sarah Daniell for reading manuscript, and Tsutomu Miyatsu for figures. We also thank Robert M. Macnab for his invaluable advice.

8.   REFERENCES

Aizawa, S.-I. (2001) (in press).
Aizawa, S.-I. (1996) Mol Microbiol 19: 1-5.
Aizawa, S.-I., Zhulin, I., Marquez-Magana, L., and Ordal, G.W., (2001) Academic Press.
Aizawa, S.-I. and Kubori, T. (1998) Genes to cells 3: 1-10.
Asakura, S., Eguchi, G., and Iino, T. (1964) J Mol Biol 10: 42-56.
Auvray, F., Thomas, J., Fraser, G. M. and Hughes, C. (2001) J Mol Biol (in press).
Bennett, J. C. Q., Thomas, J., Fraser, G. M. and Hughes, C. (2001) Mol Microbiol 39: 781-791.
Karlinsey, J.E., Chilcott, G. S. and Hughes, K. T. (1998) Mol Microbiol 30: 1029-1040.
Collazo, C. M. and Galán, J. E. (1997) Gene 192: 51–59.
González-Pedrajo, B., Ballado, T., Campos, A., Sockett, R.E., Camarena, L. and Dreyfus, G. (1997) J Bacteriol 179: 6581-6588.
Dreyfus, G., Williams, A. W., Kawagishi, I. and Macnab, R. M. (1993) J Bacteriol 175: 3131-3138.
Fan, F. and Macnab, R. M. (1996) J Biol Chem 271: 31981-31988.
Fan, F., Ohnishi, K., Francis, N. R. and Macnab, R. M. (1997) Mol Microbiol 26: 1035-1046.
Francis, N. R., Irikura, V. M., Yamaguchi, S., DeRosier, D. J. and Macnab, R. M. (1992) Proc Natl Acad Sci USA 89: 6304–6308.
Francis, N. R., Sosinsky, G. E., Thomas, D. and DeRosier, D. J. (1994) J Mol Biol 235: 1261–1270.
Fraser, G. M., Bennett, J. C. Q. and Hughes, C. (1999) Mol Microbiol 32: 569-580.
Hirano, T., Yamaguchi, S., Oosawa, K. and Aizawa, S.-I. (1994) J Bacteriol 176: 5439-5449.
Homma, M., DeRosier, D. J. and Macnab., R. M. (1990a) J Mol Biol 213: 819–832.
Homma, M., Kutsukake, K., Hasebe, M., Iino, T. and Macnab., R. M. (1990b) J Mol Biol 211: 465–477.
Hueck, C. J. (1998) Microbiol. Mol Biol Rev 62: 379-433.
Hughes, K.T., Gillen, K.L., Semon, M.J. and Karlinsey, J.E. (1993) Science 262: 1277-1280.
Ikeda, T., Asakura, S. and Kamiya, R. (1989) J Mol Biol 209: 109-114.
Iino, T. (1969) J Gen Microbiol 56: 227-239.
Iyoda, S. and Kutsukake, K. (1995) Mol Gen Genet 249: 417-424.
Iyoda, S. and Kutsukake, K. (1995) Mol Gen Genet 249: 417-424.
Lonner, J., Brown, K.L. and Hughes, K.T. (2000a) Cell 102: 487-497.

Karlinsey, J.E., Pease, A. J., Winkler, M.E., Bailey, J. L. and Hughes, K.T. (1997) J Bacteriol 179: 2389-2400.

Karlinsey, J.E., Tanaka, S., Bettenworth, V., Yamaguchi, S., Boos, W., Aizawa, S.-I. and Hughes, K.T. (2000b) Mol Micro 37: 1220-1231.

Katayama, E., Shiraishi, T., Oosawa, K., Baba, N. and Aizawa., S.-I. (1996) J Mol Biol 255: 458-475.

Kawagishi, I., Homma, M., Williams, A.W. and Macnab, R.M. (1996) J Bacteriol 178: 2954-2959.

Kawagishi, I., Müller, V., Williams, A.W. and Macnab, R.M. (1992) J Gen Microbiol 138: 1051-1065.

Khan, I. M., Reese, T. S. and Khan, S. (1992) Proc Natl Acad Sci USA 89: 5956-5960.

Kihara, M., Minamino, T., Yamaguchi, S. and Macnab, R. M. (2001) J Bacteriol 183: 1655-1662.

Komoriya, K., Shibano, N., Higano, T., Azuma, N., Yamaguchi, T. and Aizawa, S.-I. (1999) Mol Microbiol 34: 767-779.

Kornacker, M.G. and Newton, A. (1994) Mol Microbiol 14: 73-85.

Koroyasu, S., Yamazato, M., Hirano, T. and Aizawa, S.-I. (1998) Biophys J 74: 436-443.

Kubori, T., Matsushima, Y., Nakamura, D., Uralil, J., Lara-Tejero, M., Sukhan, A., Galán, J. E. and Aizawa, S.-I. (1998) Science 280: 602-605.

Kubori, T., Shimamoto, N., Yamaguchi, S., Namba, K. and Aizawa, S.-I. (1992) J Mol Biol 226: 433-446.

Kubori, T., Sukhan, A., Aizawa, S. -I. and Galán, J. E. (2000) Proc Natl Acad Sci USA 97: 10225-10230.

Kutsukake, K. (1994) Mol Gen Genet 243: 605-612.

Kutsukake, K. (1997) J Bacteriol 179: 1268-1273.

Kutsukake, K. and Iino, T. (1994) J Bacteriol 176: 3598-3605.

Kutsukake, K., Minamino, T. and Yokoseki, T. (1994) J Bacteriol 176: 7625-7629.

Kutsukake, K., Okada, T., Yokoseki, T. and Iino, T. (1994) Gene 143: 49-54.

Kutsukake, K., Ohya, Y. and Iino, T. (1990) J Bacteriol 172: 741-747.

Kuwajima, G., Kawagishi, I., Homma, M., Asaka, J.-I., Kondo, E. and Macnab, R.M. (1989) Proc Natl Acad Sci USA 86: 4953-4957.

Lloyd, S.A., Tang, H., Wang, X., Billings, S. and Blair, D.F. (1996) J Bacteriol 178: 223-231.

Macnab, R. M. (1996) In: *Escherichia coli* and *Salmonella*: Cellular and Molecular Biology 2nd edit. (Neidhardt, F. C., Curtiss III, R., Ingraham, J. L., Lin, E. C. C., Low, K. B., Magasanik, B., Reznikoff, W. S., Riley, M., Schaechter, M. and Umbarger, H. E., eds.) ASM Press Washington DC 123-145.

Macnab, R. M. (1999) J Bacteriol 181: 7149-7153.

Makishima, S., Komoriya, K., Yamaguchi, S. and Aizawa, S.-I. (2001) Science 291: 2411-2413.

Minamino, T., Chu, R., Yamaguchi, S. and Macnab, R. M. (2000a) J Bacteriol 182: 4207-4215.

Minamino, T., Doi, H. and Kutsukake, K. (1999a) Biosci Biotechnol Biochem 63: 1301-1303.

Minamino, T., González-Pedrajo, B., Yamaguchi, K., Aizawa, S.-I., and Macnab, R. M. (1999b) Mol Microbiol 34: 295-304.

Minamino, T., Iino, T., and Kutsukake, K. (1994) J Bacteriol 176: 7630-7637.

Minamino, T. and Macnab, R. M. (1999) J Bacteriol 181: 1388-1394.

Minamino, T. and Macnab, R. M. (2000a) J Bacteriol 182: 4906-4919.

Minamino, T. and Macnab, R. M. (2000b) Mol Microbiol 37: 1494-1503.

Minamino, T. and Macnab, R. M. (2000c) Mol Microbiol 35: 1052-1064.

Minamino, T., Yamaguchi, S., and Macnab, R. M. (2000b) J Bacteriol 182: 3029-3036.

Morgan, D. G., Macnab, R. M., Francis, N. R., and DeRosier, D. J. (1993) J Mol Biol 229: 79-84.

Muramoto, K., Makishima, S., Aizawa, S.-I. and Macnab, R.M. (1998) J Mol Biol 277: 871-882.

Namba, K. (2000) Genes to Cells 6: 1-12.

Namba, K., Yamashita, I. and Vondenviszt, F. (1989) Nature (London) 342: 648-654.

Nambu, T. Minamino, T., Macnab, R. M. and Kutsukake, K. (1999) J Bacteriol 181: 1555-1561.

Ohnishi, K., Fan, F., Schoenhals, G. J., Kihara, M. and R. M. Macnab. (1997) J Bacteriol 179: 6092-6099.

Ohnishi, K., Ohto, Y., Aizawa, S.-I., Macnab, R. M. and Iino. T. (1994) J Bacteriol 176: 2272-2281.

Patterson-Delafield, J., Martinez, R.J., Stocker, B.A.D. and Yamaguchi, S. (1973) Arch Mikrobiol 90: 107-120.

Pugsley, A. P. (1993) Microbiol Rev 57: 50-108.

Silva-Herzog, E. and Dreyfus, G. (1999) Biochim Biophys Acta 1431: 374-383.

Stephens, C., Mohr, C., Boyd, C., Maddock, J., Gober, J. and Shapiro, L. (1997) J Bacteriol 179: 5355-5365.

Suzuki, T. and Iino, T. (1981) J Bacteriol 148: 973-979.

Suzuki, T., Iino, T., Horiguchi, T, and Yamaguchi, S. (1978) J Bacteriol (1978) 133: 904-915.

Vogler, A. P., Homma, M., Irikura, V. M. and Macnab, R. M. (1991) J Bacteriol 173: 3564-3572.

Yamaguchi, S., Aizawa, S.-I., Kihara, M., Isomura, M., Jones, C. J. and Macnab, R. M. (1986a) J Bacteriol 168: 1172-1179.

Yamaguchi, S. Fujita, H., Ishihara, A., Aizawa, S.-I. and Macnab, R. M. (1986a) J Bacteriol 166: 187-193.

Yokoseki, T. and Kutsukake, K. (1995) Jpn. J Genet 70: 778.

Yokoseki, T. Iino, T. and Kutsukake, K. (1996) J Bacteriol 178: 899-901.

Yokoseki, T., Kutsukake, K., Ohnishi, K. and Iino, T. (1995) Microbiology 141: 1715-1722.

Williams, A.W., Yamaguchi, S., Togashi, F., Aizawa, S.-I., Kawagishi, I. and Macnab, R.M. (1996) J Bacteriol 178: 2960-2970.

Chapter 14

PROTEIN SECRETION IN GRAM-POSITIVE BACTERIA

Rob Meima[1] and Jan Maarten van Dijl[2]

[1] Department of Molecular Microbiology, Vrije Universiteit de Boelelaan 1087, 1081 HV Amsterdam, NL
Present address: DSM Food Specialties, Wateringseweg 1, Postbus 1, 2600 MA Delft, NL

[2] Department of Pharmaceutical Biology, University of Groningen, Antonius Deusinglaan 1, 9713 AV Groningen, NL

1.   INTRODUCTION

Studies on protein secretion in Gram-positive bacteria have become of particular importance since many species are of great commercial value and medical interest. Several Gram-positive bacteria have been applied successfully for cost-effective industrial production of enzymes since, due to the absence of an outer membrane, proteins are secreted directly into the growth medium from which they can be easily purified. In addition, the study of the protein secretion repertoire of Gram-positives has contributed to our fundamental understanding of notorious pathogens, such as *Staphylococcus aureus*, and may facilitate the rational design of novel drugs, specifically targeted against these organisms (Alksne *et al.*, 2000).

The present chapter will summarize the current knowledge on protein secretion mechanisms in Gram-positive bacteria. In this context, *Bacillus subtilis* will serve as a model, since extensive experimental data are available for this bacterium. Comparisons with other well-studied export machineries, such as those of *Escherichia coli* and *Saccharomyces cerevisiae* will be made, and some of the unique features of the translocation machineries of *B. subtilis* and a number of other Gram-positive bacteria will be discussed.

B. Oudega (ed.), Protein Secretion Pathways in Bacteria, 271–296.
© 2003 *Kluwer Academic Publishers. Printed in the Netherlands.*

2. *BACILLUS SUBTILIS* AND OTHER GRAM-POSITIVES

Arguably, the best-studied Gram-positive bacterium is *Bacillus subtilis*, an endospore-forming, non-pathogenic, aerobic soil bacterium that can be found in sediments throughout the world (Sonenshein *et al.*, 2001). Other representatives of the genus *Bacillus* include *Bacillus amyloliquefaciens* and *Bacillus licheniformis*, both widely used in industrial production processes besides *B. subtilis*. Although most bacilli are harmless, some pathogenic species are known, including *Bacillus anthracis*, the causative agent of anthrax, *Bacillus cereus*, which causes food poisoning, and *Bacillus thuringiensis*, which produces toxins with powerful insecticidal activities. Other well-studied Gram-positive organisms include *Lactococcus lactis*, one of the most important species in starter cultures for cheese production, and *Staphylococcus aureus*. The latter is of great importance from a medical point of view and, since the development and establishment of powerful sequencing and data managing techniques, much effort has been put in the analysis of its genome by various research centers, both public and industrial.

B. subtilis and other bacilli have a long history of safe use in industrial and food applications. In fact, strains of *B. subtilis* have been used for centuries in the production of *natto*, a soybean-based food product which is widely appreciated in Japan. This fermentation process requires secretion of several enzymes necessary for extracellular production of polymers, such as polyglutamate and levan. At present, *Bacillus* species are widely used for the large-scale production of enzymes for use in laundry detergents and food processing, as well as the production of fine chemicals, pharmaceuticals, food-additives, antibiotics and insecticides (Ferrari *et al.*, 1993; Bron *et al.*, 1999). The world production of industrial enzymes constitutes an annual market of an estimated US$ 1 billion, half of which is produced by *Bacillus* species. Powerful fermentation technologies have been developed for several bacilli, enabling production of extracellular enzymes at levels well in excess of 1 g/l (Bron *et al.*, 1999). Likewise, *L. lactis* and other lactic acid bacteria have been applied in dairy processes since ancient times. Today, *L. lactis* is used for the bulk production of cheese. With the availability of the complete genome sequence of this (Bolotin *et al.*, 2001) and other lactic acid bacteria, strategies are being developed to further optimize their use in food technology, including the production of novel bacteriocins to be used as biopreservatives in the prevention of food spoilage caused by pathogens such as *Listeria monocytogenes* (Kuipers *et al.*, 2000).

Since the isolation of naturally transformable strains (Spizizen, 1958), *B. subtilis* has become the paradigm of Gram-positive physiology and genetics. Elegant cloning and expression systems have been developed and, when published in 1997, it represented the first Gram-positive bacterium for which the complete genome sequence was established and annotated (Kunst *et al.*, 1997). In addition to previously identified genes involved in protein secretion, several new functions were discovered in the annotated version of the sequence.

As nutrients become limiting, *B. subtilis* cells undergo a series of developmental changes in response to the changing environment. These post-exponential phenomena include degradative enzyme synthesis, the development of genetic competence (the ability to take up and incorporate exogenous DNA) and sporulation. Characteristic of the early stages in the transition from exponential to stationary phase of growth is the secretion of degradative enzymes including proteases, nucleases, and phosphatases, in an attempt to process and internalize nutrients (Priest, 1977; Ferrari *et al.*, 1993). In fact, relatively few proteins are secreted during exponential growth (Antelmann *et al.*, 2001). This is exemplified by the fact that the expression of several elements of the secretion apparatus increases sharply around the transition from exponential to stationary growth phase (Herbort *et al.*, 1999; van Dijl *et al.*, 2001). A survey of the annotated ORFs in the genome sequence indicated that about 300 genes encode proteins with putative amino-terminal signal peptides, which can be classified into four distinct groups (Tjalsma *et al.*, 2000; see below). Subsequent studies on the extracellular proteome (the collection of native exoproteins of *B. subtilis*), revealed approximately 200 protein spots that could be resolved by two-dimensional gel-electrophoresis (Antelmann *et al.*, 2001; Hirose *et al.*, 2000).

3. PROTEIN SECRETION PATHWAYS IN GRAM-POSITIVE BACTERIA

At a first glance, secretion in Gram-positive eubacteria appears to be a relatively simple and efficient process since the architecture of the Gram-positive cell is far less complex than that of eukaryotic cells. Moreover, the absence of an outer membrane, such as that present in Gram-negative organisms, implies that proteins need to pass only a single phospholipid bi-layer to be secreted directly into the growth medium. However, the presence of a relatively thick negatively charged cell wall poses a serious barrier for proteins to pass on their way out of the cell and into the medium (for review, see Foster and Popham, 2001). The cell wall is a complex heteropolymer of peptidoglycan, covalently linked to anionic polymers, such as teichoic and

274

teichuronic acids. Substantial evidence has mounted suggesting that the fate of secreted proteins is largely determined by interactions at the membrane-cell wall interface (see below).

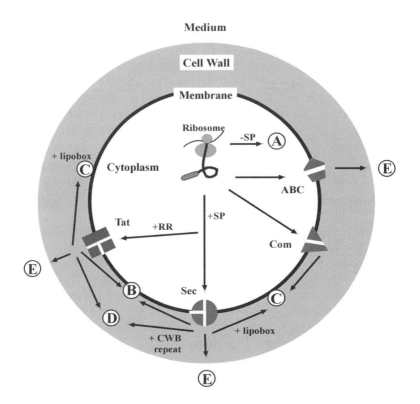

Fig. 1. Schematic cross-section of a *B. subtilis* cell with the predicted protein transport pathways. Based on the predictions of signal peptides and various retention signals, it is hypothesized that at least four different protein transport pathways exist in *B. subtilis* that can direct proteins to at least four different subcellular destinations (B-E). Ribosomally synthesized proteins can be sorted to various destinations depending on the presence (+SP) or absence (-SP) of an amino-terminal signal peptide and specific retention signals such as lipid modification or cell wall-binding repeats (CWB). (A) Proteins devoid of a signal peptide remain in the cytoplasm. (B) Proteins with one or more membrane-spanning domains are inserted into the membrane either spontaneously (not shown), via the Sec pathway or possibly via the Tat pathway (+RR). (C) Proteins which have to be active at the extracytoplasmic side of the membrane can either be lipid-modified proteins (+lipobox) exported via the Sec or (possibly) Tat pathways, or pseudopilins exported by the Com system. (D) Proteins that need to be retained in the cell wall can be exported via the Sec or Tat pathway. In order to be retained, the mature parts of these proteins contain cell wall-binding repeats (+CWB). (E) Proteins can be secreted into the medium via the Sec or Tat pathways or by ABC transporters. Adapted from Tjalsma *et al.* (2000).

As in other organisms, most secretory proteins are synthesized as precursors carrying an amino-terminal extension, the signal peptide, which is removed during or shortly after translocation. The known signal peptides can be classified according to the export pathway used or the type of signal peptidase (SPase) responsible for their proteolytic processing. The structure and cleavage sites of these different classes of signal peptides will be summarized in the following section. The mechanisms by which signal peptide-containing preproteins are targeted to and translocated across the membrane, schematically represented in Fig. 1, are subsequently discussed in following sections.

3.1   Signal peptide classification

The major class of signal peptides is cleaved by type I SPases. In a survey of the *B. subtilis* genome, a total of 179 putative SPase I-cleaved signal peptides were identified. On average, the signal peptides of this class are longer in Gram-positives than in Gram-negatives (Tjalsma *et al.*, 2000; van Dijl *et al.*, 2001). Nevertheless, they are composed of the three distinct domains that typically make up signal peptides directing protein secretion in other prokaryotes and eukaryotes. The amino terminal portion, or N-domain, contains several positively charged residues and is thought to be involved in interactions with anionic lipids and the translocation machinery (Chen and Nagarajan, 1994; Fekkes *et al.*, 1999). The central portion of the signal peptide, the H-domain, is highly hydrophobic and is supposed to adopt an α-helical structure in a lipid environment. The presence of helix breakers (Pro, Gly) in the middle of the H-domain facilitates hair pin-like insertion of the signal peptide. The H-domain usually ends with a helix breaker, which facilitates signal peptide cleavage at an SPase I recognition site in the C-domain with the consensus sequence $^{-3}$Ala-$^{-2}$X-$^{-1}$Ala$\downarrow$ (*X*, any residue except cysteine and proline; Nielsen *et al.*, 1997). A subgroup of these SPaseI-cleaved signal peptides is characterized by the presence of the so-called twin-arginine motif (R-R-X-ϕ-ϕ, where ϕ represents a hydrophobic residue; Cristóbal *et al.*, 1999) preceding the H-domain, which is typical of proteins secreted via the Tat pathway. A screen of the annotated genome sequence for ORFs containing this motif retrieved 15 putative Tat-dependent precursors (Jongbloed *et al.*, 2000; van Dijl *et al.*, 2001).

The second class of signal peptides is found in prelipoproteins, which are cleaved by SPase II or lipoprotein-specific SPase (Tjalsma *et al.*, 2001). In general, these signal peptides are somewhat shorter than SPase I-cleaved signal peptides. They are characterized by the presence of the so-called lipobox with the consensus sequence $^{-3}$L-$^{-2}$(A/S)-$^{-1}$(A/G)$\downarrow$C^{+1}. The invariable Cys residue constitutes the site of lipid modification and, after removal of the signal peptide, becomes the first amino acid of the mature protein. In *B. subtilis*, a total of 114 possible lipoprotein signal peptides were identified (Tjalsma *et al.*, 2000).

A distinct and relatively rare class of signal peptides is found in type IV pilin-like proteins, or pseudopilins. Although resembling SPase I- and SPase II-cleaved signal peptides with respect to their modular structure, they are translocated independently of the Sec-machinery. Moreover, their cleavage occurs at the cytoplasmic side of the membrane and is catalyzed by a dedicated pseudopilin SPase which, in the case of *B. subtilis*, is the ComC protein. The consensus cleavage site for pseudopilin SPases is $^{-2}$K-$^{-1}$G$\downarrow$F^{+1}-X_3-E^{+5} (where X represents any residue, and the arrow indicates the cleavage site; Chung *et al.*, 1998). Notably, this site is located between the N- and H-domains of the signal peptide. Upon processing the H-domain remains attached to the mature protein, while the Phe residue at position +1 is modified by amino-methylation (Lory, 1994).

Finally, proteins that are transported via ATP-binding cassette (ABC)-transporters contain signal peptides that lack the H-domain and cleavage sites typical for all other signal peptides described above. In *B. subtilis*, only four proteins with signal peptides of this type are known (discussed below).

3.2   Sec-dependent protein secretion

As can be inferred from the abundance of classical signal peptides, most exoproteins in *B. subtilis* and other organisms are secreted via the Sec-dependent secretion machinery. The key components of this pathway have all been identified and to a certain extent analyzed in *B. subtilis* and other Gram-positive bacteria (Table 1). The individual steps in the Sec-dependent secretion pathway, which is drawn schematically in Fig. 2, can be classified into early (targeting), middle (translocation), and late (post-translocational processing) stages of protein secretion. The following paragraph will describe these individual processes accordingly.

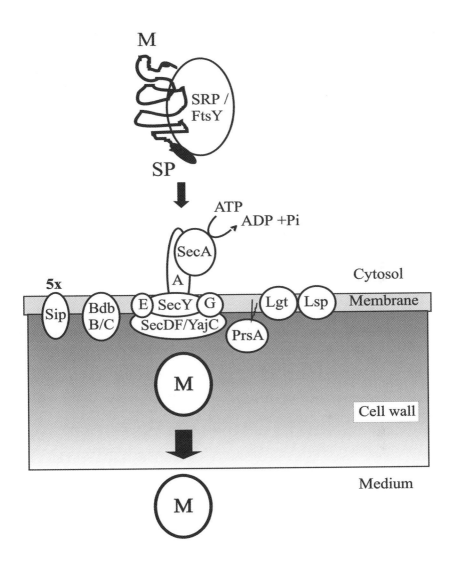

Fig. 2. Main components of the Sec-dependent *B. subtilis* protein secretion machinery. The SRP complex consists of Ffh, scRNA and HbsU. See text for further details. A, SecA; E, SecE; G, SecG; M, mature protein; SP, signal peptide.

278

Table 1. Components of the Sec-dependent protein export machineries of *B. subtilis*, *L. lactis*, *S. aureus*, and *E. coli*. Adapted and modified from Tjalsma *et al.* (2000) and van Dijl *et al.* (2001).

Component	*B. subtilis*	*L. lactis*[a]	*S. aureus*[b]	*E. coli*
Secretion-specific chaperones	Ffh, scRNA, HbsU	Ffh, HslA/B	Ffh, HbsU	Ffh, 4.5S RNA, "HupA/B" (?)
			FtsY[c], -	
	FtsY, FlhF (?)	FtsY, -	-	FtsY, -
	CsaA	-	-	YgjH (?)
	-	-		SecB
General chaperones	GroEL, GroES	GroEL, GroES	GroEL, GroES (Ohta et al., 1993)	GroEL, GroES
	DnaK, DnaJ, GrpE	DnaK	DnaK, DnaJ, GrpE (Ohta et al., 1994)	DnaK, DnaJ, GrpE
		TF[d]		
	TF[d]		TF[d]	TF[d]
Translocation motor	SecA	SecA	SecA1 (Segarra and Iandolo, 1997) SecA2 (57)	SecA
Translocation channel	SecYEG	SecYEG	SecY1 (71), SecY2 (49), SecE, SecG	SecYEG
	SecDF, YrbF	-, YwaB	SecDF, YrbF	SecD, SecF, YajC
Lipid modification	Lgt	Lgt	Lgt	Lgt
SPases	SipP/S/T/U/V	SipL	SpsB (Cregg et al., 1996)	LepB
		-	-	-
	SipW LspA	LspA	LspA	LspA
SPPases	SppA, TepA	Pi323 (?)	-, -	SppA, OdpA

Folding catalysts	PrsA, YacD (?)	PrtM, Usp45, pi314	PrsA	SurA, PpiD
			-	RotA, FkpA
	-	-	DsbA	DsbA/B/C/D/ E/G
	BdbA/B/C/D	-		

[a], annotations, based on similarity (Bolotin *et al.* 2001); [b], based on BLAST searches in the "unfinished microbial genomes" with the corresponding *B. subtilis* proteins (http://www.ncbi.nlm.nih.gov/Microb_ blast/unfinishedgenome.html); the levels of similarity (% identical residues and conservative replacements) are shown in parenthesis; [c], the putative *ffh* and *ftsY* genes were found on a single contig for all *S. aureus* strains sequenced. In addition, a homolog of *ylxM* was found between them, suggesting that their genetic organization is similar to that in *B. subtilis*; [d], trigger factor

3.2.1 Targeting of secretory precursors to the translocation apparatus

Most exported proteins are translocated across the cytoplasmic membrane in a more or less unfolded state. To maintain this export-competent conformation, precursor proteins interact with cytosolic chaperones, which in some cases are also involved in targeting of the preproteins to the translocation machinery. The only secretion-specific chaperone identified to date in *B. subtilis* is the signal recognition particle (SRP), a ribonucleoprotein complex consisting of the Ffh protein assembled onto a 7S RNA scaffold. The Ffh protein is a GTPase homologs to the 54-kD subunit of the eukaryotic SRP and P48 of *E. coli* (for a recent review, see Herskovits *et al.*, 2000). Recent studies have shown that the histon-like protein of *B. subtilis* (HbsU) is able to bind to the 7S RNA Alu-domain, forming an integral part of the SRP (Nakamura *et al.*, 1999; see also below). The SRP recognizes and binds to the amino-terminal signal peptides of presecretory proteins emerging from the ribosome during the early stages of translation. The resulting ribosome-nascent chain (RNC) complex is targeted to the SRP receptor FtsY. The RNC-FtsY complex is then directed to the translocon, an event that is thought to be stimulated by either (i) affinity of FtsY for anionic phospholipids; (ii) affinity of FtsY for the translocon (de Leeuw, 2000; de Leeuw *et al.*, 2000); (iii) affinity of the RNC for the translocon (Bibi, 1998); or (iv) affinity of Ffh for SecA (Bunai *et al.*, 1999). GTP binding by FtsY subsequently stimulates GTP binding by SRP, after which the RNC is released. Finally, the protein is translocated co-translationally through the SecYEG translocation channel (see below).

To date, limited information is available on alternative targeting pathways operating in *B. subtilis*. Recent studies suggest that most, if not all, SecA-dependent expoproteins are targeted by the SRP pathway (Hirose *et al.*, 2000). However, the existence of alternative secretion-specific chaperones cannot be ruled out. Surprisingly, an ortholog of SecB, involved in post-translational targeting and translocation of proteins in *E. coli* and several other organisms, was not detected in *B. subtilis*, neither by classical genetic and biochemical screens, nor in the annotated genome sequence (Kunst *et al.*, 1997; van Wely *et al.*, 2001). The same is true for *L. lactis* (Bolotin *et al.*, 2001) and *S. aureus*; in fact, SecB homologs have only been discovered in Gram-negative bacteria so far. A possible functional analog of SecB in *B. subtilis* is the CsaA protein, which was shown to suppress a temperature-sensitive *secA* mutation in *E. coli*. Recent studies demonstrated that CsaA has an affinity for SecA and several preproteins, implying a role of CsaA in protein targeting (Müller *et al.*, 2000; 2000a; Kawaguchi *et al.*, 2001). In addition, the general chaperones GroEL-ES, DnaK-DnaJ-GrpE, and TF (trigger factor) may facilitate secretion by preventing proteins from adopting a translocation-incompatible conformation in the cytoplasm (Fekkes *et al.*, 1999; Müller *et al.*, 2000b). However, little evidence exists to date suggesting a direct role for the general chaperone machinery in protein secretion in Gram-positive bacteria (Lyon *et al.*, 1998; Wu *et al.*, 1998).

3.2.2   The translocation apparatus

The current working model for preprotein translocation in *B. subtilis* is largely based on results obtained in *E. coli* (for reviews, see Driessen *et al.*, 1998; van Dijl *et al.*, 2001; van Wely *et al.*, 2001). Interaction of the translocation motor SecA (Sadaie *et al.*, 1991) with the translocation channel component SecY and acidic phospholipids stimulates recognition and binding of targeted preprotein nascent chains by SecA. Subsequent binding of ATP induces conformational changes in the SecA protein, causing insertion of its carboxyl-terminus in the translocation channel and, concomitantly, the insertion of a portion of the preprotein. Hydrolysis of ATP results in release of the precursor and deinsertion of SecA from the translocation channel. Further translocation is driven by repeated cycles of ATP-dependent SecA insertion and de-insertion and the proton motive force. Homologs of *E. coli* SecA have been identified in *B. subtilis*, *L. lactis*, and *S. aureus* (Table 1).

Like in *E. coli*, the translocation pore of *B. subtilis* consists of the SecY, SecE, and SecG proteins. The corresponding genes have all been identified by complementation studies or via DNA sequencing. The SecY protein is an integral membrane protein containing 10 transmembrane spanning domains and is well conserved between *E. coli* and several Gram-positive bacteria (Suh *et al.* 1990; van Wely *et al.*, 2001). Although unable to restore growth at non-permissive temperature, the *B. subtilis secY* gene was shown to functionally complement an *E. coli secY*(ts) mutant (Nakamura *et al.*, 1990). The SecE proteins of *B. subtilis, L. lactis* and *S. aureus* are considerably shorter (1 transmembrane domain) than that of *E. coli* (3 transmembrane domains) and show relatively little sequence similarity with the latter. However, based on functional complementation of a cold-sensitive, export-defective *E. coli secE* mutant, it was shown that the *B. subtilis secE* gene encodes a true SecE homologue (Jeong *et al.*, 1993). Likewise, the SecG protein of *B. subtilis* (2 transmembrane domains), encoded by the *yvaL* gene, shows relatively little overall similarity with its *E. coli* counterpart. Nevertheless, the disruption of this gene causes a reduced efficiency of translocation, and results in a cold-sensitive phenotype typical for *sec* mutants (van Wely *et al.*, 1999).

In addition to these core elements of the translocase, accessory functions have been identified in *B. subtilis* and several other Gram-positives. Interestingly, the SecDF protein of *B. subtilis* (12 transmembrane domains) represents a molecular "Siamese twin" as a result of a natural fusion between the *secD* and *secF* genes (Bolhuis *et al.*, 1998). In most organisms, the *secD* and *secF* genes are expressed separately, and their products subsequently form a heterotrimeric complex with a third protein termed YajC in *E. coli* (see below). A similar fusion between *secD/F* genes was found in representatives of both Gram-negative bacteria, including *Synechocystis* PCC6803 (Kaneko et al., 1996), *Deinococcus radiodurans* (White *et al.*, 1999) and *Chlamydia trachomatis* (Read *et al.*, 2000), as well as Gram-positive bacteria, including several *Bacillus* species (TIGR at http://www.tigr.org; UOKNOR at http://www.genome.ou.edu/bstearo.html) and *S. aureus* (TIGR at http://www.tigr.org; Sanger; http://www.sanger.ac.uk/Projects/S_aureus/; OUACGT at http://www.genome.ou.edu/staph.html). In contrast to *E. coli* SecD/F, which was shown to be involved in the cycling of SecA (see: Driessen *et al.*, 1998) and the release of mature proteins from the translocase (Matsuyama *et al.*, 1993), the role of SecDF in protein secretion in *B. subtilis* is not fully understood. Although required for a high protein secretion capacity, the *secDF* gene is not essential for growth and viability of *B. subtilis* (Bolhuis *et al.*, 1998). Strikingly, *secD/F* genes are completely absent from the genome of *L. lactis* (Bolotin *et al.*, 2001; Table 1). Homologs of the *E. coli* YajC protein, which forms a heterotrimeric complex with SecD and SecF, were identified in *B. subtilis* (YrbF), *S. aureus*, and *L. lactis* (YwaB). The latter observation is remarkable in view of the absence of SecD/F from *L. lactis*. The role, if any, of YajC and its orthologous in Gram-positive protein translocation remains to be established.

During or shortly after translocation, the signal peptide is cleaved either by SPase I or II, depending on the nature of the signal peptide. The combined use of an activity assay, PCR strategies and genome sequencing, resulted in the identification of five type I SPase-encoding genes in *B. subtilis*, denoted *sipS*, *sipT*, *sipU*, *sipV* and *sipW* (van Dijl *et al.*, 1992; Tjalsma *et al.*, 1997, 1998). For comparison, *L. lactis* contains only a single type I SPase-encoding gene (*sipL*), the product of which is highly similar to SipS, SipT, SipU, and SipV of *B. subtilis*. The four latter SPases are of the prokaryotic type (P-type), sharing similarity with SPases common to all bacteria. In contrast, SipW is highly similar to SPases found in archaea and the endoplasmic reticulum of eukaryotes, representing one of the few known prokaryotic ER-type SPases. SipW appears to be specifically involved in the processing of the exported YqxM and TasA pre-proteins, encoded by genes of the *yqxM-sipW-tasA* operon. Interestingly, mature TasA is both secreted to the medium and sorted to the inter membrane space between mother cell and forespore during the early stages of sporulation (Stöver and Driks, 1999a, 1999b, 1999c).

The gene encoding the SPase II of *B. subtilis* was cloned by complementation of a temperature-sensitive *lsp* mutant of *E. coli* (Pragai *et al.*, 1997). Unlike *E. coli* cells, *B. subtilis* cells lacking SPaseII are viable, suggesting that processing of lipoproteins is not strictly required for their functionality (Tjalsma *et al.*, 1999). This view is underscored by the fact that the lipoprotein PrsA, a peptidyl-prolyl *cis/trans* isomerase required for the efficient folding of translocated non-lipoproteins, is essential for viability of *B. subtilis* (Jacobs *et al.*, 1993; Kontinen and Sarvas, 1993; Kontinen *et al.*, 1991).

Upon processing of pre-proteins in *E. coli*, signal peptides are rapidly degraded by the concerted activities of signal peptide peptidases (SPPases). Two known SPPases of *E. coli* are the membrane-bound protease IV and the cytoplasmic OpdA protein (Novak and Dev, 1988). A homolog of *E. coli* SppA was found in *B. subtilis*. In addition, the *tepA* gene of *B. subtilis* encodes a second SppA-like protein that also shows similarity to ClpP. As TepA seems to be localized in the cytoplasm, this protein may act as a functional analog of OpdA. While SppA is merely required for efficient pre-protein processing, TepA is required both for efficient pre-protein translocation and processing (Bolhuis *et al.*, 1999). BLAST searches using the genome sequence of *L. lactis* did not reveal any genes specifying SppA homologs. Limited similarity was, however, observed with a protein similar to ClpP of *Caenorhabditis elegans*. Hence, despite the lack of overall sequence similarity, the corresponding gene (*pi323*) might encode an SPPase-analog of *L. lactis*.

3.2.3 Post-translocational processing and quality control

After completion of their translocation, secreted proteins need to adopt their native conformation on the *trans* side of the cytoplasmic membrane. Fast and accurate folding is crucial, since unfolded proteins are highly susceptible to proteolysis by one of several proteases associated with the cell wall of *B. subtilis* (Tjalsma *et al.*, 2000; van Dijl *et al.*, 2001; van Wely *et al.*, 2001). Hence, interactions at the membrane-wall interface are critical to determine the fate of secreted proteins.

Folding of translocated proteins is mediated by several extracytoplasmic folding catalysts. In addition to the peptidyl-prolyl *cis/trans* isomerase PrsA described in a previous section, thiol-disulfide oxidoreductases have been implicated in extracytoplasmic folding of proteins. In *E. coli*, where the formation of disulfide bonds occurs in the periplasm, several proteins involved in this process were identified, namely DsbA, DsbB, BsbC, DsbD, DsbE, and DsbG (Raina and Missiakas, 1997; Andersen *et al.*, 1997). In *B. subtilis*, only a limited number of exported proteins contain disulfide bridges in their native conformation. Among these are the pseudopilins ComGC and ComGG, both of which are components of the DNA binding and uptake machinery. The first *Bacillus* protein with demonstrated thiol-disulfide oxidoreductase activity was Bdb of *Bacillus brevis* (Ishihara *et al.*, 1995). Subsequently, thiol-disulfide oxidoreductases were also identified in *B. subtilis* (Bolhuis *et al.*, 1999a). Of these, BdbB and BdbC were shown to be important for the folding of translocated *E. coli* PhoA, which contains two disulfide bonds in its native conformation. In contrast, disruption of the *B. subtilis bdbA* gene did not significantly affect folding of translocated PhoA. Recently, a fourth thiol-disulfide oxidoreductase was discovered (Meima *et al.*, 2001), which is highly similar to *dsbA* of *S. aureus* (Dumoulin, 2000). The coding gene, *bdbD*, forms a bicystronic operon with the *bdbC* gene. Interestingly, BdbC and BdbD are both essential for competence, most likely because they act as a redox pair catalyzing disulfide-bond formation in the ComGC protein. Like BdbC, BdbD is essential for folding of translocated PhoA (Meima *et al.*, 2001).

In addition to the foldases described above, other folding catalysts such as Ca^{2+} and Fe^{3+} are required for proper biogenesis of secreted proteins. Due to its overall negative charge, the cell wall acts as a reservoir for these cations, which are crucial for the activity and stability of several secreted proteins (Petit-Glatron *et al.*, 1990, 1993; Stephenson *et al.*, 1998; Veltman *et al.*, 1997). By varying the growth conditions (*eg.* phosphate limitation) or by introducing specific mutations, the composition and charge of the cell wall can be modulated (Foster and Popham, 2001). For instance, *dlt* mutations inhibit the D-alanylation of teichoic acid, causing an increased net negative charge of the cell wall. As a consequence, more Ca^{2+} can be bound by the cell wall, which may explain why the folding and yield of *Bacillus licheniformis* AmyL and engineered derivatives thereof is improved in *dlt* mutants of *B. subtilis* (Hyyryläinen *et al.*, 2000). Finally, the charge properties of the cell wall also affect its interaction with translocated secretory proteins, thereby influencing their folding rates (Stephenson *et al.*, 1998).

Several exoenzymes produced by Gram-positive bacteria are synthesized as so-called pre-pro-proteins (Vasantha *et al.*, 1984). After cleavage of the signal ("pre-") peptide during translocation across the membrane, the pro-peptide directs the proper folding of the pro-protein, which is further processed to produce the mature and biologically active enzyme by removal of the pro-sequence. Notably, certain propeptides can be physically separated from the corresponding mature protein without loss of function, showing that they are genuine molecular chaperones (Wandersman *et al.*, 1989; Zhu *et al.*, 1989). The pro-peptide of the extracellular alkaline protease subtilisin of *Bacillus* sp. is autoproteolytically removed from the mature portion of the protein (Power *et al.*, 1986). Similar observations were made with several neutral proteases secreted by bacilli. Maturation of the envelope-associated serine protease PrtP of *L. lactis*, involved in the initiation of casein degradation, was also shown to be based on self-cleavage (Haandrikman *et al.*, 1991a,b). However, the activation of PrtP requires the lipoprotein PrtM, a homolog of the peptidyl-prolyl cis/trans isomerase PrsA of *B. subtilis* (Table 1). These observations imply that the pro-peptide alone is not sufficient to facilitate the folding of PrtP.

3.3   Sec-independent protein secretion

Although the majority of exoproteins in *B. subtilis* is secreted via the Sec-machinery, several proteins are exported Sec-independently. The following sections will discuss the characteristics of the three Sec-independent pathways known to exist in several Gram-positive bacteria, namely (i) the Tat pathway, (ii) a pseudopilin-specific export and assembly pathway, and (iii) ABC transporter-dependent secretion pathways.

3.3.1 Twin-arginine translocation (Tat) pathway

First discovered in plant chloroplasts, this ΔpH-dependent protein secretion pathway appears to exist in most eubacteria, certain archaea and, possibly, some plant mitochondria (Berks *et al.*, 2000; Dalbey and Robinson, 1999; Wu *et al.*, 2000). In contrast to Sec-dependent secretion, in which unfolded proteins are threaded through the translocase, the Tat pathway enables the export of fully folded proteins across the chloroplast thylakoidal membrane and the bacterial plasma membrane. Among these are several cofactor-binding proteins, such as flavin-, molybdopterin-, and iron-sulfur cluster-binding proteins. In *E. coli*, four genes encoding components of the Tat pathway were identified. Three of these are localized in an operon (*tatABC*), whereas the fifth, *tatE*, is monocistronic. Functional analysis of the *tatA* gene and its paralogs, *tatB* and *tatE*, revealed that efficient Tat-dependent secretion requires TatB in combination with either TatA or TatE. Disruption of *tatC* completely blocks Tat-specific export (Bogsch *et al.*, 1998), showing that the TatC protein plays a pivotal role in the Tat pathway.

The specific roles of TatA/E, TatB and TatC are currently debated. It has been proposed that TatC, a polytopic membrane protein with six transmembrane domains, forms the translocation channel. Alternatively, the channel may be composed of multiple copies of TatA/E and TatB, which all contain a single transmembrane domain (Berks *et al.*, 2000; Wu *et al.*, 2000). In the latter concept, TatC would constitute the signal peptide-binding component. Evidence for the latter model was recently obtained in a study on the Tat pathway of *B. subtilis* (Jongbloed *et al.*, 2000).

The individual components of the Tat machinery of *B. subtilis* have been identified through the genome sequence. The *tatAc*, *tatAd*, *tatAy* genes are homologs to the *tatA/B/E* genes of *E. coli*. Interestingly, two homologs of *tatC* are present in *B. subtilis*, namely *tatCd* and *tatCy*. Recent data showed that secretion of PhoD in *B. subtilis* is strongly dependent on TatCd, but not on the paralogous TatCy, suggesting that TatC acts as a specificity determinant (Jongbloed *et al.*, 2000). Homologs of the *tatA/B/E* and *tatC* genes have also been identified in the *S. aureus* genome sequence (van Dijl *et al.*, 2001), indicating that this bacterium has a functional Tat pathway. In contrast, the Tat machinery appears to be absent in *L. lactis*.

3.3.2 Pseudopilin synthesis

A second class of Sec-independently translocated proteins is formed by the pseudopilins. The amino-terminal signal peptides of these preproteins resemble those of Sec-dependent exoproteins, but their processing depends on a cleavage event occurring on the cytoplasmic side of the membrane. The only proteins of *B. subtilis* containing typical pseudopilin signal peptides are the ComGC, ComGD, ComGE, and ComGG proteins, which are required for DNA binding and transport during genetic competence (Tjalsma *et al.*, 2000). Processing and methylation of the ComG proteins is achieved by ComC, an integral membrane protein which is homologs to pseudopilin SPases of several other organisms. In addition to ComC, the ATPase ComGA and the integral membrane proteins ComGB and ComGF are required for the export and assembly of the four ComG pseudopilins of *B. subtilis* (Chung *et al.*, 1998; Dubnau and Lovett, 2001).

Interestingly, even though a DNA uptake machinery is not known to exist in *L. lactis*, orthologs of the *B. subtilis* ComEA, ComEC, ComFA, ComFC, ComGA, and ComGB proteins, all of which are required for DNA binding and uptake, were identified by examining the *L. lactis* genome sequence (Bolotin *et al.*, 2001). In addition, orthologs of the ComGC and ComGD proteins, containing typical pseudopilin cleavage sites, as well as the SPase ComC were detected. The function of the Com proteins in lactic acid bacteria remains to be elucidated. Although they may represent the components of a DNA uptake machinery, it is also conceivable that these Com proteins are involved in the transport of another type of macromolecule. This would be in line with the general observation that pseudopilins have evolved to facilitate different transport processes in different organisms (Dubnau and Lovett, 2001).

3.3.3 Export via ABC-transporters

The third class of proteins that are known to be secreted independently of the Sec system makes use of dedicated ABC transporters. A very well described group of proteins secreted via this route consists of ribosomally synthesized antimicrobial peptides, or bacteriocins, which are produced by a large variety of Gram-negative and Gram-positive bacteria. ABC transporter-dependent export of bacteriocins has been documented particularly well for Gram-positives, such as lactic acid bacteria. Therefore, the latter systems will not be reviewed here. Instead, this section is focused on ABC transporter-dependent protein export in *B. subtilis*.

Three dedicated ABC transporters have been implicated in the secretion of ribosomally-synthesized bacteriocins in *B. subtilis*. Firstly, the ABC transporter SunT of *B. subtilis* 168 is most likely required for the export of the lantibiotic sublancin 168 (SunA; Paik *et al.*, 1998). SunT is predicted to have a dual role in sublancin 168 secretion as it belongs to a group of ABC transporters that facilitate both signal peptide removal and translocation of mature lantibiotics across the membrane. As shown for other ABC transporters of this type, amino-terminally located conserved Cys and His residues are required for signal peptide cleavage (Havarstein *et al.*, 1995). During export of mature sublancin 168, one β-methyllanthionine bond and two disulfide bonds are formed in this protein. The thiol-disulfide oxidoreductases BdbA and BdbB have been implicated in the latter step, as the corresponding genes are located immediately downstream of the *sunA-sunT* genes (Bolhuis *et al.,* 1999a). Secondly, the lantibiotic subtilin (SpaS) of *B. subtilis* ATCC 6633 is secreted by a dedicated ABC transporter complex, composed of the SpaB, SpaC and SpaT proteins (Chung *et al.*, 1992; Kiesau *et al.*, 1997). Thirdly, the antilisterial bacteriocin subtilosin (Sbo) of *B. subtilis* 168, which is exported as a cyclic peptide with several modifications including one disulfide bond, requires the ABC transporter AlbC for translocation. The *albC* gene is part of an operon located downstream of the *sbo* gene that is essential for subtilosin production. Interestingly, this operon contains genes for two peptidases (AlbE and AlbF) that are likely to be required for subtilosin processing (Zheng *et al.*, 1999). Finally, although conclusive evidence is not documented, the competence pheromone ComX is likely to be secreted via a dedicated ABC transporter, since it is ribosomally synthesized as a precursor and modified prior to secretion (Dubnau and Lovett, 2001).

4. UNIQUE FEATURES OF PROTEIN SECRETION IN GRAM-POSITIVE BACTERIA

The recent functional genomic analyses have made it evident that *B. subtilis* is naturally equipped for the high-level secretion of proteins (Antelmann *et al.*, 2001; Tjalsma *et al.*, 2000). In the soil, the natural habitat of *B. subtilis*, starvation conditions are likely to be the rule rather than the exception. This may have provided the selective pressure that has ultimately resulted in the establishment of a high secretion capacity as an adaptation strategy to survive and compete under conditions of nutrient deprivation. The duplication of many genes, including the type I SPase (*sip*) genes and the *tatC* genes, should probably be regarded as exponents of this remarkable degree of adaptation. Similarly, the identification of plasmids bearing additional type I SPase-encoding genes (*sipP*; Meijer *et al.*, 1995) in strains selected for optimal *natto* production, supports the view that the multiplication of genes for secretion machinery components is an adaptation strategy to increase the fitness of bacilli in a given environment. Obviously, the latter consideration is of major relevance for microbial production processes.

Interestingly, the multiplication of genes for secretion machinery components is a phenomenon not only observed in *B. subtilis*, but also in other Gram-positive eubacteria. For example, duplicated copies of the *secA* and *secY* genes are present in *S. aureus*. Similarly, two copies of *secA* have been reported for *Mycobacterium smegmatis*, one of which (*secA1*) is the essential equivalent of the *E. coli* and *B. subtilis secA* genes, while the other gene (*secA2*) is not essential (Braunstein *et al.*, 2001). At present, it is unknown whether all *secA* and *secY* genes of *S. aureus* are expressed, and whether or not they specify functional SecA and SecY proteins. The analysis of upstream regions of the *secA2* and *secY2* genes of *S. aureus* revealed the presence of putative ribosome binding sites for both genes (GGAAG and AGAGG at -3 and -6 from the predicted start codons, respectively), indicating that these additional *sec* genes can be expressed.

An alternative strategy to adjust the capacity for protein secretion to environmentally determined demands is the controlled expression of genes for secretion machinery components. Strikingly, most genes for secretion machinery components of *B. subtilis* are expressed in a growth phase- and/or medium-dependent manner, the *sipU*, *sipV*, *tepA*, and *bdbC* genes being the only exceptions that appear to be transcribed constitutively (van Dijl *et al.*, 2001). This is likely to relate to the fact that the genes for many secretory proteins are expressed in a growth phase- and medium-dependent manner. Thus, secreted degradative enzymes are mainly synthesized in the stationary phase, when nutrients become limiting (Antelmann *et al.*, 2001). Similarly, the pseudopilin export machinery of *B. subtilis* is only induced upon competence development (Dubnau and Lovett, 2001). Finally, the formation of endospores is characterized by the sequential and compartmentalized expression of specific subsets of genes in the sporulating mothercell on the one hand and the prespore on the other (for review, see Piggot and Losick, 2001). Genes for secretion machinery components that are activated during sporulation include, for example, *sipW* and *ftsY*. SipW, a special type I SPase which is more closely related to ER and archaeal SPases, appears to be mainly dedicated to the secretion of spore-specific proteins, such as pre-TasA (Stöver *et al.*, 1999a,b). Consistently, eubacterial homologs of SipW have, so far, only been observed in spore forming Gram-positives (Tjalsma *et al.*, 1998, 2000; van Roosmalen *et al.*, 2001). In addition, sporulation-specific expression of *ftsY* from an alternative, σ^K-dependent promoter, was shown to be crucial for targeting and assembly of spore coat proteins during the later stages of sporulation (Kakeshita *et al.*, 2000). Together, these data demonstrate that the flexibility of the secretion apparatus enables the cell to adjust its range of secreted proteins to the specific requirements imposed on the bacterium during different stages of growth.

Another aspect of the Gram-positive secretion process that seems to differ considerably from the protein export process in Gram-negative bacteria concerns the targeting of proteins to the membrane. In addition to the fact that SecB appears to absent from Gram-positive eubacteria, the SRP-dependent targeting systems of both groups of organisms display remarkable differences. For instance, the FtsY protein of *B. subtilis* lacks the highly acidic A-domain typical of *E. coli* FtsY and yeast SRα (Fig. 3). In the latter, this domain is thought to be involved in membrane-association through interactions with the membrane-bound SRβ receptor. Furthermore, several lines of evidence suggest that the FtsY protein of *E. coli* contains a 14-amino acid N-terminal extension with a putative pseudopilin processing site, which might be involved in its targeting (Luirink *et al.*, 1994; de Leeuw, 2000). A similar pseudopilin cleavage site is not present in the FtsY protein of *B. subtilis*. As shown in the alignment of secretion-related GTPases (Fig. 3), a third SRP54-type GTP-binding protein, denoted FlhF, is present in *B. subtilis*. This protein, which is absent from *E. coli*, has been implicated in cell motility (Carpenter *et al.*, 1992).

Interestingly, the scRNA molecules of *B. subtilis* and several other *Bacillus* sp. (Nakamura *et al.*, 1992a, b) are considerably longer than all other known bacterial scRNA species, their length and structure being very similar to archeal and eukaryotic SRP RNA (for alignments, see http://psyche.uthct.edu/dbs/SRPDB/SRPDB.html). The structure of the mammalian 7S RNA *Alu* domain, involved in SRP9 and SRP14 binding and elongation arrest, was recently determined (Weichenrieder *et al.*, 2000). The equivalent *Alu* domain of the *B. subtilis* scRNA was shown to be involved in HbsU binding (Nakamura *et al.*, 1999). However, homologs of SRP9 and SRP14 have not been identified in *B. subtilis*. The role of the *Alu* domain and HbsU in the functionality of *B. subtilis* SRP remains to be further addressed. One intriguing possibility is that the presence of the *Alu* domain is critical for the proper development of endospores, since (i) mutants lacking the *Alu* domain are no longer able to sporulate (Nishiguchi *et al.*, 1994); (ii) *E. coli* 4.5S RNA can not complement the sporulation deficiency of such mutants; and (iii) among eubacteria, the *Alu* domain is found exclusively in spore-forming bacilli and *Clostridium perfringens*, another Gram-positive, spore-forming bacterium. Further analysis of the role of the *Alu* domain, as well as the identification of additional proteinaceous components of the SRP of bacilli will probably provide novel insights into the mechanisms of SRP-dependent targeting in *B. subtilis* and other Gram-positives.

290

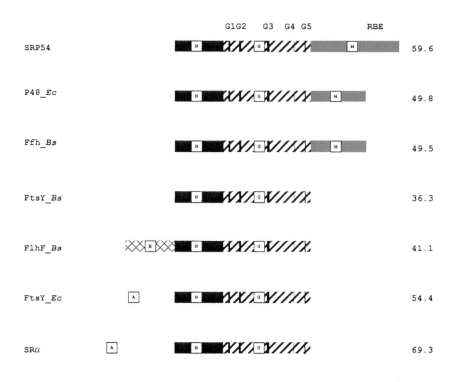

Fig. 3. Alignment of the SRP54-type GTPases of yeast (SR), *E. coli* (Ec), and *B. subtilis* (Bs). "A", acidic domain; "B", basic domain, "M", M-domain, involved in RNA and preprotein binding; "G", GTP-binding domain. G1-G5, conserved boxes in the GTPase domains; RBE, RNA-binding element in the SRP54/Ffh family (Eichler and Moll, 2001).

4. FUTURE PERSPECTIVES

Despite the extensive knowledge accumulated over the past two decades, numerous questions concerning the components of the protein targeting and secretion machinery of Gram-positive eubacteria remain to be addressed. Firstly, from the evidence available to date it cannot be ruled out that alternative secretion-specific chaperones operate in the targeting of certain preproteins in *B. subtilis*. To address this question, it will be essential to determine the natural substrates for the SRP pathway. Also, based on the structure of the scRNA of *B. subtilis* and other bacilli, it is conceivable that, in addition to Ffh and HbsU, other proteins are assembled onto the RNA scaffold. The identification of such additional components of the SRP may provide further insight in the mechanisms of SRP-dependent targeting. Secondly, a study of the relative contribution of paralogous genes in the secretion process in several Gram-positive bacteria, similar to the studies of Tjalsma et al. (1998), could provide further insights into the regulation and specificity of the translocation machineries. For instance, dissecting the roles of each of the two SecA and SecY paralogs of *S. aureus*, or the SecA paralogs of mycobacteria could lead the way to the development of highly specific drugs, which might be of great benefit in the treatment of infections with these pathogens. Finally, further studies into the mechanisms by which the Tat-pathway operates and, equally important, the determinants directing different exported proteins into either the Sec or the Tat secretion pathways are required. The information obtained from such analyses could allow the development of novel strategies for the secretion of those proteins which, due to their folding properties, can not be exported via the Sec-dependent pathway. Several lines of evidence demonstrated that fully folded proteins, such as the mouse dihydrofolate reductase (Hynds et al., 1998), the 23kDa thylakoid lumen protein (Bogsch et al., 1998) and active jellyfish green fluorescent protein (GFP; Santini et al., 2001; Thomas et al., 2001) can be transported via the Tat machinery when provided with twin-arginine signal peptides. Notably, active GFP can not be exported *via* the Sec pathway (Feilmeier et al., 2000). Similar achievements in *Bacillus* species could be of major industrial and pharmaceutical importance, since it would potentially allow for the large-scale production of a host of new therapeutics.

One aspect of protein targeting that has not been addressed so far is the biogenesis of integral membrane proteins (IMPs) in Gram-positive bacteria. In recent years, exciting new data have emerged, demonstrating the activity of proteins that mediate the proper insertion and assembly of IMPs in several organisms. The first known example of such an assembly mediator was the Oxa1p protein in the yeast mitochondrial inner membrane, which was shown to be required for cytochrome c oxidase assembly and, consequently, respiration and growth on non-fermentable carbon sources (Bauer et al., 1994). Recently, a more general role for Oxa1p was proposed (Altamura et al., 1996; Hell et al., 1998, 2001). Since the identification of Oxa1p and its role in the biogenesis of IMPs, homologs of this protein were identified and studied in various organisms. Thus, the *E. coli* YidC protein was shown to interact with several IMPs of different topology (Houben et al., 2000, Samuelson et al., 2000; Scotti et al., 2000). Similarly, the Albino III protein is involved in the biogenesis of a light-harvesting chlorophyll-binding protein in the thylakoidal membrane of chloroplasts (Moore et al., 2000; Sundberg et al., 1997). As exemplified with YidC of *E. coli*, these

292

general assembly mediators can interact with several different components of the Sec translocon. However, in the inner membrane of yeast mitochondria, which lack a conserved Sec machinery, Oxa1p itself seems to represent a novel type of translocon (Luirink *et al.*, 2001). Homologs of these genes have now been identified in the genomes of several Gram-positive bacteria, including *B. subtilis* (*spoIIIJ* and *yqjG*), *L. lactis* (*ybdC*, *yfgG*), *S. aureus*, *Listeria* sp. and *Streptomyces* sp. Disruption of *spoIIIJ* blocks sporulation of *B. subtilis* at an intermediate stage (Errington *et al.*, 1992), but it is presently unclear whether this phenotype relates to the impaired assembly of general or spore-specific membrane proteins. Notably, many of the key elements of the Sec, Tat and pseudopilin translocases themselves depend on proper targeting and assembly in the cytoplasmic membrane. Consequently, a thorough understanding of the mechanisms that underlie membrane protein insertion and assembly, and the role of Oxa1p homologs in these processes, will be required for a complete definition of the protein secretion pathways in Gram-positive bacteria.

6. ACKNOWLEDGEMENTS

The authors wish to thank Drs. A. Bolhuis, S. Bron, J.D.H. Jongbloed, J. Luirink, B. Oudega, W.J. Quax, and H. Tjalsma for stimulating discussions and support. R.M. was supported by grant VBI.4837 from the "Stichting Technische Wetenschappen; JMvD was supported by the European Union grants QLK3-CT-1999-00413 and QLK3-CT-1999-00917.

7. REFERENCES

Alksne, L.E., Burgio, P., Hu, W., Feld, B., Singh, M.P., Tuckman, M., Petersen, P.J., Labthavikul, P., McGlynn, M., Barbieri, L., McDonald, L., Bradford, P., Dushin, R.G., Rothstein, D., and Projan S.J. (2000) Antimicrob Agents Chemother 44: 1418-1427.
Altamura, N., Capitanio, N., Bonnefoy, N., Papa, S., and Dujardin, G. (1996) FEBS Lett 382: 111-115.
Andersen, C.L., Matthey-Dupraz, A., Missiakas, D., and Raina, S. (1997) Mol Microbiol 26: 121-132.
Antelmann, H., Tjalsma, H., Voigt, B., Ohlmeier, S., Bron, S., van Dijl, J.M., Hecker, M., (2001) Genome Research 11: 1484-1502.
Bauer, M., Behrens, M., Esser, K., Michaelis, G., and Pratje, E. (1994) Mol Gen Genet 245: 272-278.
Berks, B.C., Sargent, F., and Palmer, T. (2000) Mol Microbiol 5: 260-274.

Bibi, E. (1998) Trends Biochem Sci 23:51-55.

Bogsch, E., Sargent, F., Stanly, N.R., Berks, B.C., Robinson, C., and Palmer, T. (1998) J Biol Chem 273: 18003-18006.

Bolhuis, A., Broekhuizen, C.P., Sorokin, A., van Roosmalen, M.L., Venema, G., Bron, S., Quax, W.J., and van Dijl, J.M. (1998) J Biol Chem 273: 21217-21224.

Bolhuis, A., Matzen, A., Hyyrylaïnen, H.L., Kontinen, V.P., Meima, R., Chapuis, J., Venema, G., Bron, S., Freudl, R., and van Dijl, J.M. (1999) J Biol Chem 274: 24585-24592.

Bolhuis, A., Venema, G., Quax, W.J., Bron, S., and van Dijl, J.M. (1999a) J Biol Chem 274: 24531-24538.

Bolotin, A., Wincker, P., Mauger, S., Jaillon, O., Malarme, K., Weissenbach, J., Ehrlich, S.D. and Sorokin, A. (2001) Genome Res 11: 731-53.

Braunstein, M., Brown. A.M., Kurtz, S., and Jacobs, W.R. (2001) J Bacteriol 183: 6979-6990.

Bron, S., Meima, R., van Dijl, J.M., Wipat, A., and Harwood, C. (1999) In: A.L. Demain, and E. Davies (eds), Manual of industrial microbiology and biotechnology (2nd edition), ASM press,Washington, D.C., pp. 392-416.

Bunai, K., Yamada, K., Hayashi, K., Nakamura, K., and Yamane, K. (1999) J Biochem (Tokyo) 125:151-159.

Carpenter, P.B., Hanlon, D.W., and Ordal, G.W. (1992) Mol Microbiol 6: 2705-2713.

Chen, M., and Nagarajan, V., (1994) J Bacteriol 176: 5796-5801.

Chung, Y.J., Steen, M.T., and Hansen, J.N. (1992) J Bacteriol 174: 1417-1422.

Chung, Y.S., Breidt, F., and Dubnau, D. (1998) Mol Microbiol 29: 905-913.

Cregg, K.M., Wilding, I., and Black, M.T. (1996) J Bacteriol 178: 5712-5718.

Cristóbal, S., de Gier, J.W., Nielsen, H., and von Heijne, G. (1999) EMBO J 18: 2982-2990.

Dalbey, R.E, and Robinson, C. (1999) Trends Biochem Sci 24: 17-22.

de Leeuw, E. (2000) Thesis, Vrije Universiteit Amsterdam, The Netherlands

de Leeuw, E., te Kaat, K., Moser, C., Menestrina, G., Demel, R., de Kruijff, B., Oudega, B., Luirink, J., and Sinning, I., (2000) EMBO J 19:531-541.

Driessen, A.J.M., Fekkes, .P., and van der Wolk, J.P. (1998) Curr Opin Microbiol 1: 216-222.

Dubnau, D., and Lovett, C.M. (2001) In: A. L. Sonenshein, J. A. Hoch, and R. Losick (eds.), Bacillus subtilis and its closest relatives: from genes to cells. American Society for Microbiology, Washington, D.C., pp. 453-471.

Dumoulin, A. (2000) Genbank, acc.nr. AAG41993.

Errington, J., Appleby, L., Danie, R.A., Goodfellow, H., Partridge, S.R., and Yudkin, M.D. (1992) J Gen Microbiol 138:2609-2618.

Eichler, J., and Moll, R. (2001) Trends Microbiol 9: 130-136.

Feilmeier, B.J., Iseminger, G., Schroeder, D., Webber, H., Phillips, G.J. (2000) J Bacteriol 182: 4068-4076.

Fekkes, P., and Driessen, A.J.M. (1999) Microbiol Mol Biol Rev 63: 161-173.

Ferrari, E., Jarnagin, A. S., and Schmidt, B. F. (1993) In: A. L. Sonenshein, J. A. Hoch, and R. Losick (eds.), Bacillus subtilis and other gram-positive bacteria. American Society for Microbiology, Washington, D.C., pp. 917-937.

Foster, S.J., and Popham. D.L. (2001) In: A. L. Sonenshein, J. A. Hoch, and R. Losick (eds.), Bacillus subtilis and its closest relatives: from genes to cells. American Society for Microbiology, Washington, D.C., pp. 21-41.

Haandrikman, A.J., Kok, J., and Venema, G. (1991a) J Bacteriol 173: 4517-4525.

Haandrikman AJ, Meesters R, Laan H, Konings WN, Kok J, Venema G (1991b) Appl Environ Microbiol 57: 1899-1904.

Havarstein, L.S., Diep, D.B., and Nes, I.F. (1995) Mol Microbiol 16: 229-240.

Hell, K., Herrmann, J.M., Pratje, E., Neupert, W., and Stuart, R.A. (1998) Proc Natl Acad Sci USA 95: 2250-2255.

294

Hell, K., Neupert, W., and Stuart, R.A. (2001) EMBO J 20: 1281-1288.

Herbort, M., Klein, M., Manting, E.H., Driessen, A.J., and Freudl, R. (1999) J Bacteriol 181: 493-500.

Herskovits, A.A., Bochkareva, E.S., and Bibi, E. (2000) Mol Microbiol 38: 927-939.

Hirose, I., Sano, K., Shioda, I., Kumano, M., Nakamura, K., and Yamane, K. (2000) Microbiology 146: 65-75.

Houben, E.N.G., Scotti, P.A., Valent, Q.A., Brunner, J., de Gier J.-W.L., Oudega, B., and Luirink, J. (2000) FEBS Lett 476: 229-233.

Hynds, P.J., Robinson, D., and Robinson, C. (1998) J Biol Chem 273: 34868-34874.

Hyyryläinen, H.L., Vitikainen, M., Thwaite, J., Wu, H., Sarvas, M., Harwood, C.R., Kontinen, V.P., and Stephenson, K. (2000) J Biol Chem 275: 26696-26703.

Ishihara, T., Tomita, H., Hasegawa, Y., Tsukagoshi, N., Yamagata, H., and Udaka, S. (1995) J Bacteriol 177: 745-749.

Jacobs, M., Andersen, J.B., Kontinen, V.P., and Sarvas, M. (1993) Mol Microbiol 8: 957-966.

Jeong, S.M., Yoshikawa, H., and Takahashi, H. (1993) Mol Microbiol 10: 133-142.

Jongbloed, J.D.H., Martin, U., Antelmann, H., Hecker, M., Tjalsma, H., Venema, G., Bron, S., van Dijl, J.M., and Müller, J. (2000) J Biol Chem 275: 41350-41357.

Kakeshita, H., Oguro, A., Amikura, R., Nakamura, K., and Yamane, K. (2000) Microbiol 146:2595-2603.

Kaneko, T., Sato, S., Kotani, H., Tanaka, A., Asamizu, E., Nakamura, Y. Miyajima, N., Hirosawa, M., Sugiura, M., Sasamoto, S., Kimura, T., Hosouchi, T., Matsuno, A., Muraki, A., Nakazaki, N., Naruo, K., Okumura, S., Shimpo, S., Takeuchi, C., Wada, T., Watanabe, A., Yamada, M., Yasuda, M., and Tabata, S. (1996) Res 3: 109-136.

Kawaguchi, S., Muller, J.P., Linde, D., Kuramitsu, S., Shibata, T., Inoue, Y., Vassylyev, D.G., and Yokoyama, S. (2001) EMBO J 20: 562-569.

Kiesau, P., Eikmanns, U., Gutowski-Eckel, Z., Weber, S., Hammelmann, M., and Entian, K.D. (1997) J Bacteriol 179: 1475-1481.

Kontinen, V.P., Saris, P., and Sarvas, M. (1991) Mol Microbiol 5: 1273-1283.

Kontinen, V.P., and Sarvas, M. (1993) Mol Microbiol 8: 727-737.

Kuipers, O.P., Buist, G., and Kok, J. (2000) Res Microbiol 151: 815-822.

Kunst, F., Ogasawara, N., Moszer, I., Albertini, A.M., Alloni, G., Azevedo, V., Bertero, M.G., Bessieres, P., et al. (1997) Nature 390: 249-256.

Lory, S. (1994) In: von Heijne, G. (ed.) Signal peptidases. R.G. Landes Company, Austin, TX, 17-29.

Luirink, J., ten Hagen-Jongman, C.M., van der Weijden, C.C., Oudega, B., High, S., Dobberstein, B., and Kusters, R. (1994) EMBO J 13: 2289-2296.

Luirink, J., Samuelsson, T., and de Gier, J.-W. (2001) FEBS Lett 501:1-5.

Lyon, R.L., Gibson, C.M., and Caparon, M.G. (1998) EMBO J 17: 6263-6275.

Matsuyama, S., Fujita, Y., and Mizushima, S. (1993) EMBO J 12: 265-270.

Meijer, W.J., de Jong, A., Wisman, G.B.A., Tjalsma, H., Venema, G., Bron, S., and van Dijl, J.M. (1995) Mol Microbiol 17: 621-631.

Meima, R., Eschevins, C., Fillinger, S., Bolhuis, A., Hamoen, L.W., Dorenbos, R., Quax, W.J., van Dijl, J.M., Provvedi, R., Chen, I., Dubnau, D., and Bron, S. (2001) J Biol Chem 277: 6994-7001.

Moore, M., Harrison, M.S., Peterson, E.C., and Henry, R. (2000) J Biol Chem 275: 1529-1532.

Müller, J.P., Ozegowski, J., Vettermann, S., Swaving, J., van Wely, K.H.M., and Driessen, A.J.M. (2000) Biochem J 348: 367-373.

Müller, J.P., Bron, S., Venema, G., and van Dijl, J.M. (2000a) Microbiol 146: 77-88.

Müller, M., Koch, H.G., Beck, K., and Schäfer, U. (2000b) Progr Nucl Acid Res Mol Biol 66: 107-158.

Nakamura, K., Takamatsu, H., Akiyama, Y., Ito, K., and Yamane, K. (1990) FEBS Lett 273: 75-78.

Nakamura, K., Imai, Y., Nakamura, A., and Yamane, K. (1992a) J Bacteriol 174: 2185-2192.

Nakamura, K., Minemura, M., Nishiguchi, M., Honda, K., Nakamura, A., and Yamane, K. (1992b) Nucl Acids Res 20: 5227-5228.

Nakamura, K., Yahagi, S., Yamazaki, T., and Yamane, K. (1999) J Biol Chem 274: 13569-13576.

Nielsen, H., Engelbrecht, J., Brunak, S., and von Heijne, G. (1997) Protein Eng 10: 1-6.

Nishiguchi, M., Honda, K., Amikura, R., Nakamura, K., and Yamane, K. (1994) J Bacteriol 176:157-165.

Novak, P., and Dev, I.K. (1988) J Bacteriol 177: 5067-5075.

Ohta, T., Honda, K., Kuroda, M., Saito, K., and Hayashi, H. (1993) Biochem Biophys Res Commun 193: 730-737.

Ohta, T., Saito, K., Kuroda, M., Honda, K., Hirata, H., and Hayashi, H. (1994) J Bacteriol 176: 4779-4783.

Paik, S.H., Chakicherla, A., and Hansen, J.N. (1998) J Biol Chem 273: 23134-23142.

Petit-Glatron, M.-F., Monteil, I., Benyahia, F., and Chambert, R. (1990) Mol Microbiol 4: 2063-2070.

Petit-Glatron, M.-F., Grajcar, L., Munz, A., and Chambert, R. (1993) Mol Microbiol 9: 1097-1106.

Piggot, P.J, and Losick, R. (2001) In: A. L. Sonenshein, J. A. Hoch, and R. Losick (eds.), Bacillus subtilis and its closest relatives: from genes to cells. American Society for Microbiology, Washington, D.C. 483-517.

Power, S.D., Adams, R.M., and Wells, J.A. (1986) Proc Natl Acad Sci USA 83: 3096-3100.

Pragai, Z., Tjalsma, H., Bolhuis, A., van Dijl, J.M., Venema, G., and Bron, S. (1997) Microbiol 143: 1327-1333.

Priest, F.G. (1977) Bacteriol Rev 41: 711-753.

Raina, S., and Missiakas, D. (1997) Annu Rev Microbiol 51: 179-202.

Read, T.D., Brunham, R.C., Shen, C., Gill, S.R., Heidelberg, J.F., White, O., Hickey, E.K., Peterson, J., Utterback, T., Berry, K., Bass, S., Linher, K., Weidman, J., Khouri, H., Craven, B., Bowman, C., Dodson, R., Gwinn, M., Nelson, W., DeBoy, R., Kolonay, J., McClarty, G., Salzberg, S.L., Eisen, J., and Fraser, C.M. (2000) Nucleic Acids Res 28:1397-1406.

Sadaie, Y., Takamatsu, H., Nakamura, K., and Yamane, K. (1991) Gene 98: 101-105.

Samuelson, J.C., Chen, M., Jiang, F., Moller, I., Wiedmann, M., Kuhn, A., Phillips, G.J., and Dalbey, R.E. (2000) Nature 406:637-641.

Santini C.-L., Bernadac, A., Zhang, M., Chanal, A., Ize, B., Blanco, C., Wu, L.-F. (2001) J Biol Chem 276: 8159-8164.

Scotti, P.A., Urbanus, M.L., Brunner, J., de Gier, J.-W.L., von Heijne, G., van der Does, C., Driessen, A.J.M., Oudega, B., and Luirink, J. (2000) EMBO J 19: 542-549.

Segarra, R.A., and Iandolo, J.J. (1997) Genbank, acc.nr. AAB54024.

Sonenshein, A.L., Hoch, J.A., and Losick, R. (2001) In: A. L. Sonenshein, J. A. Hoch, and R. Losick (eds.), Bacillus subtilis and its closest relatives: from genes to cells. American Society for Microbiology, Washington, D.C., 3-5.

Spizizen, J. (1958) Proc Natl Acad Sci USA 44: 1072-1078.

Stephenson, K., Carter, N.M., Harwood, C.R., Petit-Glatron, M.-F., and Chambert, R. (1998) FEBS Lett 430: 385-389.

Stöver, A.G., and Driks, A. (1999a) J Bacteriol 181: 1664-1672.

Stöver, A.G., and Driks, A. (1999b) J Bacteriol 181: 5476-5481.

Stöver, A.G., and Driks, A. (1999c) J Bacteriol 181: 7065-7069.

Suh, J.W., Boylan, S.A., Thomas, S.M., Dolan, K.M., Oliver, D.B., and Price, C.W. (1990) Mol Microbiol 4: 305-314.

296

Sundberg, E., Slagter, J.G., Fridborg, I., Cleary, S.P., Robinson, C., and Coupland, G. (1997) Plant Cell 9: 717-730.

Thomas, J.D., Daniel, R.A., Errington, J., Robinson, C. (2001) Mol Microbiol 39: 47-53.

Tjalsma, H., Noback, M.A., Bron, S., Venema, G., Yamane, K., and van Dijl, J.M. (1997) J Biol Chem 272: 25983-25992.

Tjalsma, H., Bolhuis, A., van Roosmalen, M.L., Wiegert, T., Schumann, W., Broekhuizen, C.P., Quax, W.J., Venema, G., Bron, S., and van Dijl, J.M. (1998) Genes Dev 12: 2318-2331.

Tjalsma, H., Kontinen, V.P., Prágai, Z., Wu, H., Meima, R., Venema, G., Bron, S., Sarvas, M., and van Dijl, J.M. (1999) J Biol Chem 274: 1698-1707.

Tjalsma, H., Bolhuis, A., Jongbloed, J.D.H., Bron, S., and van Dijl, J.M. (2000) Microbiol Mol Biol Rev 64: 515-547.

Tjalsma, H., Zanen, G., Bron, S., and van Dijl, J.M. (2001) In: R.E. Dalbey, and D.S. Sigman (eds.), The Enzymes: Co- and posttranslational proteolysis of proteins, 3rd edition, Volume XXII. Academic Press, San Diego, CA, USA. 3-26.

van Dijl J.M., de Jong, A., Vehmaanperä, J., Venema, G., and Bron, S. (1992) EMBO J 11: 2819-2828.

van Dijl, J.M., Bolhuis, A., Tjalsma, H., Jongbloed, J.D.H., de Jong, A., and Bron, S. (2001) In: A. L. Sonenshein, J. A. Hoch, and R. Losick (eds.), Bacillus subtilis and its closest relatives: from genes to cells. American Society for Microbiology, Washington, D.C., 337-355.

van Roosmalen, M.L., Jongbloed, J.D.H., Dubois, J.-Y. F., Venema, G., Bron S., and van Dijl, J.M. (2001) J Biol Chem 276: 25230-25235

van Wely, K.H.M., Swaving, J., Broekhuizen, C.P., Rose, M., Quax, W.J., and Driessen, A.J.M. (1999) J Bacteriol 181: 1786-1792.

van Wely, K.H.M., Swaving, J., Freudl, R., and Driessen, A.J.M. (2001) FEMS Microbiol Rev 25: 437-454.

Vasantha, N., Thompson, L.D., Rhodes, C., Banner, C., Nagle, J., and Filpula, D. (1984) J Bacteriol 159: 811-819.

Veltman, O.R., Vriend, G., van den Burg, B., Hardy, F., Venema, G., and Eijsink, V.G. (1997) FEBS Lett 405: 241-244.

Wandersman, C. (1989) Mol Microbiol 3:1825-1831.

Weichenrieder, O., Wild, K., Strub, K., and Cusack, S. (2000) Nature 408: 167-173.

White, O., Eisen, J.A., Heidelberg, J.F., Hickey, E.K., Peterson, J.D., Dodson, R.J., Haft, D.H., Gwinn, M.L., et al. (1999) Science 286: 1571-1577.

Wu, S.C., Ye, R., Wu, X.C., Ng, S.C., and Wong, S.L. (1998) J Bacteriol 180: 2830-2835.

Wu, L.-F., Ize, B., Chanal, A., Quentin, Y., and Fichant, G. (2000) J Mol Microbiol Biotechnol 2: 170-189.

Zheng, G., Yan, L.Z., Vederas, J.C., and Zuber, P. (1999) J Bacteriol 181: 7346-7355.

Zhu, X.L., Ohta, Y., Jordan, F., and Inouye, M. (1989) Nature 339: 483-484.